Electromagnetic Fields and Biomembranes

Electromagnetic Fields and Biomembranes

Edited by
Marko Markov
Sofia University
Sofia, Bulgaria

and
Martin Blank
Columbia University
New York, New York

Plenum Press • New York and London

Library of Congress Cataloging in Publication Data

International School of Electromagnetic Fields and Biomembranes (1st: 1986: Pleven,
Bulgaria)
  Electromagnetic fields and biomembranes.

  "Proceedings of the First International School of Electromagnetic Fields and Bio-
membranes, held October 6–12, 1986, in Pleven, Bulgaria"—T.p. verso.
  Includes bibliographies and index.
  1. Membranes (Biology)—Congresses. 2. Electromagnetism—Physiological effect—
Congresses. I. Markov, Marko (Marko S.) II. Blank, Martin, date. III. Title. [DNLM: 1.
Biological Transport congresses. 2. Cell Membranes—physiology—congresses. 3.
Electromagnetic—congresses. W3 IN884 1st 1986e/QT 34 I662 1986e]
QH601.I524 1986   574.87′5   87-29276
ISBN-13: 978-1-4615-9509-0   e-ISBN-13: 978-1-4615-9507-6
DOI: 10.1007/978-1-4615-9507-6

Proceedings of the First International School of Electromagnetic Fields
and Biomembranes, held October 6–12, 1986, in Pleven, Bulgaria

© 1988 Plenum Press, New York
Softcover reprint of the hardcover 1st edition 1988

A Division of Plenum Publishing Corporation
233 Spring Street, New York, N.Y. 10013

PREFACE

The First International School on "Electromagnetic Fields and Biomembranes" took place in Pleven, Bulgaria on 6-12 October 1986. It was designed as an advanced course through a collaboration of the Biological Faculty of Sofia University and the Council of the Bioelectrochemical Society. In an advanced course the lecturers are specialized in particular areas, and the students are usually specialists in related areas. We have captured the expertise of both groups of participants in this volume. The longer papers prepared by the lecturers are joined with the shorter papers based on the posters presented by the "students" to provide a summary of the school as well as an indication of current research directions in the field.

The course was designed to provide the latest information about biomembrane structure and function, covering the properties of both the lipid matrix and the recently characterized proteins that function as specialized channels and receptors. Real membranes and various models were covered, with an emphasis on understanding their mechanisms of interaction with various exogenous stimuli (e.g., electric, magnetic, light, etc.). Several practical applications of this information (e.g., electroporation, electro-fusion) were also presented with indications of the possibilities for new developments in biotechnology. The mixture of basic science with practical applications, together with the intermingling of lecturers and students from many different countries produced a stimulating atmosphere and effective teaching. We hope that this volume will transmit some of this atmosphere.

We appreciate the support provided by various organizations, especially the Local Administration of the town of Pleven, who shared the enthusiasm of the organizers. We trust that this permanent record of the course will be a stimulus for continuing efforts on this subject.

Marko Markov, University of Sofia
Martin Blank, Bioelectrochemical Society

CONTENT

MEMBRANE LIPIDS AND THEIR MODEL SYSTEMS

Johannes de Gier
Laboratory of Biochemistry, State University of Utrecht
Padualaan 8, 3584 CH Utrecht, The Netherlands

INTRODUCTION

In the generally accepted concepts on the structural organization of biological interfaces these membranes are depicted as fluid lipid bilayers in which protein structures are embedded. It is well recognized that the proteins form channels, carriers, ion pumps, receptors, etc. and so add specific functions to each interface, but also the lipids may contribute to the specificity of the membrane. Whereas a fluid lipid bilayer could be formed with one suitable lipid species, nature uses quite complex mixtures of lipids for building up membranes. The composition of such mixtures appears to be characteristic for a given type of membrane. The significance of this lipid diversity is not yet fully understood, but next to specific chemical properties of individual membrane lipids the overall physical behaviour of the lipid matrix seems to be important in determining the functional proporties of the interface.

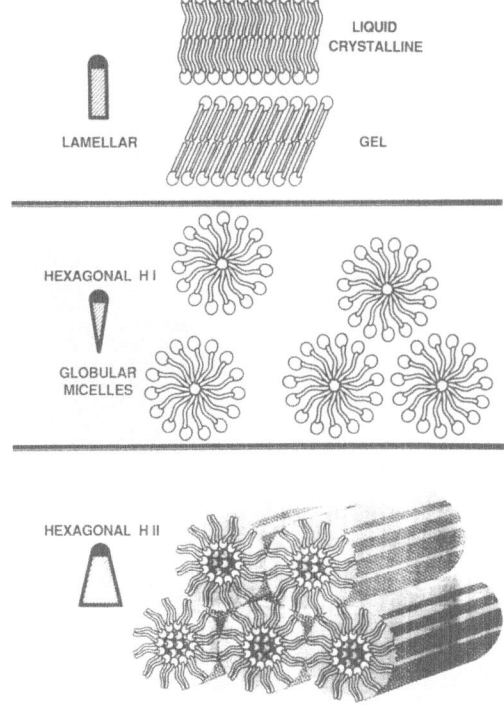

Fig. 1. Schematic presentation of different membrane lipid phases.

Lipid polymorphism

Natural occurring membrane lipids differ in size and charge of the polar headgroups and in length and unsaturation of the paraffin chains. A physical expression of these chemical differences can be observed when isolated and purified lipids are dispersed in a water phase as it appears that different association structures can be formed (compare figure 1). The type of organization which is preferred is easily understood from simple geometric considerations. Lipid molecules, like e.g. lecithins and diglycosyldiglycerides, in which there is good balance between the size of the hydrated polar headgroup and the extent of the hydrophobic part of the molecule spontaneously form a bilayer organization. Depending on the nature of the attached fatty acid chains a liquid crystalline or gel state bilayer organization can be formed. In the liquid crystalline state the lipid molecules have considerable degree of motional freedom. They can rotate around their axes and undergo rapid lateral diffusion furthermore the fluid condition of the chains allows rotation around the C-C bonds causing kinking of the chains due to trans-gauche conformational transitions. In the gel state the lipid molecules have less freedom and the paraffin chains are in a stretched, predominantly trans orientated conformation.

Membrane lipids of which the size of the polar group is dominating like as for example in lysophospholipids and gangliosides generally prefer to form a hexagonal H_I phase, which in excess water is converted into a solution of globular micelles. On the other hand, lipids of which the hydrophobic part requires relatively more space, like in unsaturated phosphatidylethanolamines and monoglucosyldiglycerides, tend to form, also in excess of water, a stable hexagonal H_{II} phase, an organization in which the lipid molecules are arranged in hexagonally packed cylinders and in contrast to the hexagonal H_I phase with the polar heads pointing to the inside of the cylinders.

Phase transitions due to changes in environmental conditions.

Lipid phases are not only dependent on the chemical composition of the lipids but also can be affected to a large extent by extrinsic factors. Temperature is an important parameter in this respect, but also the state of hydration, pH, the presence or absence of divalent cations and lipid protein interactions can have modifying consequences.

In detecting transitions from a gel state to a liquid crystalline bilayer differential scanning calorimetry has been shown to be a useful technique. The example shown in figure 2 concerns the

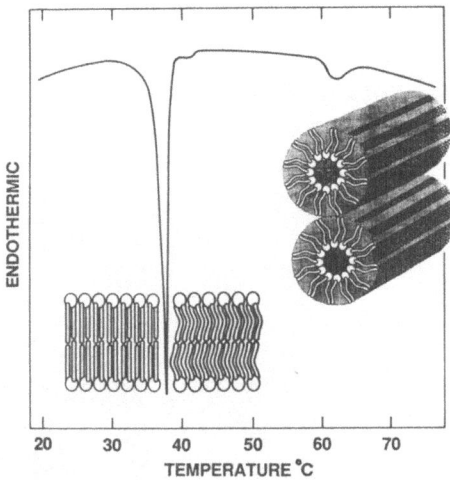

Fig. 2. Differential calorimetric scan of a liposome dispersion of dielaidoylphosphatidyl-ethanolamine

synthetic lipid dielaidoylphosphatidylethanolamine dispersed in excess of water. The main transition at 40 °C indicates the melting of the paraffin chains; below this temperature the bilayer is in the gel condition and above a liquid crystalline condition is reached. Of course the calorimetric technique by itself does not reveal any detail on the molecular organisation and additional structural analysis with X-ray or spectroscopic techniques above and below the transition temperature are required to support an interpretation. Gel to liquid crystalline transition temperatures have been analysed for many lipids and it is found that variations in chain length and unsaturation and also variations in the chemical identity of the polar head group are correlated with wide variations in transition temperature. The gel to liquid crystalline transition of many natural occurring lipids which cary unsaturated fatty acids is found in a temperature range well below the physiological values, but the transition of e.g. dipalmitoylphosphatidylcholine, which quantitatively is an important lipid species in many membranes, occurs at 41 °C. Furthermore several sphingolipids carrying long chain fatty acids have transition temperatures well above this temperature.

The analysis shown in figure 2 also reveals a minor transition at around 65 °C which with X-ray and NMR techniques has been identified as the transition from a liquid crystalline bilayer to the hexagonal H_{II} phase. This polymorphic phase transition is easily understood in terms of the shape concept. The polar headgroup of the phosphatidylethanolamine is rather small as hydration is limited due to direct hydrogen binding between the polar headgroups and with increasing temperature the increasing motion of the chains is enlarging the extent of the hydrophobic part of the molecule which promotes the conical shape of the lipid and results in formation of the H_{II} phase. The natural occurring phosphatidylethanolamine carrying unsaturated fatty acids with cis double bonds have their transition to the hexagonal phase at much lower temperature. As these unsaturated phosphatidylethanolamines are major lipid species in many biological membranes it can be concluded that at their physiological temperature biological membranes often contain significant fractions of lipids which prefer a non-bilayer organization.

Variations in pH and divalent ion concentrations are particularly important with respect to acidic membrane lipids. The gel to liquid crystalline transition of dimyristoylphosphatidic acid, for example, measured at at pH = 6 is found at around 50 °C but above pH = 7 a large decrease in transition temperature can be noticed as a consequence of increasing repulsive forces between the polar headgroups due to a second ionization on the phosphate group. Negatively charged lipids also have affinity to divalent ions such as Ca^{++} and as a result uncharged salt complexes may be formed. The complex formation reduces the hydration and also the repulsive forces between the molecules and can have quite different consequences for the phase behaviour. Ca-salts of saturated negatively charged lipids often form very efficiently packed bilayer organizations with extremely high transition temperatures. These bilayers are so rigid that a biconcave curvature is impossible and instead of closed liposomes stacked lamellar or "cochleated structures" are formed. On the other hand Ca-complex formation in case of unsaturated lipid in some cases (e.g. mitochondrial cardiolipin and dioleoylphospatidic acid) results in a transition from a lamellar to a hexagonal H_{II} phase, which again easily can be appreciated in terms of the phase concept.

For further information concerning gel to liquid crystalline and polymorphic phase transitions the reader is referred to more extensive recent reviews (1, 2).

Barrier function of the bilayer organization

When lecithins are dispersed in a water phase, depending on the dispersion technique, multilayered liposome structures or one bilayer bounded vesicles are formed. The multilayered liposomes and also the large unilamellar vesicles in salt solutions behave as ideal osmometers from which it has been concluded that the lipid bilayer is a selectively permeable membrane; permeable to water but impermeable to the hydrated ions. From swelling phenomena in isotonic solutions of small polar non-electrolytes it became clear that the bilayer also is permeable to molecules as urea, glycerol, erythritol, etc. (3). The molecular motion in the membrane barrier which enables this selective permeation is to be found in trans-gauche conformational transitions of the orientated paraffin chains.

3

The kinking of the chains, arising from such transitions results in the formation of small cavities which can be filled with single water or other small non-electrolyte molecules jumping in from the water phase. As the kinks move along the orientated chains the system works as an intrinsic carrier system (4). It is obvious that the cavities which are formed have limited size allowing only small non-electrolytes to pass the barrier. For the passage of larger non-electrolytes and of hydrated ions more drastic defects in the membrane organization will be required.

The proposed explanation of non-electrolyte permeability is supported by the finding that the rate of permeation is strongly dependent on the membrane fluidity. Effects of changes in unsaturation, chain length and cholesterol content on the membrane viscosity are clearly reflected in the permeation rate of the small non-electrolytes (4). In the gel state osmotic flow of water is extremely slow, but nevertheless a clear osmotic response can be measured indicating that in the gel state the barrier is not as rigid as could be expected (5).

Interestingly a lecithin bilayer membrane shows a permeability maximum at the temperature of phase transition so that under this particular physical condition the barrier becomes permeable to ions (6). Apparently defects in the barrier organisation which seldomly occur in the completely fluid bilayer are occurring more easely when solid and fluid patches coexist in the membrane. It can be suggested that due to disorientation of lipid molecules in the phase border region transient pores are formed as suggested in figure 3. This hypothesis is further suggested by the finding that at the phase

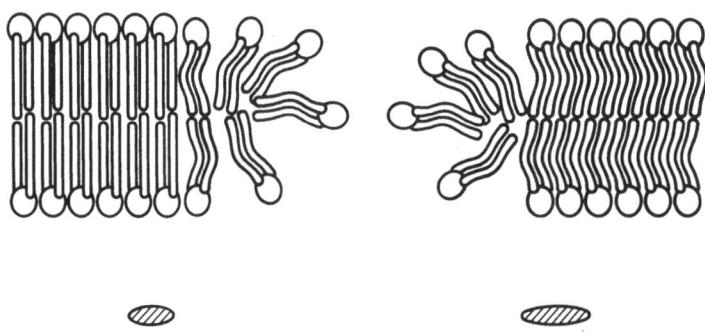

Fig. 3. Disorientation from the bilayer organization of lipid molecules at the phase border region.

transition also transmembrane diffusion of the lipid molecules can be noticed, whereas in the full liquid crystalline state such flip-flop movement is extremely low (7). To reach an estimate of the apparent pore size, trapping abilities for a series of non-electrolytes with increasing molecular weight have been studied. Using liposomes of dimyristoylphosphatidylcholine it was concluded that pores with radii up to about 8 Å can be formed (8). The extend of the permeability increase appeared to depent largely on the chemical composition of the lipid. Increase in chain length moderates the effect (6), whereas also hydrogen bonding between the polar headgroups can reduce the defect formation (9).

Extrapolating to biological membranes, barrier defects on phase border regions may explain loss of viability combined with leakage of cytoplasmic constituents, which sometimes can be noticed when micro-organisms are subjected to a cold chock. In a study on an unsaturated fatty acid auxotroph of *Escherichia coli* with variable fatty acid composition it was concluded that release of ions and other molecules from these cells upon rapid cooling could be correlated with a partial transition of the membrane to a gel state condition (10).

To maintain the barrier function of the membrane a full liquid crystalline condition of the lipid matrix is important. In this respect it is of interest to consider the high concentrations of sterol occurring in cell membranes of eukariotic cells. When such concentrations are introduced in liposomal bilayers phospholilpid-sterol interaction brings the bilayer into an intermediate state of viscosity from which transition to a gel state is impossible. Therefore an important function of cholesterol in the membrane could be to avoid formation of phase border regions (11).

Liposomes of bilayer and non-bilayer lipids

When non-bilayer lipids are introduced into the bilayer it can be expected that this destabilizes the bilayer organization and affects its barrier function. Studying mixtures of lecithins and lyso-lecithins it appeared that high concentrations of the micel forming lysophospholipid can be accomodated in the bilayer organization of the diacyl phospolipid. Dispersion of mixtures up to about 50 mole % of lyso-lecithin resulted in the formation of liposomal structures only above this concentration a mixed micellar organisation was obtained. However already at around 20 mole % of lyso-lecithin the liposomal membrane became freely permeable to K^+ ions indicating that transient defects in the bilayer frequently occur due to the presence of the non-bilayer lipid (12).

Membrane model studies with hexagonal H_{II} forming lipids are of particular interest because of their abundance in biological membranes. As discussed already in pure form these lipids can undergo a bilayer to H_{II} phase transition which then completely destroys the membrane organization. In mixtures with lipids, which form a stable bilayer an overall membrane organization can be maintained but a resulting increased molecular motion may be important for specific membrane functions. When liposomes are formed of a mixture of lecithin with an unsaturated phosphatidylethanolamine and the H_{II} character of the phosphatidylethanolamine is promoted by increase of temperature this results in fusion events. The multilayered liposome is then transfered into a sponge like structure (13). Similar results can be observed upon addition of Ca^{++} to liposomes composed of a mixture of lecithin and cardiolipin. It is proposed that at the membrane connection points inverted micelles are formed which can accommodate the cone shaped H_{II} lipids. The occurrence of these inverted micelles is nicely demonstrated in electron micrographs of freeze fracture faces of such liposomes in which they appear as lipidic particles with a diameter of about 10 nm (14).

Transient non-bilayer structures and specific transport

It is intriguing to speculate that bilayer to non-bilayer transitions could play a role in speciic membrane transport. It can be suggested that permeants which are able to perturb the bilayer organization, for example divalent ions or defined protein structures, upon specific interaction with lipids in the membrane, locally induce transition to a non-bilayer organization resulting in their own translocation over the membrane. Subsequent dissociation of the permeant lipid complex could restore the bilayer organization so that efficient barrier function is maintained. In this respect it is of interest that phosphatidic acid has been suggested to play a role in Ca^{++} transport over the plasma membrane, whereas cardiolipin could play in the electrically silent efflux of Ca^{++} mitochondria (15). A direct role of such phospholipids in the Ca^{++} translocation according to a mechanism suggested in the above hypothesis has been studied in membrane model systems. In our approach we have used large unilamellar vesicles and and improved version of an excisting translocation assay in which the inflow of Ca^{++} is detected by the Ca^{++} chelating indicator arsenazo III enclosed in the interior of the vesicles (16). The vesicles were prepared according to the reverse phase technique and were composed of dioleoylphosphatidylcholine and 20 mole % of an acidic phospholipid. When Ca^{++} was added in the outside medium of vesicles containing dioleoylphosphatidylglycerol as the acidic phospholipid a perfect barrier function of the membrane was maintained. However, when the dioleoylphosphatidylglycerol was replaced by dioleoylphosphatidate or cardiolipin an in time increasing Ca^{++} aresenazo III complex function could be measured, indicating that the phosphatidate and cardiolipin are both able to mediate in Ca^{++} translocation, whereas the phosphatidylglycerol is not. Studies on the specificity of the translocation process showed defined selectivity in the low millimolar range of Ca^{++} (17). Under these conditions clear translocation of Ca^{++} can be observed whereas a perfect barrier is maintained for solutes like K^+ and $SO_4^=$ ions and carboxyfluorescein. Upon increase of the Ca^{++} concentration (> 3 mM for dioleoylphosphatidate and > 12 mM for beafheart cardiolipin) the selectivity of the process gradually diminished. Remarkably in the low

Ca^{++} concentration range specific translocation of Ca^{++} could be accompanied by translocation of Ca^{++} chelating anions such as EDTA and the indicator ion arsenazo III. To explain this permeability behaviour we suppose that specific phase behaviour of the two negatively charged lipids enables flip-flop movement of Ca-phospholipid or Ca-phospholipid-chelator complexes over the membrane as indicated in figure 4.

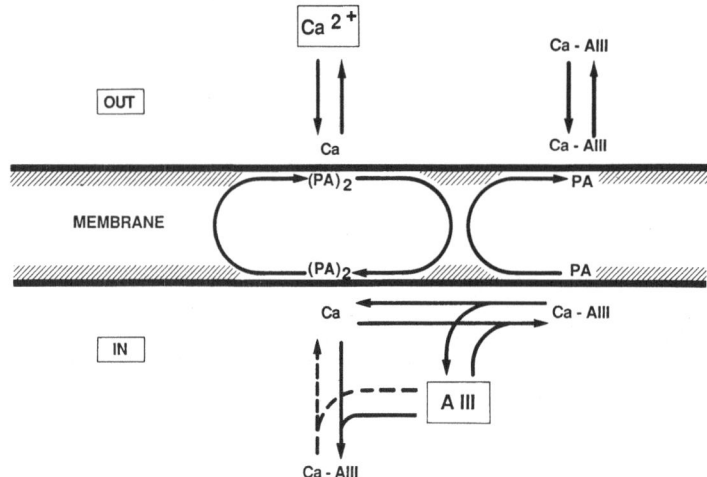

Fig.4 Schematic presentation of phospholipid flip-flop mediated translocation of Ca^{++} and arsenazo III

The results of these permeability studies could be nicely correlated with studies on the phase behaviour of the acidic phospholipids (18). Interaction of Ca^{++} with dioleoylphosphatidylglycerol did not result in disturbance of the bilayer organization. On the other hand interaction of Ca^{++} with cardiolipin is well known to result in the formation of a hexagonal H$_{II}$ phase. Furthermore studying the Ca^{++} dioleoylphosphate complex formation at pH 7.4 a new organization was defined characterized by immobolization of the headgroup whereas the acyl chains remain in fluid condition. Based on the details of these phase studies and on the structural behaviour of mixtures of the acidic phospholipids with dioleoylphosphatidylcholine it is argued that transient formation of hydrated or dehydrated inverted intramembrane complexes could be the mechanism for the in figure 4 proposed flip-flop scheme (18).

REFERENCES

1. Mc Elhaney, R.N.: The use of differential scanning colorimetry and differential thermal analysis in studies of model and biological membranes.
    Chem. Phys. Lipids 30, 229-259 (1982).
2. De Kruijff, B.; Cullis, P.R.; Verkleij, A.J.; Hope, M.J.; van Echteld, C.J.A.; Taraschi, T.F.: Lipid Polymorphism and membrane function; in Martonosi, The enzymes of biological membranes; 2nd ed. Vol 1, Membrane structure and dynamics., pp 131-204 (Plenum Press, New York 1985).
3. Bangham, A.D.; de Gier, J.; Greville, G.D.:
    Osmotic properties and water permeability of phospholipid liquid crystals.
    Chem. Phys. Lipids 1, 225-40 (1967).
4. De Gier, J.; Mandersloot, J.G.; Hupkes, J.V.; Mc Elhaney, R.N.; van Beek, W.P.:
    On the mechanism of non-electrolyte permeation through lipid bilayer and through biomembranes.
    Biochim. Biophys. Acta 233, 610-618 (1971).

5. Blok, M.C.; van Deenen, L.L.M.; de Gier, J.:
    Effect of gel to liquid crystalline phase transition on the osmotic behaviour of phosphatidyl·
    choline liposomes.
    Biochim. Biophys. Acta 433, 1-12 (1976).

6. Blok, M.C.; van der Nent-Kok, E.C.M.; van Deenen, L.L.M.; de Gier, J.:
    The effect of chain length and lipid phase transitions on the selective permeability properties ol
    liposomes.
    Biochim. Biophys. Acta 406, 187-196 (1975).

7. De Kruijff, B.; van Zoelen, E.J.J.:
    Effect of the phase transition on the transbilayer movement of dimyristoylphosphatidylcholine
    in unilamellar vesicles.
    Biochim. Biophys. Acta. 511, 105-115 (1978).

8. Van Hoogevest, P.; de Gier, J.; de Kruijff, B.:
    Determination of the size of the packing defects in dimyristoylphosphatidylcholine bilayers,
    present at the phase transition temperature.
    FEBS lett. 171, 160-164 (1984).

9. Noordam, P.C.; Killian, A.; Oude Elferink, R.F.M.; de Gier, J.:
    Comparative study on the properties of satururated phosphatidylethanolamine and
    phosphatidylcholine bilayer: Barrier characteristics and susceptibility to phospholipase A,
    degradation.
    Chem. Phys. Lipids 31, 191-204 (1982).

10. Haest, C.W.M.; de Gier, J.; van Es, G.A.; Verkley, A.J.; van Deenen, L.L.M.:
    Fragility of the permeability barrier of Escherichia Coli.
    Biochim. Biophys. Acta 288, 43-53 (1972).

11. De Kruijff, B.; Demel, R.A.:
    The function of sterols in membranes.
    Biochim. Biophys. Acta 457, 109-132 (1976).

12. Mandersloot, J.G.; Reman, F.C.; van Deenen, L.L.M.; de Gier, J.:
    Barrier properties of lecithin-lysolecithin mixtures.
    Biochim. Biophys. Acta 382, 22-26 (1975).

13. Noordam, P.C.; van Echteld, C.J.A.; de Kruijff, B.; de Gier, J.:
    Barrier characteristics of membrane model systems containig unsaturated phosphatidylethano
    lamines.
    Chem. Phys. Lipids 27, 221-232 (1980).

14. Verkleij, A.J.:
    Lipidic intramembranous particles.
    Biochim. Biophys. Acta 779, 43-63 (1984).

15. Somermeyer, G., Knauss, T.C., Weinberg, J.M. and Humes, H.D.:
    Characterization of Ca^{2+} transport in rat renal brush-border membranes and its modulation b
    phosphatidic acid.
    Biochem. J. 214, 37-46 (1983).

16. Smaal, E.B., Mandersloot, J.G., de Kruijff, B.; de Gier, J.
    Essential adaptation of the calcium influx assay into liposomes with entrapped arsenazo III for studies o
    the possible calcium translocating properties of acidic phospholipids.
    Biochim. Biophys. Acta 816, 418-422 (1985).

17. Smaal. E.B.; Mandersloot, J.G.; de Kruijff, B.; de Gier, J.:
    Consequences of the interaction of calcium with dioleoylphosphatidate containing model
    membranes: changes in membrane permeability.
    Biochim. Biophys. Acta 860, 99-108 (1986).

18. Smaal, E.B., Nicolay, K., Mandersloot, J.G., de Gier, J., de Kruijff, B.:
    ^{2}H-NMR, ^{31}P-NMR and DSC characterization of a novel lipid organization in calciumdioleoyl
    phosphatidate membranes. Implications for the mechanism of the phosphatidate calcium transmembran
    shuttle.
    Biochim. Biophys. Acta 897 (1987) 453-466.

HYDRATION FORCES BETWEEN PHOSPHOLIPID MEMBRANES AND

THE POLYETHYLENE GLYCOL INDUCED MEMBRANE APPROACH

Klaus Gawrisch[a], Klaus Arnold[b], Kerstin Dietze[a], Uta Schulze[a]

[a]Department of Physics, Karl Marx University, Linnéstr.5, Leipzig, DDR 7010,(G.D.R.)

[b]Department of Medicine, Institute of Physics and Biophysics, Karl Marx University, Liebigstr.27, Leipzig, DDR 7010,(G.D.R.)

INTRODUCTION

The fusion of two biological membranes can be divided into membrane approach, formation of a common membrane between the opposing cells and rupture of this membrane. During the PEG induced membrane approach protein free membrane areas are formed (Maul et al., 1976;Knutton, 1979; Robinson et al., 1979; Krähling, 1981a; Hui et al., 1985). Most authors assume that the primary steps of membrane fusion occur between lipid bilayers of protein free areas (Ahkong et al., 1975; Zakai et al., 1977; Cullis & Hope, 1978; Hui et al., 1981; Krähling, 1981b). For a better understanding of molecular mechanisms of membrane fusion it is reasonable to investigate the influence of PEG water solutions on lipid model membranes. The application of several spectroscopic methods such as NMR is easier in model membrane ivestigations than in investigations of natural biomembranes. Our interest was concentrated on the first step of membrane fusion, the membrane approach.
Different repulsive and attractive forces act between membranes. For distances between lamellar lipid bilayers smaller than 2 nm in most cases a repulsive force caused by hydration of lipid headgroups appears, which exceeds attractive van der Waals interaction and repulsive electric double layer forces (Cowley et al., 1978; Rand, 1981; Lis et al., 1982).
Membrane approach requires either a change in the properties of the membrane surface to reduce the magnitude of repulsive forces or to increase the magnitude of attractive forces. Another possibility is the generation of an efflux of water from the space between bilayers driven by osmotic gradients (Rand, 1981). Up to now the physical mechanism of PEG induced membrane fusion has been unknown. In some papers the fusion is interpreted as a result of direct interactions of PEG with the membrane surface (Ohno et al., 1982; Boni et al., 1984a). The formation of PEG bridges between opposing membranes on the basis of the formation of hydrogen bonds between polar parts of membrane molecules and PEG has been discussed. The existence of strong van der Waals attraction between the methylene groups of PEG and the membrane has also been considered.
On the other side, it is well known that PEG has a high water binding ability. Every ethylene oxide unit binds several water molecules via hydrogen bonds (Baran et al., 1972; Krähling, 1981b; Tilcock & Fisher, 1982; Lebovka

et al., 1983). As a result low molar concentrations of PEG in water solutions decrease the activity of water in these solutions significantly, or in other words, PEG water solutions have a high osmotic pressure. If the PEG concentration in the water between membranes is decreased, in comparison to the surrounding PEG water solution part of the water is sucked off the water layer between membrane bilayers. Similar dehydration processes are caused by the addition of high molecular weight dextran water solutions to lipid membranes (Rand, 1981; LeNeveu et al., 1976; LeNeveu et al., 1977). This view of the molecular mechanism of PEG induced model membrane approach explains also the membrane dehydration observed after addition of PEG water solutions (Knutton, 1979; Robinson et al., 1979; Tilcock & Fisher, 1979; Boni et al., 1981; Krähling, 1981b; Tilcock & Fisher, 1982; Wojcieszyn et al., 1983; Boni et al., 1984b; Wilschut & Hoekstra, 1984; Hui et al., 1985; MacDonald, 1985). At the beginning of our experiments we intended to give a more detailed description of primary PEG phospholipid interactions and of secondarily induced structural changes in phospholipid water dispersions. During our research we observed that there is no indication of a direct interaction of PEG molecules with phospholipids. The experimental results show without doubt that the PEG induced approach of phospholipid membranes is caused by the osmotic properties of PEG water solutions together with a reduced solubility of PEG molecules in the water layers near the membrane. The approach is not caused by PEG bridges between membrane surfaces.

MATERIALS AND METHODS

Poly(ethylene glycols)

Poly(ethylene glycols) with molecular weights of 20,000 - 17,000 (PEG 20,000), 7,500 - 6,000 (PEG 6,000), 420 - 380 (PEG 400) from Serva and tri(ethylene glycol) (PEG 150) from Merck were used without further purification. Double distilled H_2O or ^2H_2O (isotopic enrichment 99.7%) from Isocommerz was used for the preparations.

Lipids

Lipid water dispersions were prepared either with a total egg yolk phospholipid fraction (TEPF) or with egg yolk lecithin (EYL) isolated on an aluminium oxide (Merck) column with a chloroform, methanol solution (9 : 1 v/v). The composition of TEPF was checked by thin layer chromatography on HPTLC plates (Merck) with a chloroform, methanol, ammonia (25 % in water) solution (65 : 35 : 3 v/v/v) and by high resolution 31-P NMR spectroscopy of the lipids in a micellar Triton X-100 solution in water. The mole ratio of phosphatidylcholine (PC) to phosphatidylethanolamine (PE) was 3.2 : 1. Further trace amounts of sphingomyeline, lysophosphatidylcholine, lysophosphatidylserine and neutral lipids, altogether about 5 mol%, were detected. The lipid purity was checked during the manipulations by thin layer chromatography. After one week of storage trace amounts of lysolecithin (less than 5 wt.%) appeared in some of the EYL samples.

Osmometry

Osmotic properties of PEG water solutions were determined by water vapour pressure measurements over the PEG water solutions with a homebuilt apparatus. The water vapour pressures were measured relative to the vapour pressure of pure water. The vessels containing the PEG water solution and pure water were temperature stabilized by a water bath (stability better than 0.1 K). Before the measurements the distilled water and the PEG water solution were degassed by at least three freeze-thaw cycles under vacuum.

<u>NMR</u>

For ^{2}H NMR measurements approximately 100 mg of dry lipid were placed in sample tubes of 10 mm diameter and water or the PEG water solution of the appropriate concentration was added in excess (at least 1 ml). The tubes were filled with nitrogen and sealed. The contents of the sample tubes were mixed by shaking. Before the NMR measurements the lipid phase was concentrated by centrifugation, and the samples were stored at room temperature for at least 24 h.

In some experiments dry lipid was placed in a dialysis package (Serva) impermeable to substances with a molecular weight higher than 15,000. The package was placed in a tube which contained the appropriate PEG water solution. The tubes were filled with nitrogen and sealed for at least 24 hours. Then the tubes were opened, the dialysis package was transferred to another sample tube which was also filled with nitrogen and sealed.

2-H- and 31-P NMR measurements were performed on a Bruker HX-90 spectrometer in the Fourier transform mode at 6.4 MHz and 36.4 MHz, respectively.

EXPERIMENTAL RESULTS

<u>NMR measurements as functions of water concentration</u>

We prepared lipid water dispersions with increasing water concentrations from 2 to 100 water molecules per lipid. The phase state of lipids was checked by 31-P NMR. All samples showed an anisotropy of chemical shift characteristic of a liquid crystalline lamellar phase (Seelig, 1978; Arnold et al., 1981). In a previous publication (Gawrisch et al., 1985) we showed that the structure of EYL and TEPF water dispersions changes with increasing water concentrations. Up to 15 moles water per mole of phospholipid the samples consist of flat bilayers separated by water layers. At higher water concentrations multilamellar liposomes with diameters in the μm range and smaller are formed spontaneously. Because of lateral diffusion of water molecules over the curved bilayer surface the quadrupole splitting of water in liposomes with diameters smaller than 2 μm is averaged out and a more or less broadened isotropic 2-H NMR signal appears. Similar averaging processes of the anisotropy of chemical shift in the 31-P NMR spectra were not detected because the lateral diffusion of lipid molecules is not as fast as that of water molecules.

The measured 2-H NMR lineshape and quadrupole splitting are a reproducible indicator for the number of water molecules per lipid between bilayers. At water concentrations higher than 15 waters per lipid the fraction of multilamellar liposomes with diameters greater than 2 μm gives a quadrupole split signal which is superimposed on a broadened isotropic water signal.

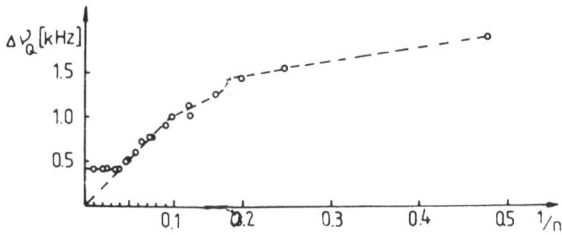

Fig. 1. ^{2}H NMR quadrupole splitting as a function of the reciprocal number of water molecules per lipid molecule (TEPF).

With increasing water concentration the width of the isotropic signal decreases and the detection of the quadrupole split signal in the base of the resonance line becomes more pronounced.
In Fig. 1 the dependence of the quadrupole splitting on the inverse number of waters per lipid is given. As shown by Finer & Darke (1974) every straight line in such a representation corresponds to a hydration shell of lipid molecules. For both EYL and TEPF molecules, about 12 waters per lipid are influenced in their parameters by the lipid surface. About five waters form a first hydration shell and the remaining seven waters a second one. At a water concentration of 23 + 2 waters per lipid the quadrupole splitting reaches a minimum value which does not decrease further with increasing water concentration, indicating that at this water concentration the incorporation of water between lipid bilayers is finished.

NMR measurements as functions of PEG concentration

Dispersions of lipids in PEG water solutions which contain from 0 to 60 wt.% PEG of different molecular weights were prepared. At all concentrations of PEG's of different molecular weights, there was no indication for hexagonal or cubic lipid phases in the NMR spectra. 2-H NMR spectra typical of lipid samples containing PEG water solutions of different concentration or having contact to these solutions via a dialysis membrane are given in Fig. 2. Water which interacts with PEG shows an isotropic NMR signal. The quadrupole split signal caused by water trapped between lipid bilayers is superimposed on it.

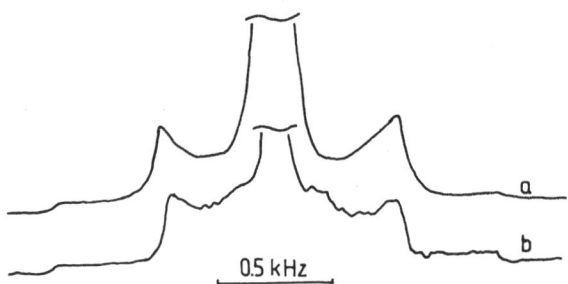

Fig. 2. ^{2}H NMR spectra of heavy water.
a) 100 mg TEPF in 1 ml PEG 20,000 heavy water solution (50 wt.% PEG)
b) 100 mg TEPF in a dialysis package in contact with a PEG 20,000 heavy water solution (50 wt.% PEG).

Within experimental error the quadrupole splitting of samples having direct contact with PEG water solutions is identical to samples which had contact with these solutions via a dialysis membrane. The quadrupole splitting of water increases with increasing PEG concentrations indicating that the hydration of lipids in the presence of PEG water solutions is limited. Each quadrupole splitting corresponds to a definite thickness of the water layer between bilayers which can be calculated if the area per lipid molecule is known. The areas per egg-PC and egg-PE molecule were taken from the paper of Lis et al., 1982. Possible changes in the area of pure lipids caused by mixing of PC and PE in the TEPF fraction were neglected.
From Fig. 2 it is seen that for membrane dehydration caused by the addition of PEG water solutions, a direct contact of PEG with lipids is not necessary. From visual and microscope observation we have additional evidence suggesting an immiscibility of PEG water solutions and lipid water dispersions. The penetration of PEG 20,000 molecules into the water layer

between lipid bilayers seems to be a very unlikely event.
In a second approach we investigated the influence of the molecular weight
of PEG on the quadrupole splitting of heavy water between lipid bilayers.
The PEG water solutions were added to the lipid water dispersions directly
(at least 1 ml). As seen in Fig. 3 the differences in $\Delta \nu_Q$-values for
comparable PEG concentrations in wt.% are small for PEG's with molecular
weights from 20,000 - 400. A significantly smaller quadrupole splitting was
observed for PEG 150. For PEG's of lower molecular weight we have to
consider the possiblity of a penetration of PEG into the space between
bilayers. Unfortunately the penetration of low moleular weight PEG into
this space cannot be prevented by a dialysis membrane. To answer the question whether low molecular weight PEG enter that space an analysis of
the forces acting between bilayers and of the osmotic properties of the
different PEG water solutions is necessary.

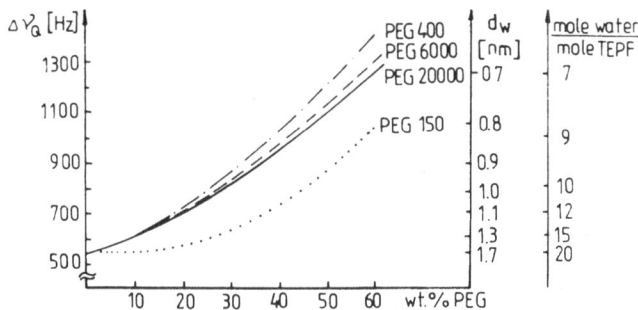

Fig. 3. Quadrupole splitting of heavy water between TEPF bilayers as a
function of PEG concentration for PEG's of different molecular
weights.

Osmotic properties of PEG water solutions

The osmotic proerties of PEG water solutions were investigated by different
authors with different methods. Unfortunately, there is a wide spread of
experimental results from different sources. For solutions of PEG in heavy
water we found no osmotic pressure data in the literature.
The activity of water a_w in the solutions was calculated by the water
vapour pressure difference between pure water and PEG water solutions. The
water activities can be well approximated by the formula

$$a_w = X_1 \exp(-K\, X_2^2) \tag{1}$$

where X_1 and X_2 are the mole fraction of water and PEG in the PEG water
solution, respectively. Chirife & Fontan (1980) demonstrated that equation
(1) follows from the Flory Huggins theory of polymeric solutions. The
osmotic pressure of the solutions can be calculated by

$$\Pi = \frac{R\,T}{V_w^\Theta} \ln a_w \tag{2}$$

13

R_o is the absolute gas constant, T the temperature in K and V_w^θ the molar volume of water. Within experimental error the osmotic pressures of PEG water
and PEG heavy water solutions are identical if the solutions contain the
same mole fraction of PEG.

Tab. 1. Experimentally determined parameters K for the calculation of
water activities in PEG water solutions (see equ. (1)).

molecular weight of PEG		K
20,000 - 17,000	(18,500)	72,500
7,500 - 6,000	(6,750)	10,800
420 - 380	(400)	28.5
150		2.8

Determination of force parameters for interactions between lipid bilayers

For PEG 20,000 water solutions membrane dehydration occurred also via a
dialysis membrane, indicating that this process is driven by osmotic effects.

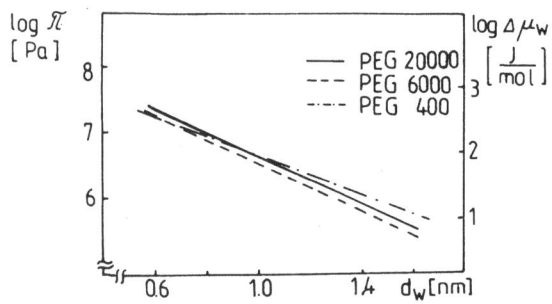

Fig. 4. Thickness of the water layer between TEPF bilayers as a function
of the osmotic pressure of PEG's with the molecular weights
20,000, 6,000 and 400.

An increase in the osmotic pressure of the solution corresponds to a decrease of the chemical potential of water in these solutions. The process of
bilayer dehydration is stopped if the chemical potential of water in the
middle between bilayers is equal to the chemical potential of water in the
surrounding PEG water solution. In thermodynamic equilibrium the repulsive
forces between bilayers are identical to the osmotic pressure of the PEG
water solution. As seen from Fig. 4 the repulsive forces decay exponenti-
ally

$$P_h = P_h^0 \exp(-d_w / \lambda)\qquad(3)$$

P_h - repulsive pressure
P_h^0 - repulsive pressure for $d_w = 0$

d_W- thickness of the water layer between lipid bilayers
λ - decay length of repulsive pressure

The magnitude of attractive van der Waals forces can be calculated by the formula (Verwey et al., 1948)

$$P_a = \frac{H}{6\,\pi} \left(\frac{1}{d_W^3} - \frac{2}{(\,d_W + d_1\,)^3} + \frac{1}{(\,d_W + 2\,d_1\,)^3} \right) \quad (4)$$

where H is the Hamaker constant and d_1 the thickness of the lipid bilayer. H is calculated from the equilibrium between attractive and repulsive forces.

Tab. 2. Parameters of forces between lipid bilayers

lipid	P_h (Pa)	λ (nm)	H (Nm)
EYL + PEG 20,000	$10^{9.0}$	0.20	$10^{-20.4}$
TEPF + PEG 20,000	$10^{8.4}$	0.24	$10^{-20.1}$
TEPF + PEG 20,000	$10^{8.4}$	0.23	$10^{-20.4}$
TEPF + PEG 400	$10^{8.2}$	0.28	$10^{-19.9}$

Down to molecular weights of 400 there is a rather good agreement with the results obtained with PEG 20,000 indicating that the PEG concentration in the water layer between bilayers is zero or negligible. PEG 150 has the highest osmotic pressure at a given concentration in wt.% but the increase in the quadrupole splitting after addition of PEG 150 water solutions is obviously smaller than for the other PEG's (see Fig. 3). This behaviour indicates that PEG 150 penetrates partly into the water layer between lipid bilayers. Nevertheless the PEG concentration in the water between bilayers is still reduced in comparison to the PEG concentration in the outer solution. As a result the bilayers are partly dehydrated.

DISCUSSION

From our investigations we conclude that cell aggregation observed after addition of PEG could be explained by the high osmotic pressure of PEG water solutions together with a reduced solubility of PEG molecules in the water layers near the membrane surface. The osmotic pressure of PEG water solutions of the same concentration in wt.% increases with decreasing molecular weight of PEG. On the other side the solubility restrictions for PEG molecules are stronger for higher molecular weight PEG. This is indicated by the measurable influx of PEG 150 into the water layers between lipid bilayers. The reduced solubility could be connected with a decrease of conformational entropy of PEG molecules near the membrane surface (Machor & van der Waals, 1952; Asakura & Oosawa, 1958; Hesselink et al., 1971; Vrij, 1976) which becomes more significant for molecules of higher molecular weight. The opposite dependence of osmotic pressure and solubility on the molecular weight of PEG could be the reason for the existence of an optimum molecular weight of PEG for cell fusion.
As shown by MacDonald (1985), dextran and sucrose solutions caused more extensive fusion of PC liposomes when applied indirectly by dialysis than when applied directly to the liposome dispersion. In contrast PEG fuse liposomes by acting on them directly and indirectly. This behaviour was explained by an exclusion of membranes from the PEG phase. Our results

confirm the conclusion that there is no direct interaction between phospholipids and PEG.

This result is further supported by measurements of the lamellar repeat spacing in PC water dispersions after addition of small amounts of polyethylene oxides (Arnold et al., 1986; Klose et al., 1986). It was observed that these surfactants are incorporated into the bilayer system with the ethylene oxide part extending into the water layer. After that incorporation the experimentally measured repeat spacing increased which was explained by the introduction of an additional repulsive force exerted by the polymer coated surface. From these experimental results we conclude that an interaction of PEG molecules with the membrane surface would generate additional repulsion between bilayers and not a reduction in repeat spacing as found in our investigations.

Within experimental error the force parameters of repulsive forces between phospholipid bilayers agree rather well with parameters of hydration repulsion determined for EYL bilayers (Lis et al., 1982). The hydration forces are connected with the structure of liquid water in the hydration layers near the membrane surface. The chemical potential of water molecules in the hydration layers is reduced in comparison to the chemical potential of free water. As a result there is a water influx into the space between lipid bilayers which acts as a repulsive force between the bilayers. The hydration repulsion for distances between bilayers smaller than 2 - 3 nm exceeds repulsive electrostatic forces and attractive van der Waals forces significantly. The hydration forces are the main barrier which has to be overcome to get membrane fusion (Rand & Parsegian, 1986). Because of osmotic pressures of PEG water solutions up to 20 MPa it is possible to dehydrate the space between bilayers and to get close contact between opposing bilayers. Changes in the hydration properties of lipids influence the repulsive hydration forces and therefore have significant influence on the approach of membranes.

The strong membrane dehydration caused by PEG influences intermolecular interactions between lipid molecules. This is indicated by the decrease in the area per lipid molecule (Lis et al., 1982) at lower hydration levels and the increase of order parameters and rotational correlation times of spin probes incorporated in model and biological membranes after addition of PEG (Surewicz, 1983; Boss, 1983; Herrmann et al., 1983; Arnold et al., 1986). Probably these changes are connected with local fusion events of opposing membranes which came into close contact after addition of PEG water solutions. This could be a so called trilamellar structure as observed in model experiments with black lipid membranes (Liberman & Nenashev, 1972). In the case of plant protoplasts and phospholipid liposomes the trilamellar structure ruptures spontaneously and a bigger cell is formed. In the case of erythrocytes the trilamellar structures seem to be more stable. After dilution of the PEG water solution the chemical potential of water in the surrounding solution increases. The water enters the space between aggregated cells and strong repulsive forces are generated. Cells without local fusion points disaggregate. Because of the increase of the inner cell pressure the trilamellar structure ruptures and fusion occurs. The real mechanism of fusion of the opposing membranes is still a matter of investigation. Because we have no indication for a direct presence of PEG at the points where fusion occurs we assume that PEG induced dehydration of membrane constituents plays a key role also in subsequent fusion steps.

References

Ahkong, Q.F., Fisher, D., Tampion, W., Lucy, J.A., 1975, Mechanisms of cell fusion, Nature, 253:194-195

Arnold, K., Löbel, E., Volke, F., Gawrisch, K., 1981, ^{31}P-NMR investigationsof phospholipids. III. Influence of water on the motion of the phosphate group, studia biophysica, 82:207-214

Arnold, K., Lvov, Y.M., Szögyi, M., Györgyi, S., 1986, Effect of poly-
  (ethylene oxide)-containing surfactants on membrane-membrane interaction, studia biophysica, 113:7-14
Arnold, K., Borin, M.L., Azizova, O.A., 1986, Influence of poly(ethylene
  glycol) on the partition of a charged spin probe. Colloid & Polymer Sci., 264:248-253
Asakura, S., Oosawa, F., 1958, Interaction between particles suspended in
  solutions of macromolecules, J. Polymer Sci., 33:183-192
Baran, A.A., Solomentseva, I.M., Mank, V.V., Kurilenko, D.D., 1972, On the
  role of solvation agent in the stabilization of disperse systems
  containing water soluble polymers, Dokl. Akad. Nauk U.S.S.R.,
  207:363-366
Boni, L.T., Stewart, T.P., Alderfer, J.L., Hui, S.W., 1981, Lipid-poly-
  ethylene glycol interactions: I. Induction of fusion between liposomes, J. Membr. Biol., 62:65-70
Boni, L.T., Stewart, T.P., Alderfer, J.L., Hui, S.W., 1981, Lipid-poly-
  ethylene glycol interactions: II. Formation of defects in bilayers, J. Membr. Biol., 62:71-77
Boni, L.T., Hah, J.S., Hui, S.W., Mukherjee, P., Ho, J.T., Jung, C.Y.,
  1984a, Aggregation and fusion of multilamellar vesicles by poly-
  (ethylene glycol), Biochim. Biophys. Acta, 775:409-418
Boni, L.T., Stewart, T.P., Hui, S.W., 1984b, Alterations in phospholipid
  polymorphism by polyethylene glycol, J. Membr. Biol., 80:91-104
Boss, W.F., 1983, Poly(ethylene glycol)-induced fusion of plant protoplasts. A spin label study, Biochim. Biophys. Acta, 730:111-118
Chirife, J., Fontan, C.F., 1980, A study of the water activity lowering
  behaviour of polyethylene glycols in the intermediate moisture
  range, J.Food Sci., 45:1717-1719
Cowley, A.C., Fuller, N., Rand, R.P., Parsegian, V.A., 1978, Measurement
  of repulsive forces between charged phospholipid bilayers, Biochemistry, 17:3163-3168
Cullis, P.R., Hope, M.J., 1978, Effects of fusiogenic agents on membrane
  structure of erythrocyte ghosts and the mechanism of membrane
  fusion, Nature, 271:672-674
Finer, E.G., Darke, A., 1974, Phospholipid hydration studied by deuteron
  magnetic resonance spectroscopy, Chem. Phys. Lipids, 12:1-16
Gawrisch, K., Richter, W., Möps, A., Balgavý, P., Arnold, K., Klose, G.,
  1985, The influece of water concentration on the structure of
  egg yolk phospholipid/water dispersions, studia biophysica,
  108:5-16
Herrmann, A., Pratsch, L., Arnold, K., Laßmann, G., 1983, Effect of poly-
  (ethylene glycol) on the polarity of aqueous solutions and on
  the structure of vesicle membranes, Biochim. Biophys. Acta,
  733:87-94
Hesselink, F.T., Vrij, A., Overbeek, J.T., 1971, On the theory of the stabilization of dispersions by adsorbed macromolecules. II. Interaction between two flat particles, J. Phys. Chem., 75:2094-2103
Hui, S.W., Stewart, T.P., Boni, L.T., Yeagle, P.L., 1981, Membrane fusion
  through point defects in bilayers, Science, 212:921-923
Hui, S.W., Isac, T., Boni, L.T., Sen, A., 1985, Action of polyethylene
  glycol on the fusion of human erythrocyte membranes, J. Membran
  Biol., 84:137-146
Klose, G., Brückner, S., Bezzabotnov, V.Yu., Borbely, S., Ostanevich, Yu.M.
  Hydration and swelling of the total lipid fraction of egg yolk
  and the influence of some additives studied by small-angle neutron scattering, Chem. Phys. Lipids, 41:293-307
Knutton S., 1979, Studies of membrane fusion. III. Fusion of erythrocytes with polyethylene glycol, J. Cell Sci., 36:61-72
Krähling, H., 1981a, Investigations on polyethylene glycol-induced cell
  fusion - freeze fracture observations, Acta Histochem. Suppl.,
  23:219-233

Krähling, H., 1981b, Discrimination of two fusogenic properties of aqueous polyethylene glycol solutions, Z. Naturforsch., 36c:593-596

Lebovka, N.I., Ovcarenko, F.D., Mank, V.V., 1983, Proton NMR investigations of hydration properties of polyethylene oxides, Dokl. Akad. Nauk U.S.S.R., 268:123-125

LeNeveu, D.M., Rand, R.P., Parsegian, V.A., 1976, Measurements of forces between lecithin bilayers, Nature, 259:601-603

LeNeveu, D.M., Rand, R.P., Parsegian, V.A., Gingell, D., 1977, Measurement and modification of forces between lecithin bilayers. Biophys. J., 18:209-230

Liberman, E.A., Nenashev, V.A., 1972, Fusion of bimolecular membranes as a model of cell contact, Biophysica U.S.S.R., 17:1017-1023

Lis, L.J., McAlister, M., Fuller, N., Rand, R.P., Parsegian, V.A., 1982, Interactions between neutral phospholipid bilayer membranes, Biophys. J., 37:657-666

MacDonald, R.I., 1985, Membrane fusion due to dehydration by polyethylene glycol, dextran, or sucrose, Biochemistry, 24:4058-4066

Machor, E.L., van der Waals, J.H., 1952, The statistics of the adsorption of rod-shaped molecules in connetion with the stability of certain colloidal dispersions, J. Colloid Sci., 7:535-550

Maul, G.G., Steplewski, Z., Weibl, J., Koprowski, H., 1976, Time sequence and morphological evaluations of cells fused by polyethylene glycol 6000, In Vitro, 12:787-796

Ohno, H., Sakai, T., Tsuchida, E., Honda, K., Sasakawa, S., 1982, Interaction of human erythrocyte ghosts or liposomes with PEG detected with fluorescence polarization, Biochem. Biophys. Res. Commun., 102:426-431

Rand, R.P., 1981, Interacting phospholipid bilayers: Measured forces and induced structural changes, Ann. Rev. Biophys. Bioeng., 10:277-314

Rand, R.P., Parsegian, V.A., 1986, Mimicry and mechanism in phospholipid models of membrane fusion, Ann. Rev. Physiol., 48:201-212

Robinson, J.M., Roos, D.S., Davidson, R.L., Karnovsky, M.J., 1979, Membrane alterations and other morphological features associated with polyethylene glycol-induced cell fusion, J. Cell Sci., 40:63-75

Seelig, J., 1978, Phosphorus-31 nuclear magnetic resonance and the polar group structure of phospholipids in membrane, Biochim. Biophys. Acta, 515:105-140

Surewicz, W.K., 1983, ESR study on the mechanism of PEG-membrane interaction, FEBS Lett., 151:228-232

Tilcock, C.P.S., Fisher, D., 1979, Interaction of phospholipid membranes with poly(ethylene glycol)s, Biochim. Biophys. Acta, 577:53-61

Tilcock, C.P.,S., Fisher, D., 1982, The interaction of phospholipid membranes with vesicle aggregation and lipid exchange, Biochim. Biophys. Acta, 688:645-652

Verwey, E.J.W., Overbeek, J.T., van Nes, K., 1948, "Thoery of the stability of lyophobic colloids," Elsevier Publ. Comp. Inc. N.Y., Amsterdam, London, Brussels

Vrij, A., 1976, Polymers at interfaces and the interactions in colloidal systems, Pure Appl. Chem., 48:471-483

Wilschut, J., Hoekstra, D., 1984, Membrane fusion: From liposomes to biological membranes, Trends in Biochem. Sci., 9:479-483

Wojcieszyn, J.W., Schlegel, R.A., Lumley-Sapanski, K., Jacobson, K.A., 1983, Studies on the mechanism of polyethylene glycol mediated cell fusion using fluorescent membrane and cytoplasmic probes, J. Cell Biol., 96:151-159

Zakai, N., Kulka, R.G., Loyter, A., 1977, Membrane ultrastructural changes during calcium phosphate induced fusion of human erythrocyte ghosts, Proc. Natl. Acad. Sci. U.S.A., 74:2417-2421

ELECTRICAL DOUBLE LAYERS IN MEMBRANE TRANSPORT

AND NERVE EXCITATION

Martin Blank

Dept. of Physiology and Cellular Biophysics
Columbia University, College of Physicians & Surgeons
630 West 168th Street, New York, NY 10032, USA

SURFACE PROCESSES IN MEMBRANE TRANSPORT

Because of the historical linkage of nerve excitation to the electrical stimulation of cells, electrophysiology has always appeared to offer the possibility of explaining a basic biological process in terms of electrochemical and physical properties. Thus, each new development in the physical chemistry of electrolyte solutions and membranes, has been used to contribute to a deeper insight into the properties of the biological system. Even the 200 year old controversy over Galvani's "animal electricity" was resolved when Volta was able to show parallels between the structures and functions of the natural electric organ and a man-made inorganic battery (1).

At the beginning of this century, Bernstein (2) used the newly emerging concepts of diffusion potential and selective permeability of membranes to propose that excitation was due to a transient reversible breakdown of the membrane permeability barrier. Electrical double layer theory, which had been formulated earlier by Helmholtz, started to develop rapidly soon after Bernstein's theory was published. For some reason it did not have a direct influence on membrane theory even though it indicated that membrane surfaces had different electrical potentials, ion concentrations, capacitance, etc. Thirty years later, Teorell (3) proposed that two surface potentials at the membrane/solution interfaces were needed along with the diffusion potential, to describe the membrane potential. However, even this development did not influence the main stream of electrophysiology, i.e., Goldman's (4) derivation of the steady state relation between membrane potential and the ionic mobilities and concentrations, and the Hodgkin-Huxley (5) formulation of ionic currents during an action potential. In recent years the situation has changed and there have been many applications of electrical double layer theory to steady state problems in membranes (6,7,8).

The critical role of electrical double layers in ion transport, especially during the rapid transients of excitation, can be seen from their effect on the ion concentration gradients across membranes. In figure 1 we see the sodium and potassium ion gradients across an uncharged squid axon membrane as would be expected from the bulk concentrations. Because of the surface charge (6), there are large changes in the ionic concentrations at the membrane surfaces, and the trans-membrane gradients change markedly. Since the outer face of the membrane is more highly charged, the sodium ion gradient increases while

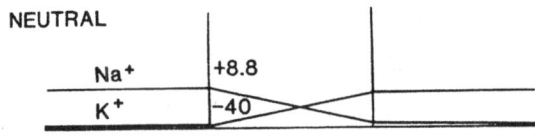

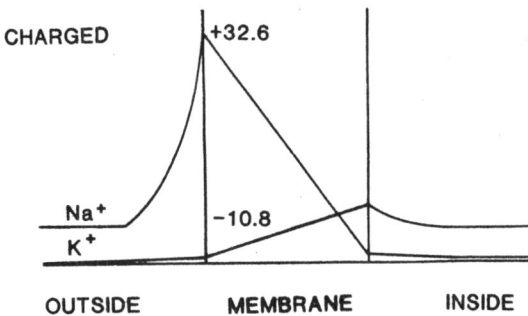

OUTSIDE   MEMBRANE   INSIDE

Figure 1. The sodium and potassium ion concentration profiles across a
squid axon membrane and the regions at the membrane/solution
interfaces are shown under two sets of conditions. The upper
diagram shows the profiles if the membrane surfaces were
uncharged, with the gradients across the membrane indicated.
In the lower diagram, the ion concentration profiles are
calculated taking into account the measured surface charge
densities (6). Both ion gradients across the membrane are
changed because of the surface charges, and the effect on the
sodium and potassium are in opposite directions because of the
charge asymmetry. If the charges on the membrane were
discharged instantaneously, as in a rapid depolarization, the
ions would diffuse into the solutions to reestablish the
gradients appropriate for an uncharged membrane. If channels
were to open as a result of the depolarization, part of the
transient ion flux would be through the channels. (Reproduced
with permission from Biochem. Biophys. Acta.)

the potassium gradient decreases. In the steady state each ion
is in equilibrium with ions in the bulk of the solution, because the
change in the concentration (i.e., chemical potential) is balanced by the
change in the electrical potential. However, the electrical force can be
changed rapidly by imposing an electric field, while the chemical force
adjusts by the much slower process of diffusion. For this reason,
unusual concentration gradients are generated from electrical double
layers when electrical depolarizations cause the steady state gradients
to relax.

    To study these ideas and determine the impact of electrical double
layer theory on membrane phenomena, we have used the simplest and
earliest model of the electrical double layer due to Helmholtz. The
build-up of counter ions and the depletion of co-ions has been described
by straight line gradients, and the entire electrical double layer region
has been treated as a compartment. These two simplifications form the

basis of the Surface Compartment Model (SCM), and channel function has been incorporated operationally as a voltage dependent permeability (9).

In this paper, we shall review the development of the SCM theory from first principles, and show how the system of equations responds to both voltage clamp and oscillating electric fields. The SCM theory can explain the ion flows in excitable membranes in terms of electrodiffusion, and can also account for the apparent selectivity of ion channels. The insights afforded by these ideas appear to be useful in understanding the effects of electromagnetic stimuli on living systems.

THE SURFACE COMPARTMENT MODEL

Theoreticians usually start their analysis of a problem by applying first principles (i.e. the fundamental laws of physics and chemistry) to a simplified model of the system of interest. In trying to elucidate the mechanisms of ion transport across biological membranes, the complexity of biological structure at the sub-microscopic level and the existence of specialized membrane proteins involved in catalysis (i.e., enzymes) and transport (i.e. channels) are often cited as obstacles to this approach. Although the properties of membrane proteins are not well understood, it has been possible to develop a more realistic description of membrane processes by using the SCM to take into account ionic processes at membrane surfaces.

The driving force for ion transport is the difference in electrochemical potential, $\Delta\mu e$, and the ion flux, J, can be described by a Nernst-Planck expression, where J is directly proportional to $\Delta\mu e$. Since the ion concentrations and electrical potentials in a surface can vary with time as a result of changes in polarization, we can calculate instantaneous values of J for each ion and use these values to determine the transient responses of the whole system. The set of non-linear, independent differential equations that we use for this purpose are given below, in FORTRAN notation, in terms of the SCM membrane system defined in figure 2. The SCM diagram applies to any charged surface through which there are ion fluxes, including the channel proteins of membranes.

Applying the principle of conservation of mass to the surface compartments, we can write six equations to describe the time variation of the ionic concentrations in the two surface compartments:

$$\dot{N}2 = (1/L2)*(JN1 - JN - PN - \dot{N}22) \tag{1}$$

$$\dot{K}2 = (1/L2)*(JK1 - JK - PK - \dot{K}22) \tag{2}$$

$$\dot{A}2 = (1/L2)*(JA1 - JA - PA) \tag{3}$$

$$\dot{N}3 = (1/L3)*(JN + PN - JN3 - \dot{N}33) \tag{4}$$

$$\dot{K}3 = (1/L3)*(JK + PK - JK3 - \dot{K}33) \tag{5}$$

$$\dot{A}3 = (1/L3)*(JA + PA - JA3). \tag{6}$$

Applying the principles of chemical kinetics to ion binding at the surfaces, we can write the following four equations to describe the changes in the bound cations at the two surfaces:

$$\dot{N}22 = BF * X2 * N2 - BR * N22 \tag{7}$$

$$\dot{K}22 = BF * X2 * K2 - BR * K22 \tag{8}$$

$$\dot{N}33 = BF * X3 * N3 - BR * N33 \tag{9}$$

21

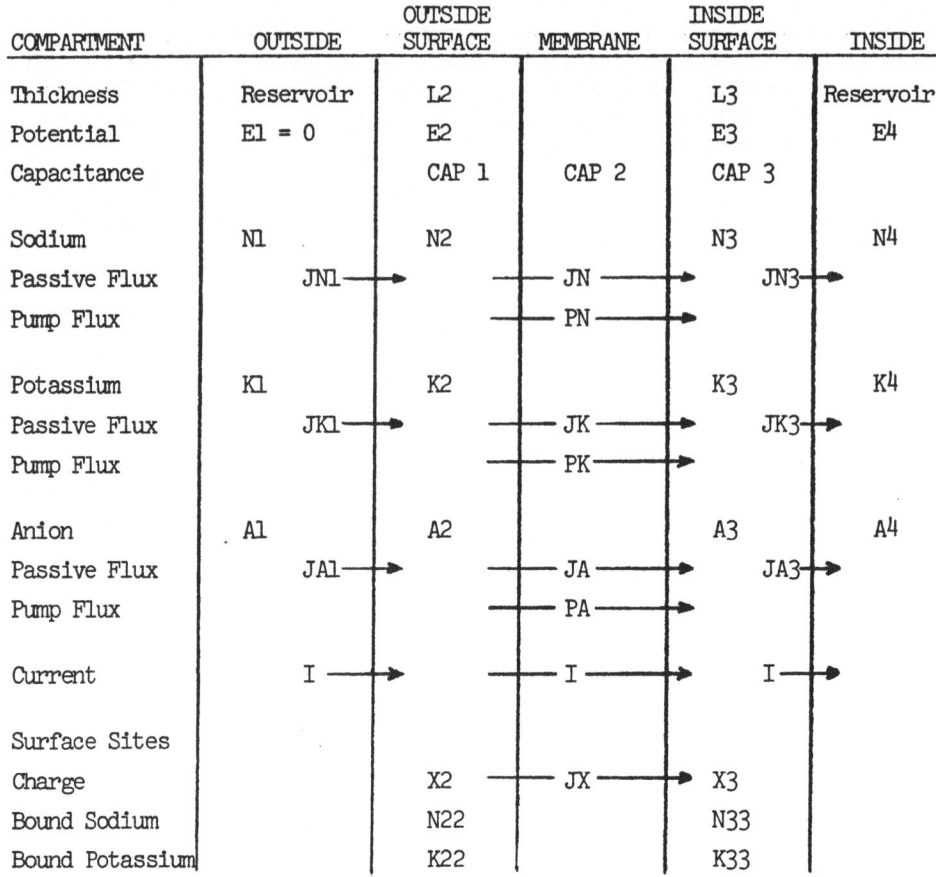

COMPARTMENT	OUTSIDE	OUTSIDE SURFACE	MEMBRANE	INSIDE SURFACE	INSIDE
Thickness	Reservoir	L2		L3	Reservoir
Potential	E1 = 0	E2		E3	E4
Capacitance		CAP 1	CAP 2	CAP 3	
Sodium	N1	N2		N3	N4
Passive Flux	JN1 →		— JN — →	JN3 →	
Pump Flux			— PN — →		
Potassium	K1	K2		K3	K4
Passive Flux	JK1 →		— JK — →	JK3 →	
Pump Flux			— PK — →		
Anion	A1	A2		A3	A4
Passive Flux	JA1 →		— JA — →	JA3 →	
Pump Flux			— PA — →		
Current	I →		— I — →	I →	
Surface Sites					
Charge		X2	— JX — →	X3	
Bound Sodium		N22		N33	
Bound Potassium		K22		K33	

Figure 2. A diagram of the Surface Compartment Model defining the various regions of the membrane system and the symbols used for the different physical properties in these regions. Fluxes between regions are shown as arrows pointing in the positive direction. (Reproduced with permission from Bioelectrochem. Bioenerg.)

$$\dot{K}33 = BF * X3 * K3 - BR * K33. \tag{10}$$

The ratio of the forward (BF) and reverse (BR) kinetic constants, BF/BR = BEQ, is the binding equilibrium constant.

Applying the principle of conservation of (negative) charge to the two sides of the membrane, we obtain:

$$\dot{X}2 = -(\dot{N}22 + \dot{K}22) - JX \tag{11}$$

$$\dot{X}3 = -(\dot{N}33 + \dot{K}33) + JX, \tag{12}$$

where the JX is the gating current associated with changes of polarization. In these equations, the gating charges are assumed to be negative.

Conserving charge during the flow current in the surface compartments and the membrane we obtain:

$$I = FA * (JN1 + JK1 - JA1) - CAP1 * \dot{E}2 \tag{13}$$

$$I = FA * (JN3 + JK3 - JA3) + CAP3 * (\dot{E}3 - \dot{E}4) \tag{14}$$

$$I = FA * (JN + JK - JA - JX + PN + PK - PA) +$$
$$CAP2 (\dot{E}2 - \dot{E}3), \qquad (15)$$

where FA is the Faraday and the three capacitances are assumed to be constant as a first approximation.

The conservation equations (eqs. 1-6 and 11-15) are derived from first principles, and the equations describing the kinetics of ion binding (eqs. 7-10) are based on chemical kinetics principles, with the assumption of a bimolecular forward rate of ion binding, and a monomolecular reverse rate of ion release. The ion fluxes are given by Nernst-Planck expressions, the fundamental equations of electrodiffusion, where the driving forces are the electrochemical potential differences. The gating currents (i.e., JX) that are assumed to arise from the redistribution of negative surface charges following changes in membrane polarization, are also calculated in that form. The channel opening mechanism, described elsewhere (10), is triggered by the gating currents.

The parameters of the model represent physical properties of the system, most of which can be found in the literature. For example, the conductances across the membrane, denoted GN, GK and GA for the ions and GX for the gating charge, are calculated from published values of resting fluxes for the ions and gating fluxes for the surface charge. Only the ionic mobility in the surface compartment and the rate constant for binding or release must be assumed. The conductance of an open channel is also fixed on the basis of measured currents.

ELECTRODIFFUSION DURING EXCITATION

The ionic flows during excitation in nerve have been well characterized (5), and they appear not to follow the equations of electrodiffusion. The squid axon, the prototype excitable cell, has two different types of channels with different kinetics that show an apparent specificity for sodium and potassium ions, respectively. The faster responding channel is selective for sodium ions and the slower one for potassium. The absolute selectivity is difficult to explain because other ions can pass through with ease (lithium in the sodium channel and thallium in the potassium channel), and there is no relation between the selectivity and the conductance of a channel as would be expected with a bottleneck type of structure.

The usual analysis of the ion flows in the squid axon is based on the simple barrier model of a membrane, with no consideration of surface processes. When the SCM is used to describe the ionic processes in nerve, it is possible to show that the unusual ion flows characteristic of electrical excitation phenomena arise from the electrodiffusion equations. A typical result showing the "voltage clamp" response of SCM voltage gated channels is given in figure 3. The ionic currents in both fast ("sodium") and slow ("potassium") channels can be described by electrodiffusion equations. Both cations pass through each channel, but because the ion diffusion rate and the rate of channel opening have different kinetic constants, the observed fluxes can be very different. A sodium channel is characterized by rapid gating currents and a potassium channel by slower gating currents, as observed in excitable membranes. It is possible that channel kinetics may be the basis of selectivity in other systems.

The success of the SCM in accounting for the ion flows during excitation has demonstrated the essential validity of the electrodiffusion description of processes at the membrane level. Both the ion flows during excitation and the apparent selectivity of different channels can be explained in terms of electrochemistry. We therefore have a useful membrane model for approaching the problem of transduction of externally applied electric and electromagnetic fields.

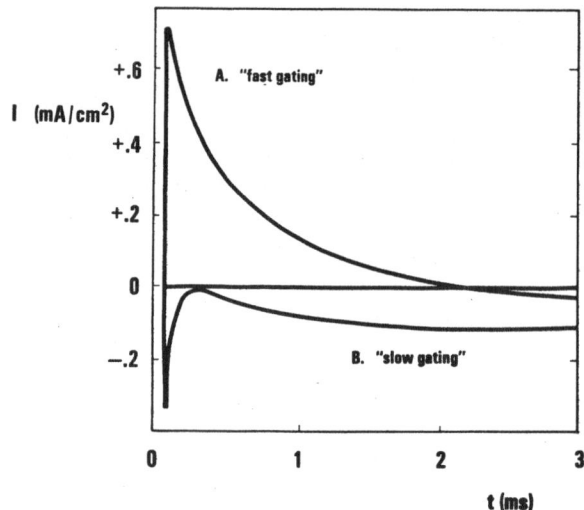

Figure 3. The ionic current I (in mA/cm^2) vs. the time t (in ms.) for different gating rates. Curve A for fast gating is 15 times faster than curve B for slow gating. The parameters of curves A and B are identical except for the one that controls the gating current. (Reproduced with permission from Bioelectrochem. Bioenerg.)

OSCILLATING ELECTRIC FIELDS

The excitation problem has shown how DC electric fields can cause charges to shift, channels to open and close, and ions to flow. Similar effects have been demonstrated with AC fields using the same theoretical membrane model. It is possible to show that oscillating electric fields lead to periodic changes in the electrical double layers that are functions of the frequency. All of the ionic concentrations change, but the relative change is greatest in those concentrations with the lowest steady state values. In particular, the sodium on the inner surface and the potassium on the outer surface show maximal effects at 100-200 Hz (11). The two ionic concentrations normally control the activity of the ion pump enzyme, the Na-K ATPase of cell membranes, and the calculated increases at particular frequencies would stimulate the enzyme and result in enhanced ion transport. A similar frequency dependence has been observed in the ouabain sensitive accumulation of rubidium by erythrocytes (12). Recent calculations show an optimal amplitude for ion accumulation, also in line with observations on erythrocytes. It should be noted that the same mechanism that leads to low frequency optima in terms of ion transport processes between electrical double layer regions can also account for the additive effects of very low amplitude signals over time.

Since the largest changes occur in those cations with the lowest steady state concentrations, one would expect relatively large changes in both protons and calcium ions. Both ions are known to affect membrane (and cellular) properties as a result of very small changes in ambient concentrations. This idea opens up possibilities for linking the physical-chemical model for ion fluxes with biochemical processes associated with the Na-K ATPase and various calcium stimulated protein kinases. It appears likely that the ion concentration changes stimulated by externally imposed electric fields could be acting as second messengers in the stimulation of biochemical reactions.

SUMMARY

     The Surface Compartment Model (SCM) emphasizes the role of electrical double layers in ion transport. When the SCM is applied to the membrane of an excitable cell, the calculated voltage clamp currents are similar to those observed in the sodium and potassium channels of excitable membranes. The difference in the selectivity of the two types of ion channels appears to be determined by the difference in gating current, suggesting a kinetic basis for the ion selectivity of voltage gated channels. Oscillating voltages applied across the SCM cause periodic changes in ion concentrations with frequency optima in the range where biological effects of alternating currents have been demonstrated. The calculated changes in internal sodium and external potassium ion concentrations appear able to account for observed ion accumulation involving stimulation of the Na-K ATPase.
     Generally, the analysis of natural membranes as physical-chemical systems is meant to establish the non-biological framework for a consideration of the biological processes associated with cell function. However, as we learn more about the physical chemistry of ion flow across membranes, especially the role of surface processes, we appear to explain many of the biological properties as well.

REFERENCES

1. C.H. Wu, Electric fish and the discovery of animal electricity. Amer. Sci. 72:598-607 (1984).
2. J. Bernstein, Untersuchungen zur thermodynamik der bioelectrischen strome. Pflugers Arch. 92:521-562 (1902).
3. T. Teorell, An attempt to formulate a quantitative theory of membrane permeability. Proc. Soc. Exp. Biol. Med. 33:282-285 (1935).
4. D.E. Goldman, Potential, impedance and rectification in membranes. J. Gen. Physiol. 27:37-60 (1943).
5. A.L. Hodgkin and A.F. Huxley, A quantitative description of membrane current and its application to conduction and excitation in nerve. J. Physiol. (London) 117:500-544 (1952).
6. D.L. Gilbert, Fixed surface charges, in: "Biophysics and Physiology of Excitable Membranes", Adelman, W.J., ed., Van Nostrand Reinhold, New York (1971).
7. S. McLaughlin, Electrostatic potentials at membrane-solution Interfaces. Curr. Top. Membr. Transp. 9:71-144 (1977).
8. M. Blank, ed. "Electrical Double Layers in Biology", Plenum, New York (1986).
9. M. Blank, The surface compartment model: a theory of ion transport focused on ionic processes in the electrical double layers at membrane protein surfaces. Biochem. Biophys. Acta 906 (2) in press (1987).
10. M. Blank, The surface compartment model (SCM) - Role of surface charge in membrane permeability changes. Bioelectrochem. Bioenerg. 9:615-624 (1982).
11. M. Blank, Theory of frequency-dependent ion concentration changes in oscillating electric fields. J. Electrochem. Soc. 134:1112-1117 (1987).

AN ALTERNATIVE ELECTROSTATIC MODEL FOR MEMBRANE ION CHANNELS

D.T. Edmonds

The Clarendon Laboratory
Parks Road
Oxford OX1 3PU, U.K.

INTRODUCTION

There are at least two reasons for making mathematical models of real biological systems. The first is to accurately reproduce the output of the given system in order to include it as part of a model of a larger system. The second is to construct models with real physical components capable of independent experimental verification in the hope of gaining a better understanding of the physical processes that underlie the operation of the system. The famous Hodgkin and Huxley m^3 model (Hodgkin and Huxley 1952) of ion channel gating in excitable tissue is an example of the first kind. They wished to simulate the time variation of the action potential and needed a model for the time dependence of the sodium and potassium channel currents. Although later researchers attempted to put flesh on the mathematical model by trying to find the m and n particles as real physical entities, the original authors were careful to state that agreement between the predictions of the model and experiment was no guarantee that the particular model proposed was correct. It is well known that many different models may produce the same output just as a tape recorder and a flute can produce the same audible output while having nothing in common in their methods of sound production.

Since that time the trend is for even more abstract compartmental models, often consisting of a long string of abstract and physically undefined states connected by arbitrarily assigned rate constants. A kinetic model consisting of a cascade of 6 such states may well reproduce the complicated kinetics of sodium channel gating but it has little to say about the underlying mechanism, except that to some approximation the output may be represented by 5 linear first order differential equations.

Here I outline a model of the second kind with its components described in terms of known physical systems interacting in a calculable and physically plausible manner. A central assumption of the model is that the trans-membrane gating charge transfer that is measured (Keynes 1983) when a channel is in the process of opening is a necessary but preliminary step in the gating of a channel. The actual opening of the channel is assumed to be a subsequent and distinct step which does not involve any transfer of charge across the membrane and is thus electrically silent as far as the measurement of gating charge transfer is concerned. Evidence for such a separation is obtained by two sets of experiments. When deuterated squid axons are studied, the rise of the sodium current is delayed in time by a factor of about 1.4 in comparison with normal protonated preparations (Conti and Palmieri 1968) but the time dependence of the gating charge transfer is not changed (Meves 1974, Schauf and Bullock 1979). Secondly in high pressure experiments the activation volume associated with channel opening (Conti et al. 1982) is about twice that to be associated with the gating charge transfer (Conti et al. 1984). Both these type of experiments are consistent with an electrically silent final step in channel gating that can be modified independently of the gating charge transfer.

A second assumption of the model is that the response time of the gate itself is very short in comparison with that of the preliminary gating charge transfer. As we shall see this leads to the most powerful prediction of this model which is that the kinetics of gating are fully determined by the measured equilibrium and static characteristics of the channel together with the kinetics of the gating charge transfer. The predictions of this model will be compared here with those measured for the sodium channel in the squid giant axon, in the frog Node of Ranvier and in the frog node treated with Batrachotoxin (BTX) which are known to have very different sodium conductance kinetics.

THE SILENT GATE MODEL

The model was first described (Edmonds 1983) as a much simplified system with only four states in which a single gating particle with only two states represents the measured gating charge transfer. Subsequently the model has been much refined (Edmonds 1987a, 1987b) so that the gating charge is assumed to transfer continuously, representing the cummulative effects of many microscopic dipole reversals or charge migrations in the vicinity of the channel. I have also attempted previously to show (Edmonds 1985) by simplified but quantitative calculations that at least one physical realization of the model, which employs the known large electric dipole moment of the alpha-helix, is physically realistic.

The model has two components called the Q and N systems. The Q system represents an assembly of charges or electric dipoles that respond to the voltage applied across the membrane and which would form part of the experimentally measured gating charge transfer detected when excitable channels are switched. In Fig. 1a this is represented by a collection of charged particles capable of moving across the membrane. More realistically in Figs. 1b and 1c it is represented by a polarizable cylinder spanning the membrane composed of the rotatable dipoles and mobile charges referred to above. The N-system consists of a charged group in Fig. 1a or a neighbouring alpha-helical rod of the channel protein in Fig. 1b or a pair of oppositely directed helices in Fig. 1c. If the inside of the membrane is made more positive then, in Fig. 1a, charge will tend to migrate across the membrane and in cases b and c the electric dipole moment of the Q-system will tend to increase in an outward pointing direction. This in turn will increase the repulsive force acting on the charged group or nearest helix dipole. In each case the motion assumed for the N-system, a sideways motion of the charged group or helix in Figs. 1 a and b or a rotation of the pair of helices about their common axis in Fig. 1c, signals the opening of the channel. Because this motion does not change the electrostatic energy of the charged group or the helices in the transmembrane electric field the movement gives rise to no gating charge and is thus electrically silent. The same mathematics can represent any one of the physical realizations shown in Fig. 1 but for definiteness I will describe in detail the polarized cylinder and alpha-helix shown in Fig. 1b. For simplicity I will also assume that there is only one alpha-helix and that it can exist in only two states represented by n=0 as the ground state which represents the channel

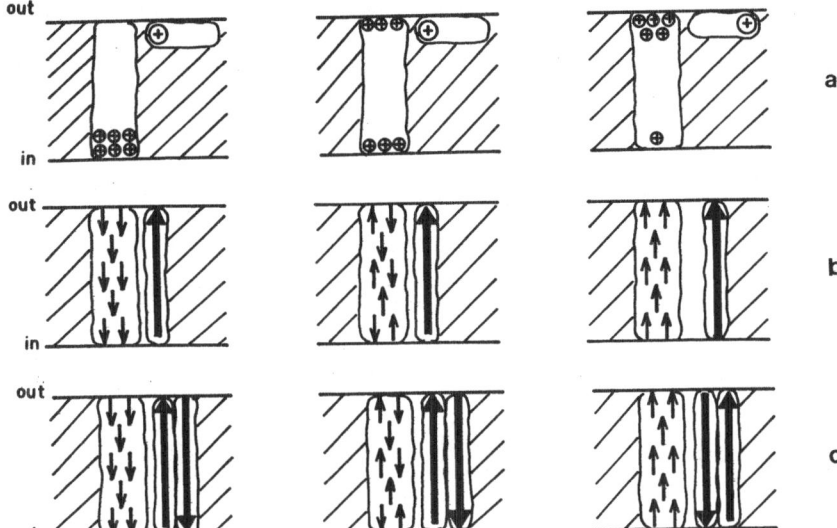

Fig. 1. Possible physical realizations of the silent gate model.
In a the Q-system consists of discrete charges that move
across the membrane and the N-system consists of a
neighbouring charged group. In b and c the Q-system is
represented more generally by a polarizable cylinder and
the N-system is either a single alpha-helix with its
large electric dipole moment in b or a pair of helices
that may rotate in c. In each figure the lefthand
diagram shows the resting state with the inner side of
the membrane negative. The righthand diagram shows the
situation after enough charge transfer has occurred to
ensure gating and the centre diagram shows an intermediate
situation when some charge transfer has occurred but not
sufficient to initiate opening.

closed and n=1 which represents the channel open and hence conducting ions
across the membrane. Unlike the Hodgkin and Huxley model the number of
identical N-systems assumed does not change (Edmonds 1987b) the kinetics
predicted by the model.

The basic operation of the model may be seen from the following
illustration. In the resting state with the inside of the membrane negative,
the core (the Q-system) is polarized inward and its dipole moment thus
attracts the dipole moment of the N-system helix tending to hold the channel
closed. If the inside is now suddenly made positive the core will gradually
reverse its polarization with some characteristic time constant until its
outward pointing electric dipole moment is sufficiently large to repel the
N-system dipole and change it from its resting n=0 state to its open n=1
state. It is thus clear that the influence of the membrane voltage on the
N-system is indirect, it does not directly change the energy of the N-system
but it does alter the dipole moment of the Q-system which in turn interacts electrostatically with the N-system causing the channel to open. As
mentioned before, calculations of the interaction energies involved, which
include an attempt to allow for the strong screening of the electric dipole
moment of the alpha-helix by the high dielectric constant fluid which
bathes the membrane, yield values several times KT, the prevailing thermal
energy and enough to ensure the desired effect (Edmonds 1985).

Channel opening is seen to be a threshold phenomenon in which a critical amount of reversal of the Q-system dipole moment must occur before the channel will open. This ensures that much of the gating charge flows prior to the channel opening and that the ion flow through the channel is S-shaped with time following a sudden depolarization of the membrane.

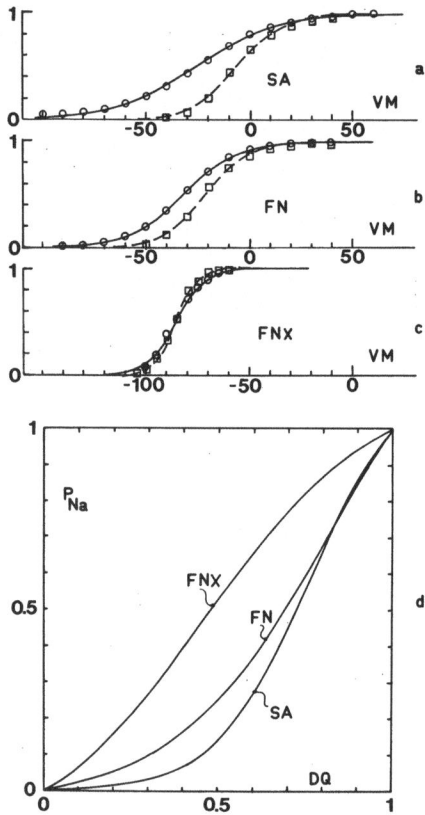

Fig. 2. a, b and c show the fit of Equations (1) and (2) to the experimentally determined equilibrium gating charge transfer (circles) and the probability of channel opening (squares) for the squid axon (SA), the frog node (FN) and the frog node treated with BTX (FNX) respectively. In Figure 2 d is shown the curves obtained by plotting the predicted normalized probability of channel opening P(n=1) directly against the predicted normalized gating charge transfer q.

Furthermore because the dipole dipole interaction energy falls off rapidly with distance (approximately like R^4) in a complex protein the interaction of the nearest alpha-helix to the channel will dominate.

Another feature of this model is that it is the cummulative electrostatic effect of the change in the dipole moment of the Q-system that increases the probability of an N-system transfer. This change can be due to the rotation of water molecules, the transfer of hydrophobic ions cross the membrane or the motion of charged groups attached to the membrane spanning proteins and there is no need for any of these movements to be associated with specific molecular groups to ensure that the channel will open.

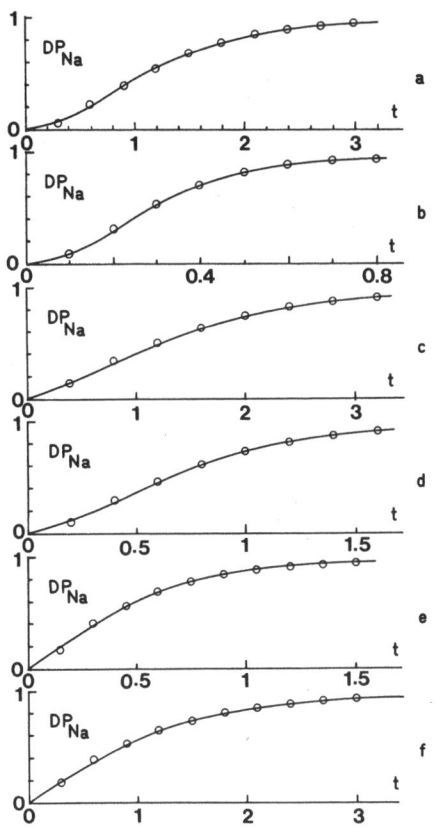

Fig. 3. A comparison between the predictions of the silent gate model (lines) and experiment (circles) for the kinetics of sodium channel activation in the squid axon (a) and (b), the frog node (c) and (d) and the frog node treated with BTX (e) and (f). The details of the comparison are given in the text.

To obtain quantitative results I will assume that the membrane voltage dependent charge transfer in the Q-system is represented by the continuous variable q so that q=0 represents the state with the inside of the membrane very negative and q=1 represents the saturated transfer that occurs for very positive trans-membrane voltage differences. I will further assume that the characteristic of equilibrium gating charge transfer as a function of applied membrane voltage VM has the form

$$q(VM) = 1/\{1+\exp[-ZQ(VM - VQ)/VT]\} \qquad (1)$$

where $VT = KT/|e| = 25$ mV with K is the Boltzmann constant and T the absolute temperature. The constants ZQ and VQ give the "effective valence" of the transferred charge and the membrane voltage for which q(VM) = 0.5.

In Fig. 2 a is shown as circles the experimentally measured gating charge transfer for the squid giant axon (SA) as a function of membrane voltage (Keynes 1983). The full line through these points is the prediction of Equation (1) with ZQ = 1.33 and VQ = -25 mV. Similarly in Figs. 2 b and c are shown as circles the experimental results for the frog node (FN) (Dubois and Schneider 1982) and for the frog node treated with BTX (FNX) (Dubois, Schneider and Khodorov 1983) with the full line representing Equation (1) with ZQ and VN given by 1.93 and -32 mV and by 3.82 and -85.6 mV. respectively. It can be seen that Equation (1) does represent the shape of these equilibrium characteristics well. I am most grateful to Profs. Keynes and Dubois for providing me with these data in numerical form.

For simplicity I will assume that the N-system has only two states, shut represented by n = 0 and open represented by n = 1. The excess in energy of the state n = 1 over the state n = 0 may be written as DU(q) in

$$DU(q) = |e| (VN - q*VNQ)$$

so that $DU(q=0) = |e|*VN$ and $DU(q=1) = |e|*(VN-VNQ)$, and VN and VNQ are constants which are expressed in mV. As q changes from 0 to 1, DU(q) change from DU(0) to DU(1) so that $|e|*VN$ may be identified with the interaction energy of the N-system with its environment excluding the Q-system and the term $q*|e|*VNQ$ may be identified with the energy of interaction between the N and Q system in a state defined by q. The normalized (0 to 1) Boltzmann probability that the N-system will be in the state represented by n=1 is then given by P(n=1) in

$$P(n=1) = [N(q) - N(q=0)]/[N(q=1) - N(q=0)], \qquad (2)$$

where

$$N(q) = 1/\{1+\exp[DU(q)/KT]\} = 1/\{1+\exp[(VN-q*VNQ)/VT]\}.$$

In Figs. 2 a, b and c are plotted as squares the experimentally measured equilibrium normalized probabilities that a given sodium channel will be open as a function of membrane voltage, corrected where appropriate for effects of inactivation, obtained for the squid axon (Keynes 1983), the frog node (Bergman and Meves 1981) and the frog node treated with BTX (Dubois, Schneider and Kodorov 1983). The frog node results have been corrected for the marked curvature of the "instantaneous" current v. voltage characteristic of this preparation by the authors concerned. Once again I am indebted to Profs. Keynes, Dubois and Meves for sending me these data. The dotted lines through the experimental points represents the best fit of Equation (2) to these data and results in sets of constants VN and VNQ for the three preparations of 145.4 mV. and 192.4 mV. , 91.84 mV. and 111.56 mV. and 49.85 mV. and 106.9 mV. respectively.

Fitting Equations (1) and (2) to the equilibrium characteristics measured experimentally has determined all of the four constants ZQ, VQ, VN and VNQ contained in the model so that there remain no adjustable constants. The kinetic predictions of the model for the time variation of channel opening is given by $P(n=1)$ in Equation (2) with the time variation provided by the time variation of q. A graphic way of understanding this point is to plot for each different preparation the normalized equilibrium probability of the channel conducting shown as broken lines in Figs. 2 a, b and c, directly against the normalized gating charge transfer shown in these same figures as full lines. This has been done in Fig. 2 d. and shows that each preparation is represented by a curved line. Because of the assumption of the model that the kinetics of the N-system are very fast in comparison with those of the Q-system the relationships shown in Fig. 2 d, which were derived from the static equilibrium characteristics, will also be obeyed kinetically as the N-system will always be in equilibrium with the instantaneous state of the Q-system. Thus knowing the kinetics of the gating charge transfer tells us about the kinetics of moving along the horizontal axis of Fig. 2 d and the model predicted kinetics for the channel opening probability is simply obtained by reflecting the variation of DQ (identical to q) in the appropriate curve and reading the resultant time variation from the vertical axis.

To illustrate this I will assume that the time variation of q is simply given by

$$q(t) = q_i + (q_f - q_i)*[1-\exp(-t/TQ)] \tag{3}$$

where q_i and q_f are the initial and final values of the total normalized gating charge transfer and

$$TQ = 2*TMAX/\{\exp[ZQ*(VM-VQ)*f/VT]+\exp[-ZQ*(VM-VQ)*(1-f)/VT]\} \tag{4}$$

where TMAX is the maximum value of the relaxation time TQ which occurs at a membrane voltage VM=VQ and f is a constant $0 < f < 1$. Equations (1), (3) and (4) represent what is expected (Adrian 1978) if a number of charges of valence ZQ move independently between two states of energy

$$EQ(0) = ZQ*(VM-VQ)*f \quad \text{and} \quad EQ(1) = -ZQ*(VM-VQ)*(1-f)$$

over an energy barrier of height B such that

$$2*TMAX = 1/[R*\exp(-B/KT)]$$

where R is the universal rate constant. I make no such detailed assumptions here but merely use Equations (3) and (4) as a convenient empirical representation of the experimental results. In fact it is an accurate description of the kinetics of gating charge motion for the frog node (Meves and Rubly 1986) and for the frog node treated with BTX (Dubois and Schneider 1985). It is only an approximation for the squid axon where the gating charge is known to be heterogeneous with at least two components (Khodorov 1979, Peganov 1979, Greef et al. 1982).

The experimental data for the growth of the sodium current with time are usually presented assuming a Hodgkin and Huxley (1952) model with a time dependent parameter m(t) given by

$$m(t) = m_i + (m_f - m_i)*\{1-\exp[-(t-d)/TM]\},$$

where m_i and m_f are the initial and final values of m and TM is a relaxation time. The change in sodium channel permeability is then given by DP_{Na} in

$$DP_{Na} = m(t)^X,$$

where X is some integer depending on the preparation. The time delay d, often called the Cole and Moore delay (Cole and Moore 1960) has been found necessary in order that such a model fit the experimental results obtained. For the squid axon such a model has been found to give good agreement with experiment with X approximately 3 (Keynes and Kimura 1983). In Fig. 3 a and b the predictions of such a Hodgkin and Huxley model are given by the open circles for sudden excursions of VM from a resting value of −50 mV. to 0 and +50mV. respectively. The characteristic markedly sigmoid curve is seen. The full line in Figs. 3 a and b are the predictions of Equation (2) with the time dependence of q(t) given by Equations (3) and (4) using the values of ZQ, VQ, VN and VNQ determined from the static characteristics with f = 0.5. The agreement is seen to be excellent. In a previous paper (Edmonds 1987a) the comparison of the predictions of the silent gate model and the experiments for the squid axon are more fully described including a comparison between the values of TM and d obtained under various conditions when the experiments on the one hand and the predictions of the silent gate model on the other hand are interpreted with a Hodgkin and Huxley model with X=3.

In Figs. 3 c and d are shown as circles the experimental shape of the sodium current switch-on for the frog node for excursions of VM from a resting value of −100 mV. to values of −30 mV. and 0 mV. , which in this case is found to fit with good accuracy a Hodgkin and Huxley model with X = 2 (Meves and Rubly 1986). The curves are seen to be less sigmoid as compared with those of the squid axon. The full lines are the predictions of the silent gate model under the same conditions and once again the agreement is seen to be excellent. Finally in Figs. 3 e and f are shown as circles the experimentally observed time dependent switch-on for the frog node treated with BTX for excursions of VM from a resting value of −120 mV. to values of −70 mV. and −90 mV. in which a simple exponential behaviour, corresponding to X = 1 is found (Dubois and Schneider 1985). Once more the silent gate model prediction, shown as a full line is able to predict the time dependence using the parameters determined by the static characteristics. A fuller comparison between the model and experiment in this case is given in a recent paper (Edmonds 1987b). In all these comparisons the times and thepredicted relaxation times and delay times are scaled by taking TMAX=1 in Equation (4).

Although not displayed here the silent gate model is equally accurate at reproducing the time dependence of the sodium current during switch-off. For example in the squid axon although the switch-on is markedly sigmoid the switch-off is much less so (Keynes and Kimura 1983). This is reproduced by the present model essentially because the critical value of q that leads to channel gating is much closer to a value 1 than to a value 0. This means that the latency period for switch-on from a resting value of q near 0 is much longer than that for switch-off from a resting value of q near 1.

CONCLUSIONS

Using only the experimentally measured equilibrium characteristics and assuming simple exponential gating charge transfer the silent gate model is seen to be capable of reproducing the very different kinetic behaviour of the sodium channel permeability in the squid axon, the frog node and the frog node treated with BTX. Besides these predictions for the collective behaviour of many channels the model may be shown to be in accord with the measured stochastic behaviour of individual single channels (Edmonds 1987a).

REFERENCES

Adrian, R.H., 1978, Charge movement in the membrane of striated muscle, Annu. Rev. Physiol. Bioeng, 7:85.

Bergman, C. and Meves, H., 1981, The excitable membrane of nerve fibres. in: "Membranes and intercelluar communication," Balian, R., Chabre, M. and Devaux, P.F., eds. North Holland, Amsterdam.

Cole, K.S. and Moore, J.W. 1960, Potassium ion current in the squid giant axon:dynamic characteristic, Biophysics J., 1:1.

Conti, F. and Palmieri, G. 1968, Nerve fibre behaviour in heavy water under voltage clamp, Biophysik, 5:71.

Conti, F., Fioravanti, R., Segal, J.R. and Stuhmer, W., 1982, J. Membr. Biol. 69:35.

Conti, F., Inoue, I., Kukita, F. and Stuhmer, W., 1984, Pressure dependence of sodium gating currents in the squid giant axon, Eur. Biophys. J., 11:137.

Dubois, J.M. and Schneider, M.F., 1982, Kinetics of intramembrane charge movement and sodium current in the frog Node of Ranvier, J. Gen. Physiol., 79:571.

Dubois, J.M., Schneider, M.F. and Khodorov, B.I., 1983, Voltage dependence of intramembrane charge movement and conductance activation of Batrachotoxin-modified sodium channels in frog Node of Ranvier, J. Gen. Physiol., 81:829.

Dubois, J.M. and Schneider, M.F., 1985, Kinetics of intramembrane charge movement and conductance activation of Batrachotoxin-modified sodium channels in frog Node of Ranvier, J. Gen. Physiol., 86:381.

Edmonds, D.T., 1983, A model of sodium channel inactivation based upon the modulated blocker, Proc. R. Soc. Lond. B219:423.

Edmonds, D.T., 1985, The alpha-helix dipole in membranes: a new gating mechanism for ion channels. Eur. Biophys. J. 13:31.

Edmonds, D.T., 1987a, A physical model of sodium channel gating, Eur. Biophys. J., 14:195.

Edmonds, D.T., 1987b, A comparison of sodium channel kinetics in the squid axon, the frog node and the frog node with BTX using the silent gate model, Eur. Biophys. J., in press.

Greef, N.G., Keynes, R.D. and Van Helden, D.F., 1982, Fractionation of the asymmetry current in the squid giant axon into inactivating and non-inactivating components, Proc. R. Soc. Lond., B215:375.

Hodgkin, A.L. and Huxley, A.F., 1952, A quantitative description of membrane current and its application to conduction and excitation in nerve, J. Physiol. Lond. 117:500.

Keynes, R.D., 1983, Voltage gated ion channels in the nerve membrane, Proc. R. Soc. Lond. B220:1.

Keynes, R.D. and Kimura, J.E., 1983, Kinetics of the sodium conductance in the squid giant axon, J. Physiol., 336:621.

Khodorov, B.I., 1979, Inactivation of the sodium gating current, Neuroscience, 4:865.

Meves, H., 1974, The effect of holding potential on the asymmetry currents in squid giant axons, J. Physiol., 243:847.

Meves, H. and Rubly, N., 1986, Kinetics of sodium current and gating current in the frog nerve fibre, Pflugers Arch. 407:18.

Peganov, E. 1979, A study of the inactivating component of the asymmetrical current in the frog nerve fibre. Neuroscience 4:539.

Shauf, C.L. and Bullock, J.O., 1979, Modifications of sodium channel gating in Myxicola giant axons by deuterium oxide, temperature and internal cations, Biophys. J. 27:193.

ELECTROORIENTATION STUDIES ON ELECTROPHYSICAL PROPERTIES OF ERYTHROCYTES IN RADIO-FREQUENCY RANGE OF ALTERNATING ELECTRIC FIELD

A.I.Miroshnikov, A.Yu.Ivanov, V.M.Fomchenkov, and A.N.Shirokova

Institute of Biological Physics, Acad.Sci.USSR
Pushchino, Moscow Region, USSR, 142292

INTRODUCTION

In recent years electroorientation methods based on optical registration of the orientation effect produced by alternating electric field /1,2/ and various electrophysical models /2,3,4,5/ have been intensively used to study electrophysical properties of cells, in particular permittivity (P), electroconductivity (EC) and polarizability.

The purpose of the present paper was to determine quantitative characterization of permittivity and electroconductivity of fixed erythrocytes by comparing theoretical and experimental electroorientation spectra (EOS). On the basis of the data obtained we attempted to explain the phenomenon of the erythrocyte orientation perpendicular with the lines of the field applied.

THEORETICAL SECTION

The orientation of nonspherical cells under uniform alternating electric field is determined by the interaction of the field applied with the electric dipole moment arising from the polarization of the object in the external field. The torque, M, leading to a rotation of the ellipsoidal particle with semiaxes a,b and c is given in Eqn.1 /2,6/:

$$\bar{M} = Re\,(\gamma_{\shortparallel} - \gamma_{\perp})\,E_0^2\,Sin\,2\theta \qquad (1)$$

where θ is the angle between the long axis of the particle (here it is oriented along a or b) and field lines (E is the field strength); $\gamma_{\shortparallel}$ and $\gamma_{\perp}$ are the longitudinal and transverse polarizabilities, respectively. For the cells of a given shape

$$\gamma_{\parallel} = \frac{\pi}{3} a^2 c \, \mathcal{E}_0 \, \frac{\mathsf{S}^*_{md} \, (\mathsf{S} - \mathsf{S}_{md})}{\mathsf{S}_{md} + n_x \, (\mathsf{S} - \mathsf{S}_{md})} \qquad (2)$$

$$\gamma_{\perp} = \frac{\pi}{3} a^2 c \, \mathcal{E}_0 \, \frac{\mathsf{S}^*_{md} \, (\mathsf{S} - \mathsf{S}_{md})}{\mathsf{S}_{md} + n_z \, (\mathsf{S} - \mathsf{S}_{md})} \qquad (3)$$

where n_x and n_z are the ellipsoid depolarization coefficients which characterize the extent of attenuation of the
external electric field inside of the ellipsoid /7/.

$$n_z = \frac{1-k^2}{k^3} \, (k - arctg \, k); \qquad k = \sqrt{\left(\frac{a}{c}\right)^2 - 1}$$

$$n_x = 1/2 \, (1 - n_z)$$

S-are the complex dielectric factors of the external medium
and cell interior /8/. For the medium $\mathsf{S}_{md} = \mathcal{E}_{md} - j\sigma_{md}/2\pi f \mathcal{E}_0$,
where $\mathcal{E}_{md}$ and σ_{md} are the permittivity and electroconduc-
tivity, respectively; f is the field frequency; $\mathcal{E}_0$ is the
permittivity of free space. For cell interior $\mathsf{S} = \mathcal{E}_e - j\sigma_e/2\pi f \mathcal{E}_0$

, where $\mathcal{E}_e$ and σ_e are the effective P and EC. When estimating the torque, the electroosmo-
tic slippage /1/ was neglected because of the insignifican-
ce of erythrocyte zeta-potential.

Effective P and EC were assumed to be frequency-dependent with regard to the main relaxation mechanisms of
electric polarization of cells and their α and β -disper-
sions /3,9/

$$\mathcal{E}_e^{\parallel,\perp} = \mathcal{E}_h + \frac{\mathcal{E}_s^{\parallel,\perp} - \mathcal{E}_m^{\parallel,\perp}}{1 + (f/f_{01}^{\parallel,\perp})^2} + \frac{\mathcal{E}_m^{\parallel,\perp} - \mathcal{E}_h}{1 + (f/f_{02})^2} \qquad (4)$$

$$\sigma_e^{\parallel,\perp} = \sigma_h + \frac{\sigma_s^{\parallel,\perp} - \sigma_m^{\parallel,\perp}}{1 + (f/f_{01}^{\parallel,\perp})^2} + \frac{\sigma_m^{\parallel,\perp} - \sigma_h}{1 + (f/f_{02})^2} \qquad (5)$$

where, $\mathcal{E}_s, \sigma_s, \mathcal{E}_m, \sigma_m, \mathcal{E}_h, \sigma_h$, , are the low-, middle-
and high-frequency values of effective P and EC; f_{01} and
f_{02} are the characteristic frequencies of α - and β-dis-
persions. On the basis of equations (1), (2) and (3), the
formula derived in /4/, as applied to the model presented,
has the form:

$$M = \frac{\pi}{3} a^2 c \, \mathcal{E}_0 \left[\frac{S_1^{\parallel} S_3^{\parallel} + S_2^{\parallel} S_4^{\parallel}}{\gamma^2 (S_1^{\parallel})^2 + (S_2^{\parallel})^2} - \frac{S_1^{\perp} S_3^{\perp} + S_2^{\perp} S_4^{\perp}}{\gamma^2 (S_1^{\perp})^2 + (S_2^{\perp})^2} \right] \qquad (6)$$

where

$$S_1^{\parallel,\perp} = \varepsilon_{md} + n_{x(z)}\left(\varepsilon_e^{\parallel,\perp} - \varepsilon_{md}\right),$$

$$S_2^{\parallel,\perp} = \sigma_{md} + n_{x(z)}\left(\sigma_e^{\parallel,\perp} - \sigma_{md}\right),$$

$$S_3^{\parallel,\perp} = \gamma^2 \varepsilon_{md}\left(\varepsilon_e^{\parallel,\perp} - \varepsilon_{md}\right) + \sigma_{md}\left(\sigma_e^{\parallel,\perp} - \sigma_{md}\right),$$

$$S_4^{\parallel,\perp} = \varepsilon_{md}\left(\sigma_e^{\parallel,\perp} - \sigma_{md}\right) - \sigma_{md}\left(\varepsilon_e^{\parallel,\perp} - \varepsilon_{md}\right),$$

$$\gamma = 2\pi f \varepsilon_0$$

Frequency dependencies of M/M_o (M_o is equal to M at $f=4\cdot 10^5$ Hz and model parameters preset) were computed by Eqn.6 with regard to Eqns.4 and 5. The following parameters of the model used were preset according to the literature data /10,11/ and by comparing theoretical and experimental results:

$a=b=3\cdot 10^{-6}$ m; $\quad c=10^{-6}$ m; $\quad \sigma_{md} = 3\cdot 10^{-4}$ S/m;
$\varepsilon_o = 80$; $\quad f_{01}^{\parallel} = 2\cdot 10^2$ Hz; $\quad f_{01}^{\perp} = 18\cdot 10^2$ Hz;
$f_{02} = 4\cdot 10^5$ Hz; $\quad \sigma_s^{\parallel} = 1.5\cdot 10^{-2}$ S/m; $\quad \sigma_s^{\perp} = 1.65\cdot 10^{-2}$ S/m;
$\sigma_m^{\parallel} = 1.8\cdot 10^{-2}$ S/m; $\quad \sigma_m^{\perp} = 2.96\cdot 10^{-2}$ S/m; $\quad \sigma_h^{\parallel,\perp} = 4\cdot 10^{-2}$ S/m;
$\varepsilon_s^{\parallel} = 3\cdot 10^3$; $\quad \varepsilon_s^{\perp} = 8\cdot 10^3$; $\quad \varepsilon_m^{\parallel,\perp} = 5\cdot 10^2$; $\quad \varepsilon_h^{\parallel,\perp} = 50$.

The characteristic frequency of α-dispersion is known to depend on cell size. This dependence was taken into account when f_{01} prescribing the different values of parallel ($f_{01}^{\parallel}$) and perpendicular ($f_{01}^{\perp}$) erythrocyte polarizabilities. The value of f_{02} was preset according to the literature data /1,10/.

MATERIALS AND METHODS

Sheep erythrocytes fixed with formalin and lyophilized were used in the experiments. Just before measurements, erythrocytes were twice washed in physiological solution (0.145 M) and once in 11% isotonic sucrose by centrifugation at 300 g for 10 min. The cells were then resuspended in isotonic sucrose. In all the experiments the working concentration of erythrocytes was 6-8 10^5 per I ml, that excluded the possibility of influence of the dipole-dipole interactions on the process of cell orientation /12/. In studying EOS, EC and pH of the erythrocyte suspension were varied by adding 0.1 M KCl, KOH, tris and citric acid. The suspension parameters were measured with a conductometer OK-102/1 and pH-meter pH-340.

The values of EOE of erythrocytes was measured as relative change of the optical density of the suspension due to orientation of particles under the action of a uniform alternating electric field. The techniques used were identical to those described in /12/ except for the measuring chamber design. To create a uniform electric field, the interior of the measuring thermostated chamber was equipped with two flat parallel platinum electrodes $(5\cdot 20)\cdot 10^{-3}$ m in size, spaced at $5\cdot 10^{-3}$ m. The measuring chamber was placed in one of the two photometer light channels and the reference one in the other. Voltage on the electrodes of the measuring chamber was 8 and 30 V. The EOS were recorded and interpreted as described in detail in /2,12/.

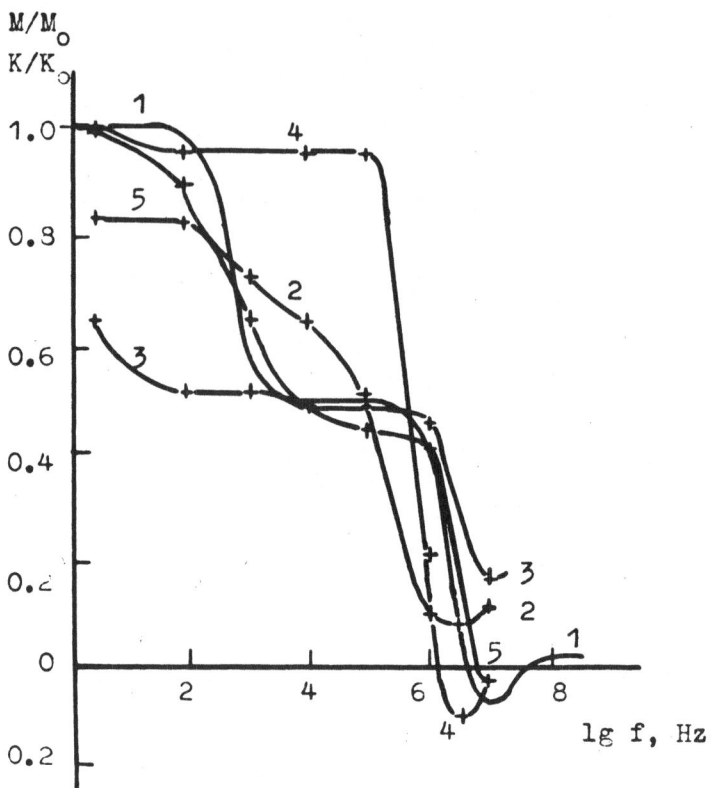

Fig. 1. Theoretical EOS M/M_0 at the model parameters
preset, where M_0 is the value of the torque
at f=20 Hz (curve I). Experimental EOS K/K_0
at various EC of the suspension (S/m): pH 5.3,
EC=5·10^{-4} (curve 2); EC=1·1·10^{-2} (curve 3):
pH 7.6, EC=8·10^{-4}(curve 4);EC=1.1·10^{-2}(curve 5).
K_0 is the change of optical density at f=20 Hz,
EC=510^{-4} and 8·10^{-4}, U = 8 V.

RESULTS AND DISCUSSION

Theoretical (curve I) and experimental (curves 2-5)
EOS obtained at various values of pH and EC of erythrocyte
suspension are presented in Fig.1. It is readily seen that
the shape of EOS changes markedly both at low and high
frequencies depending on the medium parameters. As known,
at low frequencies the value of EOE is defined by the
surface polarization including the polarization of a double
electric layer. Relaxation of surface polarization is
observed in the range from 0 to 10^4 Hz and related to α -
-dispersion. In radio-frequency range the main contribution
to cell polarization is made by polarization of cell structures, mainly of cell membrane. Relaxation of the structural polarization is observed in the range from 10^4 to 10^{8}Hz
and called β -dispersion.

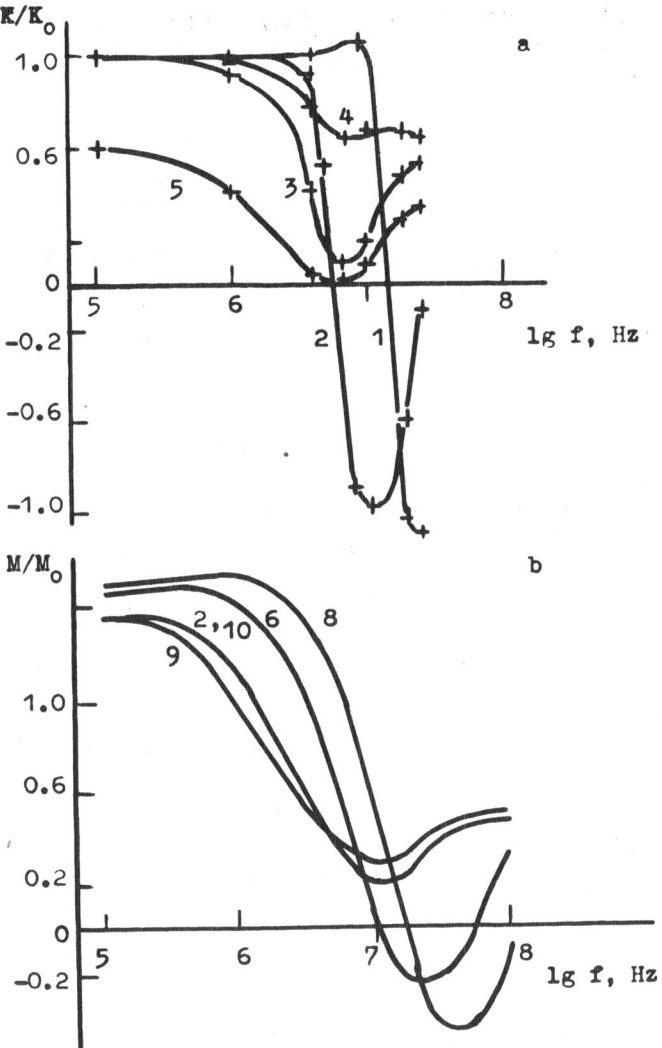

Fig. 2. a) Experimental EOS at various pH of the suspension: curve I is obtained at pH 9.6;
curve 2 at pH 7.5; curve 3 at pH 6.6;
curve 4 at pH 5.3; curve 5 at pH 3.8.
K_0 is the change in optical density of the
suspension at $f=10^5$ Hz; pH 9.6; EC=8.5$\cdot 10^{-3}$S/m;
U=30 V.
b) EOS calculated according to the data of
Table 1.

It should be noted that at some frequencies within the
range of 10^6–10^8 Hz the value of M becomes negative. This
phenomenon is consistent with the fact that the orientation
of long axis of the particle is perpendicular with the
field lines, i.e. the particle is oriented "across the
field" /13/. Experimental EOS indicate that this effect and
frequencies at which it is observed are dependent on the
medium parameters.

41

Table 1. Variations of longitudinal and transverse permittivity and electroconductivity values for the erythrocyte model

	1	2	3	4	5	6	7	8
$\sigma_h^{\parallel}$	0.03	0.04	0.06	0.08	0.10	0.12	0.15	0.20
$\sigma_h^{\perp}$	0.03	0.04	0.06	0.08	0.10	0.12	0.15	0.20
$\varepsilon_h^{\parallel}$	250	200	150	100	80	50	20	10
$\varepsilon_h^{\perp}$	130	100	50	30	25	20	15	10
		5.3				7.5		9.6pH

	9	10	11	12	13	14	15	16
$\sigma_h^{\parallel}$	0.03	0.04	0.06	0.08	0.10	0.12	0.15	0.20
$\sigma_h^{\perp}$	0.03	0.04	0.06	0.08	0.10	0.12	0.15	0.20
$\varepsilon_h^{\parallel}$	100	130	180	200	220	250	280	300
$\varepsilon_h^{\perp}$	8	10	20	23	27	30	35	40
	3.8	6.6						pH

For a better understanding of the changes of EOE value at high frequencies, EOS were measured at a frequency range of 0.1-30 MHz and various parameters of erythrocyte suspension. Fig.2 illustrates experimental spectra of erythrocytes at various pH of the suspension. For data analysis, two families of curves were plotted for high-frequency portion of the EOS (Table 1). We tended to obtain theoretical curves as closely as possible to the experimental ones in shape and position by varying effective values of high-frequency P and EC. The values of pH were then transferred from experimental to theoretical curves.

Theoretical EOS obtained according to the data of Table 1 are presented in Fig.2,b. Both the table and figure indicate that the change of high-frequency P markedly affects the value of EOE of the erythrocyte in "across the field" orientation position. In our calculations, the longitudinal and transverse P were allowed to change independently. It appears that the best agreement between the theoretical curves and experimental results was provided under such P variations. The frequency range at which the orientation of the erythrocyte is perpendicular with the field lines is also influenced by EC of the particle.

High-frequency portion of EOS as a function of EC of the erythrocyte suspension is presented in Fig.3,a. The family of curves was plotted assuming that 1) high-frequency EC of the erythrocyte is higher than that of the medium;

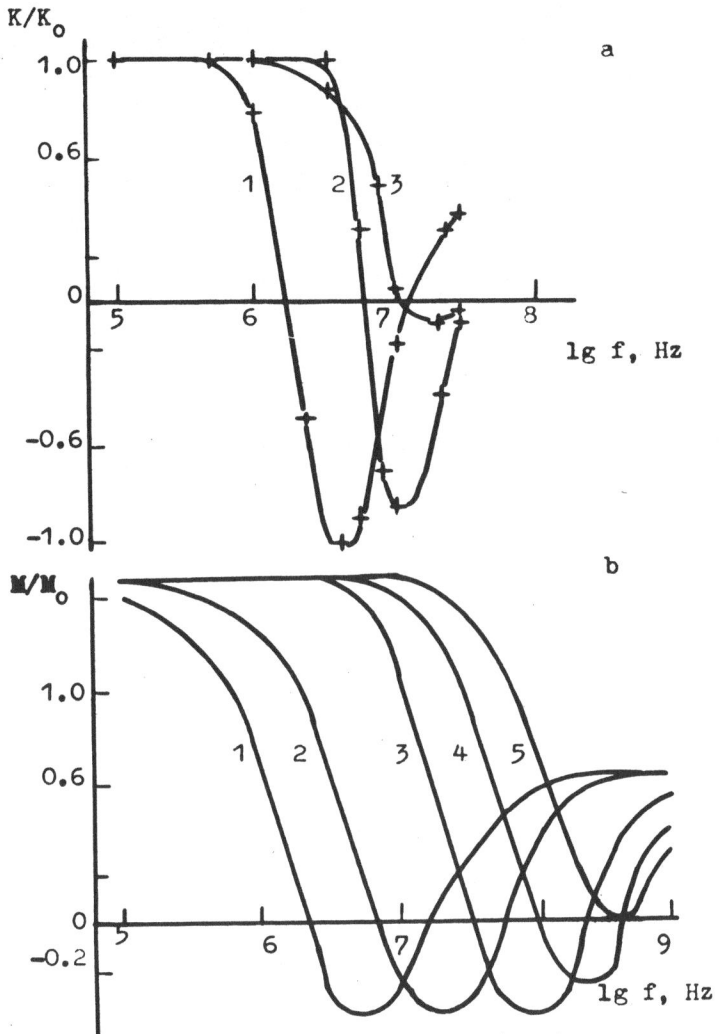

Fig. 3. a) Experimental EOS at various EC of the suspension (S/m): pH 7.6, EC=8·10⁻⁴ (curve I);
EC=1.1·10⁻² (curve 2); EC=2.6·10⁻² (curve 3).
K_0 is the change in optical density of the
suspension at f=10⁵ Hz; EC=8·10⁻⁴; V=30 V.
b) EOS calculated according to the data of
Table 2

2) low- and middle-frequency EC of the erythrocyte change
with frequency dispersion of EC. The curves were constructed (Fig.3,b) according to the date of Table 2.

Comparing experimental and theoretical EOS revealed
that the model proposed enables to obtain a good correlation between the theoretical and experimental curves. However, a discrepancy in the EOE values for the erythrocyte
in across the field orientation position can be explained
by the peculiarities of the optical method used.

Table 2. Changes of model parameters depending on variations of medium electroconductivity

	1	2	3	4	5
$\sigma_s^{\|}$	0.015	0.04	0.215	0.55	1.50
$\sigma_s^{\perp}$	0.0165	0.044	0.236	0.605	1.65
$\sigma_m^{\|}$	0.018	0.048	0.258	0.66	1.80
$\sigma_m^{\perp}$	0.0296	0.079	0.424	1.09	2.96
$\sigma_h^{\|}$	0.03	0.08	0.43	1.10	3.00
$\sigma_h^{\perp}$	0.05	0.12	0.645	1.50	4.00
$\varepsilon_h^{\|}$	30	40	60	80	150
$\varepsilon_h^{\perp}$	5	6	10	15	40
σ_{md}	$3 \cdot 10^{-4}$	$8 \cdot 10^{-4}$	$4.3 \cdot 10^{-3}$	$1.1 \cdot 10^{-2}$	$3 \cdot 10^{-2}$

Analysis of experimental and modelling results allows for estimation of the values of high-frequency P and EC as well as their changes depending on medium parameters. Depicted in Fig.4 is the dependence of high-frequency P and EC on pH of the medium. The curves were plotted according to the data of Table 1. It is seen that effective high--frequency EC of the erythrocyte increases with pH. EC in this case is compounded with the membrane EC and EC of cell interior, the latter being determined mostly by hemoglobin properties. The cellular membrane acts as a capacitor and its EC strongly depends on the field frequency.

Since the orientation of erythrocytes under the alternating electric field was shown to depend on the field frequency and medium parameters, it can be assumed that the mechanism of erythrocyte orientation is based on the change of the ratio between erythrocyte EC and EC of the medium. This change causes the redistribution of the field lines. Depending on whether the field lines pass arround the erythrocyte or across it, the particle has perpendicular orientation across the field lines or parallel orientation with the field lines.

At low frequencies, the membrane capacitance is high, and the erythrocyte behaves as a nonconducting particle in a conducting medium. In this case the field lines pass around the particle and the erythrocyte is oriented "with the field".

When the membrane capacitance decreases with increasing the field frequency, the transverse EC of the erythrocy-

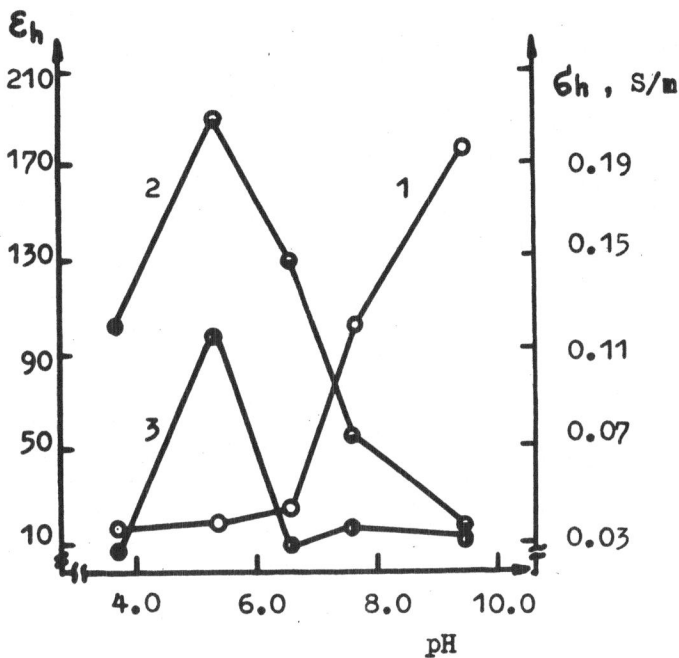

Fig. 4. High-frequency EC $\sigma_h^{\parallel,\perp}$ (I), parallel $\varepsilon_h^{\parallel}$ (2) and perpendicular $\varepsilon_h^{\perp}$ (3) permittivity of the erythrocyte model as a function of pH of the suspension. The calculations are made according to the date of Table 1

te becomes, at certain frequencies, greater than longitudinal EC, and the field lines pass across the erythrocyte, the latter being oriented "across the field". This is supported by theoretical analysis according to which the best agreement between theoretical and experimental curves is reached on the assumption that $\sigma_h^{\perp} > \sigma_h^{\parallel}$ (see Table 2).

At last, at high frequencies, the membrane capacitance may be neglected, then the erythrocyte should be considered as a well-conducting particle in a poorly conducting medium. In this case, the electric field is distributed mainly inside of the particle which in turn is oriented "with the field".

After fixation, the integrity of the erythrocyte membrane as an ion barrier is destroyed that results in the change of the properties of intracellular content. In particular, changing pH of the medium causes the alteration of this parameter inside of the erythrocyte and thus the change of ionizing degree of the intracellular hemoglobin. It is well-known that pH of hemoglobin isopoint is 6.8-7.2. After treating hemoglobin with formalin as well as with osmium tetroxide solution, the blocking of hemoglobin amino groups shifts the isopoint of it towards the more acidic region /14/. The number of dissociated ionogenic groups of hemoglobin increases with pH of the suspension. This

results in the increase of the level and mobility of the
counterions and hence in the rise of the effective EC /15/.
The preliminary studies on the electroorientation effect of
fixed erythrocytes depending on the fixation conditions
have suggested that the blocking of various numbers of hemoglobin amino groups causes the changes in high-frequency EC
of the erythrocyte and therefore in the degree and frequency range of the "across the field" orientation of the particle. Thus, studying EOE of the cells exposed to high-
-frequency electric field enables to estimate the isopoint
of intracellular content.

Both experimental and theoretical curves show that
increase of the medium EC shifts the frequency of the
"across the field" orientation towards the higher frequencies (Fig.3,a,b). At high EC of the medium, the "across
the field" orientation is not observed. This can be explained by the fact that the level of ions inside of the erythrocyte increases with EC of the medium. As a result, the
difference between longitudinal and transverse EC decreases
while the value of P increases, that makes the "across the
field" orientation of the particle impossible. Curves in
Fig.2,a,b shows that the increase of medium pH also shifts
the frequency of the "across the field" orientation towards
the higher frequencies, but the value of EOE in such case
is high. There is reason to believe that external ions do
not enter the erythrocyte, whereas the level and mobility
of the ions inside of the particle increases due to dissociation of the ionogenic groups. In other words, the
difference between longitudinal and transverse polarizabilities of the erythrocyte increases due to the increase in
the difference between EC of the particle and that of the
external medium.

REFERENCES

1. S. Stoilov, V.N. Shilov, S.S. Dukhin, V. Petkanchin, Electrooptics of colloids, Naukova dumka,
   Kiev, 200 (1977).
2. A.I. Miroshnikov, V.M. Fomchenkov, A.Yu. Ivanov,
   Electrophysical analysis and separation of
   cells, Nauka, Moscow, 198 (1986).
3. K. Asami, T. Hanai, N. Koizumi, Dielectric analysis of E.coli suspensions in the light of the
   theory of interfacial polarizability, Biophys.
   J., 31: 215-228 (1980).
4. V.M. Fomchenkov, A.K.Azhermachev, G.A. Chugunov,
   P.B. Babaev, Influence of polyelectrolites on
   the superficial electrical properties of micro-
   organisms, Kolloidnyi zhurnal, 45: 273-280
   (1983).
5. A.I. Miroshnikov, A.Yu. Ivanov, V.M. Fomchenkov,
   A.N. Shirokova, Electrophysical   model of
   erythrocyte in studying the cell polarization,
   Depoted in VINITI, N 2365-B86, 37 (1986).
6. F.D. Ovcharenko, V.V. Malyarenko, V.N. Shilov,
   Studies on electroconductivity of the suspensions and electroorientation of colloid particles, Kolloidnyi zhurnal, 39: 73-79 (1977).

7. L.D.Landau, E.M.Livshitz, Electrodynamics of continuous media, GIFML, Moscow, 532 (1959).

8. H. Pohl, Dielectrophoresis, Cambridge University Press, Cambridge, 579 (1978).

9. T. Hanai, N. Koizumi, A. Krimajiri, A method for determining the dielectric constant and the conductivity of membrane-bounded particles of biological relevance, Biophys. Struct. Mechanism, 1: 285-294 (1975).

10. K. Asami, T. Hanai, N. Koizumi, Dielectric approach to suspension of ellipsoidal particles covered with a shell in particular reference to biological cells, Japanese J.of Appl. Physics, 19: 359-365 (1980).

11. E. Bielicz, J. Terlecki, J. Flutak, J. Krupa, Electric admittance and dielectric constant of membrane coated ellipsoidal particle suspension, Studia biophysica, 51: 145-154 (1975).

12. A.Yu. Ivanov, V.M.Fomchenkov, A.I.Miroshnikov, Method for registration of cell electroorientation, Elektronnaya obrabotka materialov, 4: 53-57 (1984).

13. A.D. Gruzdev, On orientation of microparticle in electric fields, Biofisika, 10: 1091-1093 (1965).

14. J. Tooze, Measurements of some cellular changes during the fixation of amphibian erythrocytes with osmium tetroxide solutions, J. Cell Biol., 22: 551-556 (1964).

15. E.L. Carstensen, W.G. Aldridge, S.Z. Chield, P. Sullivan, H.H. Brown, Stability of cells fixed with glutaraldehyde and acrolein, J. Cell Biol., 50: 529-532 (1971).

CHARGE MOTION INSIDE PROTEINS

L. Keszthelyi

Institute of Biophysics, Biological Research Center
Szeged, Hungary

INTRODUCTION

The functioning of many proteins is connected with charge motion inside them. As examples ion channels, pumps, heme proteins may be mentioned. During these functions two different charge motions occur: a) charged parts of proteins or b) external charges through proteins can move.

Proteins are considered as dielectrics. The moving charges induce displacement current externally which may be measured with appropriate methods.

For the time behaviour of these currents to be observed, two require- ments have to be fulfilled: (i) the pumps or other proteins should be syn- chronized by starting them simultaneously in as short a time as possible; (ii) the system must be asymmetric with respect to the current-measuring electrodes. The first requirement has been fulfilled until now by exciting the systems with pulses of electric field, as in the case of neurons, and with short light pulses. The electric field excitation needs electrodes located in or on the surface of the cells. With electrodes (one in the cell and one outside) the second requirement is automatically fulfilled, but this is possible only for large cells. We deal in the following with small units not suitable for electrode insertion. In this case three dif- ferent methods are known for the production of asymmetry:

(a) Light gradient method[1-3]. Closed cells or vesicles containing light-absorbing pigments uniformly distributed on their surface absorb slightly less light at the side more distant from the light source than at the side nearer to the source, because of the absorption in the near side. The pump currents flowing at the two sides are therefore not completely compensated. Electrodes positioned in the direction of the exciting light in the suspension of the vesicles pick up the current.

(b) Membrane-bound systems[4-7]. Closed vesicles and membrane frag- ments can be adjoined to artificial membranes or layered between water/oil interfacial layers. The electrodes are in the two compartments separated by the membrane of high resistance and capacity. The displacement current caused by the moving charges is coupled capacitatively through the mem- brane and conductively through the bathing solution to the electrodes.

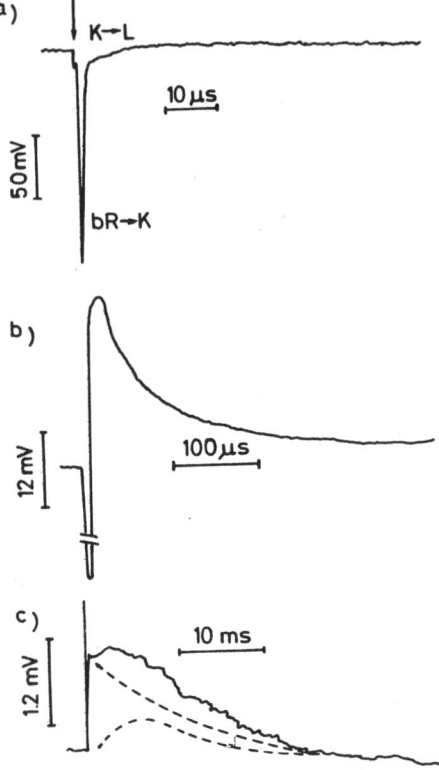

Fig. 1. Electric signals from bR-s in pm suspension oriented by electric field. a) Electric signals associated with bR→K and K→L transition (at 5°C). b) L→M transition (at 20°C). c) The electric signal is composed from two components, the single exponential corresponds to M→0, the other to 0→bR transition (at 20°C).

Asymmetry in the case of adhered vesicles is a consequence of the advantageous sensing of charge movement from the adjoining surface or of the oriented attachment in the case of membrane fragments.

(c) Oriented membrane fragments in suspension[8]. Purple membranes containing the proton pumping bacteriorhodopsin (bR) molecules from Halobacterium halobium were oriented in suspension by a quasistatic electric field. The oriented system could be used to observe the current due to charge displacement during the photocycle. The displacement current was attributed to protons pumped after laser light pulses. This system is very advantageous because the quantity of the working unit is so large that simultaneous electrical and optical measurements can be performed: events in the proton pumping can be correlated with the photocycle.

Later two other methods were worked out to have oriented systems. The pm-s were immobilized during orientation by polymerization in gel or deposited in oriented way to transparent electrodes and dried. These two methods produce samples which are easy for experimentation because electric signals may be produced without orienting electric fields. The suspension, gel and dried oriented sample methods have different possibilities to obtain important information on proton translocation in bR[9].

The proteins in crystals - under favourable conditions - may also be oriented.

In this paper some results obtained by the above methods with bR, halorhodopsin (which translocates Cl^- ions) and with myoglobin crystals are reviewed.

MEASUREMENTS

Bacteriorhodopsin

By measurement of the electric signals in suspension and gel five components with different lifetimes and amplitudes were found (Fig. 1). Comparing these lifetimes with the lifetimes of the different transitions in the bR photocycle the components of the electric signals were assigned to them. It was stated that the protons are translocated in five steps. From the amplitudes of the signal the distances the protons make in the five steps were estimated[9].

Here the main ideas of the calculations are only described, the details may be found in ref. 10. On the effect of light a charge Q takes a distance d from point 1 to 2 between two electrodes (Fig. 2).

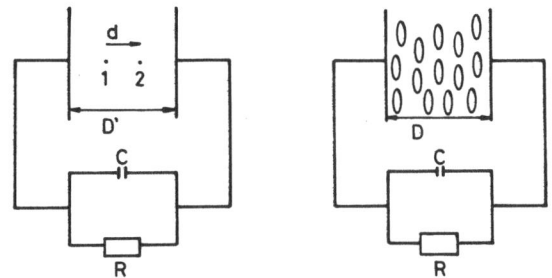

Fig. 2. The measuring system. Left: sketch of the motion of a single charge, right: the real measuring system.

The induced current i(t) is:

$$i(t) = \frac{Q\ v(t)}{\varepsilon\ D'}, \tag{1}$$

where $v(t)$ is the time depending velocity, ε is the dielectric constant, and D' is the distance of the electrodes. We assume that $v(t)$ is very large, i.e. the charge jumps the distance d. Integrating Eq. (1) with respect to time:

$$Q_{ind} = \int_0^\infty i(t) \, dt = \frac{Q}{\varepsilon D'} \int_0^\infty v(t) \, dt = \frac{Q \, d}{\varepsilon D'} . \qquad (2)$$

The induced charge Q_{ind} is proportional to d. Q_{ind} charges the external C capacitance (Fig. 2), which discharges through the resistance R. The time dependence of the voltage on the RC circuit is

$$V(t) = \frac{Q_{ind}}{C} e^{-t/RC} . \qquad (3)$$

In the real case the membranes are suspended in an oriented way containing a large number of excited states N in which charges move according to the time behaviour of the decay of the excited state. We take a simple exponential decay of the N states excited at t=0 then the number of decaying states in unit time is:

$$N'(t) = k \, N \, e^{-kt} , \qquad (4)$$

where k is the rate constant.

We assume the conductivity of the suspension is very small. This approximation means that the hypothetical electrodes in Fig. 2 are now far from each other in a distance of D to have the oriented membrane fragments containing bR between them. In this case D' may be simply replaced by D in Eq. (2).

We may now calculate the voltage form arising from the motion of N charges. It means a folding of Eqs. (3) and (4):

$$V_N(t) = \frac{N \, Q \, d \, k}{\varepsilon \, D \, C} \int_0^t e^{-kt'} e^{-(t-t')/RC} \, dt' = \qquad (5)$$

$$= \frac{N \, Q \, d \, k}{\varepsilon D} \frac{R}{1 - k R C} (e^{-kt} - e^{-t/RC}) .$$

Eq. (5) may be approximated in two limiting cases by

$$V_N(t) = \frac{N \, Q \, d}{\varepsilon \, D \, C} e^{-t/RC} , \qquad \text{if } k \gg 1/RC \qquad (6)$$

and

$$V_N(t) = \frac{N \, Q \, d}{\varepsilon \, D} R \, k \, e^{-kt} , \qquad \text{if } k \ll 1/RC . \qquad (7)$$

Eqs. (5)-(7) show that measuring $V_N(t)$ we may determine the distance d the charge makes between points 1 and 2. N is determined from separate light absorption measurement, Q is assumed to be elementary charge (a single proton in case of bR), R, C and D are known, k comes from the time measurements.

In deriving the equations we made some assumptions. The most important is to have a homogeneous dielectric charactized by ε. This is surely not correct. Therefore we consider the equations to be good only in order of magnitude. Experiments determine the $V_N(t)$ functions for all the components of charge motion. We may add the individual values of d_i and normalize to the thickness of the membrane (~ 5 nm). This way the distance of all the five steps may be determined (Table 1). Without normalization taking $\varepsilon = 2$ inside the protein the equations should be multiplied by 4 to get the correct sum of distances. Calculations based on electrolytes give essentially similar results[10].

Table 1. Displacements of protons
during the bR photocycle

d	Distance (nm)	Assignment
d_1	-0.13	bR - K
d_2	-0.02	K - L
d_3	+0.5	L - M
d_4	+3.1	M - O
d_5	+1.5	O - bR
Σd_i	+4.95	

By correlating the rate constants with the rate constants of the light absorption charges which determine the bR photocycle (Table 1) we see that during bR - K transition the charges move backward d_1= -0.13 nm. It is known that this step is a trans-cis isomerization of the retinal bound via a protonated Schiff-base[11]. Simple geometry gives distance of d_{theor}= -0.16 nm. Our value (with K - L together) is d_{exp} = -0.15 nm.

The distance d_3 was checked by a quite different method. External electric fields change the activation enthalpy (ΔH) of a transition and this way the rate constant according to:

$$\frac{1}{k(V)} = \tau(V) = \frac{1}{f} \exp \frac{\Delta H \pm FdV/2D'}{RT}, \qquad (8)$$

where V is the potential applied on one membrane with thickness D', f is the frequency factor, F the Faraday, R the universal gas constant, T the absolute temperature and d the thickness of the barrier[12]. k values of two components of the L - M transitions were measured depending on V (between ± 70 mV) and d'_3 = 0.6 nm and d''_3 = 0.4 nm in good accord with d_3=0.5 nm in Table 1 were found[13].

The three short motions occur deep inside the proteins and the two longer ones lead the protons to the external and from the internal side of the membrane. These long distances locate the binding side of the retinal about 1/3 from the internal surface of the membrane which is in accord with the structural data[14].

Our group performed a number of studies on bacteriorhodopsin in different environments, with differently modified amino acid side chains and even after different protein digestions. A review of the results is found elsewhere[15].

Halorhodopsin

A second pigment found in halobacteria is the halorhodopsin (hR) which translocates Cl^- ions on the effect of light absorption[16]. We succeeded to orient membrane patches containing hR and immobilized them in gel. Electric signals measured after laser flash excitation[17,18] are shown in Fig. 3.

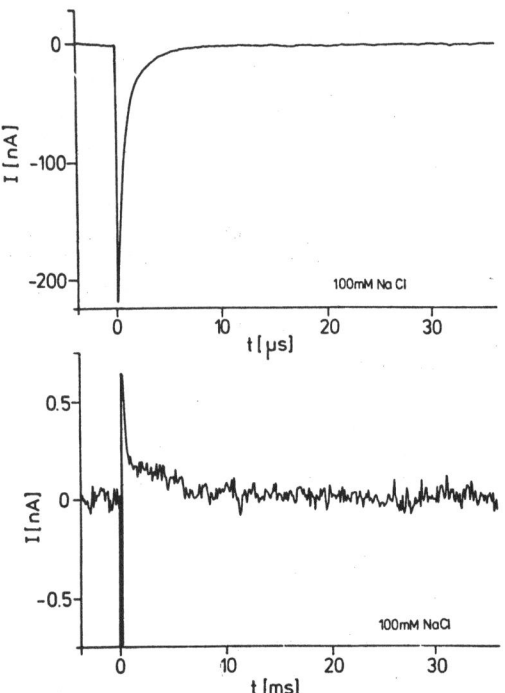

Fig. 3. Electric signals from oriented hR molecules. The membrane pellets containing hR were oriented and immobilized in gel.

Four components were found: a large negative signal which is faster than the time resolution of the measuring system, a small negative signal with 2 μs lifetime at room temperature and two positive signals with 0.5 ms and 4 ms lifetimes. The first and second negative signals are interpreted as corresponding to the trans-cis isomerization as in bR, the positive signal should be caused by the motion of the Cl^- ions because they are absent if the solution does not contain Cl^- ions.

Myoglobin crystals

The bond of carbonmonoxide-ferrous myoglobin (MbCO) may be broken by light flash. The ligand CO moves away from the Fe^{++} ions to which it is bound. This process is very fast ($\sim$ps). The time constant of spontaneous rebinding of CO in the ms region[19]. During these processes the Fe^{++} ions move out and in from the heme plane.

Fig. 4. The elementary cell of MbCO crystal contains two MbCO molecules with hemes oriented as the arrows show. The resulting diplacement of Fe^{3+} ion has an uncompensated component in the direction of the b axis.

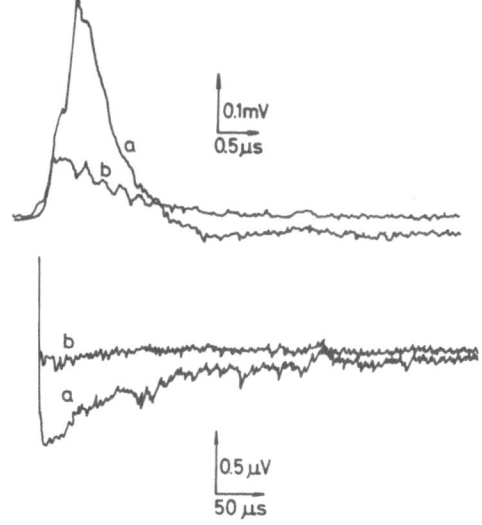

Fig. 5. Electric signals measured after laser flash in MbCO crystal. (a) signal + background, (b) background.

The elementary cell of the MbCO crystals contains two MbCO molecules as sketched in Fig. 4. The resultant of the displacements of the two Fe^{++} ions points in the direction of the b axis of the crystal which is easy to recognize from the macroscopic shape of the crystal. An electric signal is therefore expected if the crystal is between two electrodes and is illuminated by a laser flash. Fig. 5 shows the observed signals which correspond to the expectation. A fast signal is associated with the fast deligation of CO, while the rebinding causes a much slower signal with opposite sign.

SUMMARY

In this paper we tried to demonstrate that charge motions associated with different functions of proteins is measurable under favourable conditions. Oriented systems, fast excitation and sensitive electronic instrumentation are the requirements. The data obtained are valuable to understand the mechanisms of the electric phenomena intimately associated with protein functions.

REFERENCES

1. C.F. Fowler and B. Kok, Direct observation of a light induced electric field in chloroplasts, Biochim. Biophys. Acta 357:308 (1974).
2. H.T. Witt and A.Zickler, Electrical evidence for the field indicating absorption change in bioenergetic membranes, FEBS Letters 37:307 (1973).
3. P. Graber and H.W. Trissl, On the rise time and polarity of the photovoltage generated by light gradients in chloroplast suspension, FEBS Letters 123:95 (1981).
4. H.W. Trissl and M. Montal, Electrical demonstration of rapid light-induced conformational changes in bacteriorhodopsin, Nature 266:655 (1977).
5. L. Drachev, A.D. Kaulen and V.P. Skulachev, Time resolution of the intermediate steps in the bacteriorhodopsin linked electrogenesis, FEBS Letters 87:161 (1978).
6. H.W. Trissl, Novel capacitive electrode with a wide frequency range for measurements of flash induced changes of interface potential at oil water interface, Biochim. Biophys. Acta 595:82 (1980).
7. A Fahr, P. Lauger and E. Bamberg, Photocurrent kinetics of purple membrane sheets bound to planar bilayer membranes, J. Membrane Biol. 60:51 (1981).
8. L. Keszthelyi and P. Ormos, Electric signals associated with the photocycle of bacteriorhodopsin, FEBS Letters 109:189 (1980).
9. L. Keszthelyi, Intramolecular charge shifts during the photoreaction cycle of bacteriorhodopsin. in "Information and Energy Transduction in Biological Membranes", L. Bolis, E.J. Helmreich and H. Passov, eds., Alan R. Liss, Inc., New York, 1984, pp. 51-71.
10. G. Váró and L. Keszthelyi, Photoelectric signals from dried oriented purple membranes of Halobacterium halobium, Biophys. J. 43:47 (1983).
11. S.O. Smith, J. Lugtenburg and R.A. Mathies, Determination of retinal chromophore structure in bacteriorhodopsin with resonance Raman spectroscopy, J. Membrane Biol. 85:95 (1985)
12. P. Lauger, R. Benz, G. Stark, E. Bamberg, P.C. Jordan, A. Fahr and W. Brock, Relaxation studies of ion transport systems in lipid bilayer membranes, Ann. Rev. Biophys. 14:513 (1981).

13. G. Váró and L. Keszthelyi, Arrhenius parameters of the bacterio-rhodopsin photocycle in dried oriented samples, Biophys. J. 47:243 (1985).
14. W. Stoeckenius, R.H. Lozier and R.A. Bogomolni, Bacteriorhod-opsin and the purple membrane of Halobacteria, Biochim. Biophys. Acta, 505:215 (1979).
15. L. Keszthelyi, Charge motion inside the bacteriorhodopsin mol-ecule, Bioelectrochem. Bioenerg. 15:437 (1986).
16. J.K.Lányi, Halorhodopsin: a light driven chloride ion pump, Ann..Rev. Biophys. Chem. 1986, in press.
17. A.Dér, K. Fendler, L. Keszthelyi and E. Bamberg, Primary charge separation in halorhodopsin, FEBS Letters 187:233 (1985).
18. A.Dér, K. Fendler, L. Keszthelyi, D. Oesterhelt and E. Bamberg, Charge displacement during the photocycle of halorhodopsin Bio-chim. Biophys. Acta, (in press)
19. R.H. Austeen, K.W. Beeson, L. Eisenstein, H. Frauenfelder and I.C. Gunsalus, Dynamics of ligand binding to myoglobin, Biochem-istry, 14:5355(1975).

ELECTROCONFORMATIONAL COUPLING AND THE EFFECTS OF STATIC AND DYNAMIC ELECTRIC FIELDS ON MEMBRANE TRANSPORT

R. Dean Astumian[1], P. Boon Chock[1], Francoise Chauvin[2], and Tian Yow Tsong[2]

[1]Laboratory of Biochemistry
National Heart, Lung, and Blood Institute
National Institutes of Health
Building 3, Room 202
Bethesda, Maryland 20892, U.S.A.

[2]Department of Biological Chemistry
The Johns Hopkins University School of Medicine
Baltimore, Maryland 21205, U.S.A.

INTRODUCTION

Life itself depends on the maintenance of electric fields across cell membranes, and on the propagation of electric currents. Even unicellular organisms have been shown to generate, and in turn respond to, minute electric signals. Larger electric perturbations, applied in an experimental context, can activate membrane enzymes (Johnstone and Laris, 1980; Serpersu and Tsong, 1983, 1984), be utilized as an energy source for the biosynthesis of ATP or the creation of chemical gradients across membranes (Witt et al., 1976; Serpersu and Tsong, 1983; Knox and Tsong, 1984), or lead to the formation of "pores" in the membrane through which small molecules can passively diffuse into and out of the cell (Kinosita and Tsong, 1977 a,b; Neumann et al., 1982; Tsong, 1983). This latter process is known as electroporation.

These electric field effects involve interaction with structures that allow transport of material from one side of the membrane to the other. In this chapter, we will develop a number of concepts important in understanding the interaction between electric fields and membrane transport. The principle of electroconformational coupling (ECC), on which we shall focus our attention, stresses the influence of the membrane potential, $\Delta\psi$, on the conformational equilibria of membrane proteins and other structural elements. This is in addition to the better appreciated influence of $\Delta\psi$ as a "part" of an ionic electrochemical gradient.

We show that modulation of the "static" value of $\Delta\psi$ may play an important regulatory role in governing the activity of many membrane transporters. The inherent stochastic dynamicism of the local membrane potential provides a mechanism by which membrane enzymes can capture energy contained in ion electrochemical gradients for doing useful work. Additionally, as explained by Offner (1980, 1984), stochastic fluctuations may be of great importance in imbuing membrane processes with a phenomenal sensitivity to changes in the environment.

EXPERIMENTAL MANIPULATION OF $\Delta\psi$ IS EASILY ACCOMPLISHED

Unlike in homogeneous solution, where it is technically very difficult to generate field strengths on the order of 10^5 V/cm, such intense electric fields normally exist across cell membranes <u>in vivo</u>, and can be easily obtained experimentally due to field amplification across the membrane. This amplification (see Fig. 1) arises from the fact that the membrane bilayer is much less conductive than the surrounding aqueous media, and thus in an applied field, ions of opposite signs accumulate at the interior and exterior surfaces of the membrane. In fact, from Maxwell's relation, we can calculate that an external field of only 50 V/cm applied to a suspension of cells with radius of 5 μm would result in a maximum $\Delta\psi$ of 50 mV. This represents a field strength across the membrane of 100,000 V/cm (membrane thickness of 50 Å). An instrument suitable for generating sinusoidally oscillating or repetitively pulsed electric fields of this magnitude has been described elsewhere (Kinosita and Tsong, 1977 a,b; Serpersu and Tsong, 1983).

As seen in Fig. 1, the sign of the resulting membrane potential is (+)in, (-)out on the pole of the cell facing the negative electrode, and opposite this on the other pole. It might thus be imagined that any field effect occurring at one pole would be undone at the other. This, however, is not true for the highly nonlinear (with respect to the electric field) processes occurring within the bilayer.

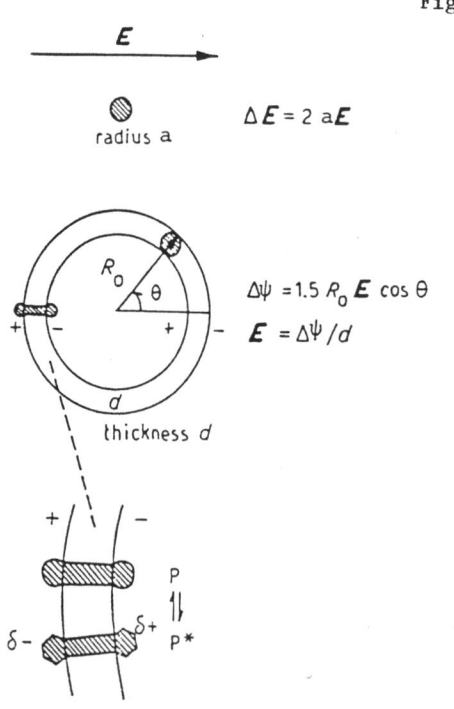

Fig. 1. Amplification of an electric field across the membrane of a spherical vesicle. A protein molecule with a radius of r will experience a potential drop across its dimension of 2 r E when it is exposed to an electric field of E. The same molecule when integrated into a membrane vesicle of radius R_0 will experience an effective field strength of $\Delta\psi/d$, d being the thickness of the lipid bilayer. $\Delta\psi$ is the field-induced transmembrane potential. The field the protein is exposed to is amplified by 1.5 R_0/d fold when it is integrated in the bilayer as compared to when it is in the homogeneous aqueous phase. This intense electric field can trigger a shift in conformational equilibrium between two states P and P[*] of a protein towards the * state if the overall dipole moment of P[*] is higher than that of P. See text for details.

THE MEMBRANE POTENTIAL DETERMINES THE POSITION OF AN ION ELECTROCHEMICAL GRADIENT

Because an ion bears an electrical charge, its chemical potential at any point in space is influenced by the presence of an electric field. If the electric potential in the bulk phase on one side of the membrane is different than that in the other, there will, at equilibrium, be a concentration difference (for some substance S) sufficient to counterbalance the electric component such that the chemical potentials of S on the two sides of the membrane are equal. Of course, if S represents a neutral substance, the condition for equilibrium is met at arbitrary $\Delta\psi$ when $[S_{in}] = [S_{out}]$.

In general, the difference in chemical potential of S between "in" and "out" (i.e., the electrochemical gradient = $\Delta\mu_S$) is given by

$$\Delta\mu S = z_S \cdot F \cdot \Delta\psi + 2.3 \cdot R \cdot T \cdot \log([S_{in}]/[S_{out}]) \tag{1}$$

with $\Delta\psi = (\psi_{in} - \psi_{out})$, z_S is the charge (including sign) on S, F is Faradays constant, R is the universal gas constant (molar Boltzmanm's constant), and T is the Kelvin temperature. $\Delta\mu_S$ is an energy, and if $\Delta\psi$ is written in mV and T=300 K, (1) becomes

$$\Delta\mu_S(kJ/mole) = .096 \cdot z_S \cdot \Delta\psi + 5.7 \cdot \log([S_{in}]/[S_{out}]) \tag{2}$$

THE MEMBRANE POTENTIAL ALSO INTERACTS WITH PROTEIN CONFORMATIONAL EQUILIBRIA

The proteins that mediate ion translocation are themselves more often than not sensitive to an electric field. The mechanisms by which these proteins accomplish their function involve conformational transitions. If intramolecular charge transfer or dipole moment changes occur during these transitions, the conformational equilibria will vary as a function of the electric field. Let us consider a simple conformational equilibrium

$$P(\mu,\alpha) \rightleftharpoons P^*(\mu^*,\alpha^*) \tag{3}$$

$\mu(\mu^*)$ and $\alpha(\alpha^*)$ are the permanent dipole moment and polarizability of the states $P(P^*)$, respectively. In general, the influence of an electric field on such an equilibrium can be written in terms of a generalized van't Hoff equation (Tsong and Astumian 1986, 1987, 1988):

$$(d\ln K/dE)_{T,p} = \Delta M/RT \tag{4}$$

If we consider that P in Eq. (3) is a membrane protein, where its rotational degrees of freedom are constrained (it can't rotate in response to the field, ΔM), the macroscopic molar polarization, may be written

$$\Delta M = (\mu^* - \mu) + (\alpha^* - \alpha) \cdot E = \Delta\mu + E \cdot \Delta\alpha \tag{5}$$

Integration of Eq. (5) yields

$$K_E = K_0 \cdot \exp(x \cdot E^2 + y \cdot E) \tag{6}$$

with $x = \Delta\alpha/(2RT)$ and $y = \Delta\mu/RT$. This equation implies that at low field strength, a positive membrane potential will favor the P^* state, and negative $\Delta\psi$ will favor the P state (relative to the zero field condition) since the magnitude of $\Delta\mu$ is usually much larger than that of $\Delta\alpha$. At very high field strengths, however, the quadratic term due to the polarizability will take over, and the P^* state will be favored irrespective of the sign of the field. Although this quadratic dependence on the field may be of importance in explaining the optimal field strengths for various phenomena observed

61

experimentally (Tsong and Astumian, 1986), for the remainder of this paper we shall consider that $x \cdot E^2 \ll y \cdot E$. Then Eq. (6) may be simply rewritten

$$K_E = K_o \cdot \exp(y \cdot E) = K_o \cdot \phi \tag{7}$$

and thermodynamically consistent rate constants are

$$k_{f,E} = k_{f,o} \cdot \phi^{\delta}$$
$$k_{r,E} = k_{r,o} \cdot \phi^{(\delta-1)} \tag{8}$$

The physical meaning of these parameters may be better understood in reference to a hypothetical plot of G (Gibbs free energy) versus the "reaction coordinate", as shown in Fig. 2.

If P^* repressnts an active state of an enzyme and P an inactive one, it can immediately be seen that such an enzyme will be subject to regulation by modulation of the electric field. This is illustrated in Figs. 3a & b. Additionally, the rate coefficients, and hence the rate of the transition, will be significantly (by orders of magnitude) changed from the zero field case, as shown in Fig. 3c. For the case in which the electrically sensitive conformational transition is within an enzyme catalytic cycle, see the section below on electrostatic regulation of membrane transport.

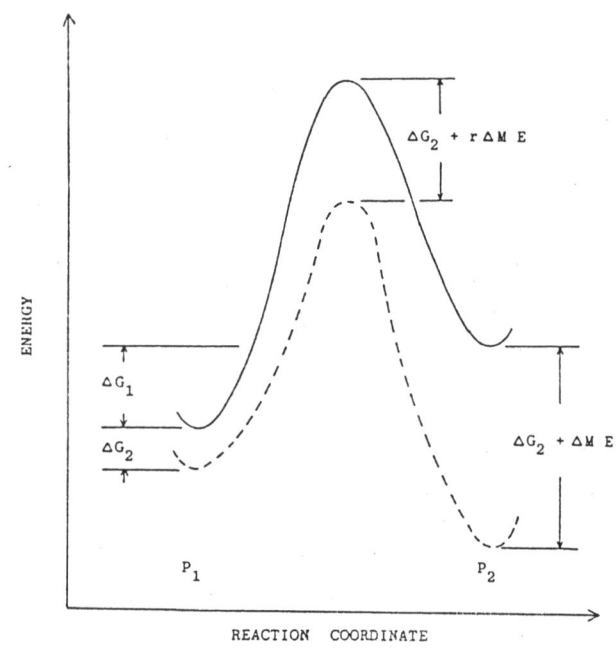

Fig. 2. Free energy diagram of $P \rightleftharpoons P^*$ reaction, under zero-field condition (solid curve) and under the influence of an electric field, E (dashed curve). P^* state has a greater molar electric moment than P state by ΔM. At zero-field, P state is more stable than P^* by ΔG_1. When the system is exposed to the electric field, P is stabilized by ΔG_2, but P^* state is stabilized by $[\Delta G_2 + \Delta ME]$. This shifts the equilibrium in favor of P^* state. Under the electric field, the transition state is stabilized by $[\Delta G_2 + \delta \Delta ME]$ when compared to the zero-field condition, where r is the apportion ment constant and has a value between 0 and 1.

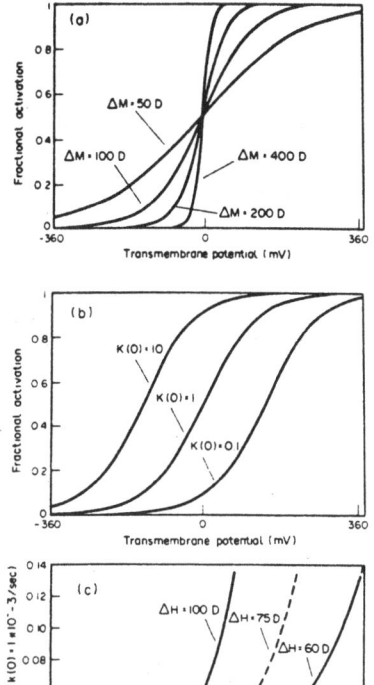

Fig. 3. Effects of membrane potential on the equilibrium and kinetic properties of a simple conformational transition of Eq. (10). P^* is assumed to be the active form. (a) Plot of fractional activation versus membrane potential, with $K_O = 1$, and different ΔM. For $\Delta M = 100$ D, transition between 0.1 and 0.9 fractional activation occurs in the range of -100 mV to +100 mV. (b) Changing K_O changes the midpoint of transition, but not the steepness of the transition. (c) Plot of the forward rate constant for the reaction, with $\delta = 1/2$, as a function of the membrane potential. Note that for ΔM of 60 D, k_f increases almost 100 fold by changing the membrane potential from 150 mV to 450 mV.

ELECTROPORATION IS AN EXAMPLE OF DRASTIC ELECTROCONFORMATIONAL CHANGES OF MEMBRANE PROTEINS AND LIPID STRUCTURES

The application of moderately high electric pulses (sufficient to induce membrane potentials of around 600 mV or higher) to biological membranes results in a dramatic increase in the permeability of the bilayer to ions and small molecules. This increased permeability, if allowed to persist, will eventually lead to cell lysis due to osmotic effects. This is a secondary result of the electric field, since lysis can be prevented by the addition of large moelecules which balance the colloidal osmotic pressure of the cytoplasmic macromolecules. At 37°C, if lysis is prevented for a sufficiently long time, the pores will spontaneously reseal, although it is impossible to entirely prevent loss of the intracellular ionic contents immediately after the pulse. At the instant of electric exposure, the pore size can be quite large, as evidenced by the incorporation of latex beads into the cell. However, within milliseconds, the pores shrink significantly to attain a size at which they are stable. Depending on the conditions, this size can be small enough to allow for the passage of Na^+ but not sucrose, or large enough for the tetrasaccharide stachyose to get through (Kinosita and Tsong, 1977 a,b). Pores thus formed can be utilized to facilitate the introduction of various materials into the cell, such as pharmaceuticals or DNA (Kinosita and Tsong, 1978; Neumann et al., 1982). While the former has not been as successful as originally anticipated due to the fact that many drugs are fairly membrane permeant and leak out rapidly, the latter is becoming increasingly important as a tool in allowing transfection of cells with various genetic materials.

Another factor of great importance is the fact that electroporation greatly enhances the susceptibility of cells to fusion. This has allowed for the development of a method of electrically selecting cell fusions for antibody formation (Lo et al., 1984). Although correlation has been noted for some time, only recently has Sowers (1983) demonstrated the relationship between electroporation and electrofusion in an unequivocal manner. In this experiment, he first electroporated the cells, and then brought them into contact by dielectrophoresis, where fusion was observed.

A number of questions repeatedly recur concerning the mechanism of electroporation. First, it is often imagined that this phenomenon is due primarily to Joule heating of the solution, or to thermally induced osmotic shock. These have been ruled out by experiments done at different ionic strengths. It was shown (Kinosita and Tsong, 1977a) that electroporation occurs at a constant field strength and is independent of the current density. Additionally, if the isotonicity of the solution is maintained constant, but with either NaCl (which gives a large Joule heating) or with sucrose (which yields a small Joule heating), the results are unaffected.

A second question has been "where in the membrane does electroporation occur?". Since pure lipid bilayers are even more susceptible to electroporation than are protein containing biological membranes, it might be imagined that predominately lipid material only is involved in pore formation. That this is not so has been proved by the demonstration of a oubain sensitive, electrically induced conductance in erythrocyte membranes (Tessie and Tsong, 1980; Serpersu and Tsong, 1983, 1984) which makes up approximatel 30% of the total conductance caused by the field. Since the K_i for oubain inhibition is the same as that for inhibition of the Na,K ATPase, we believe that this enzyme may be involved. Perhaps the "pore" formation is preceded by an electroconformational change in the Na,K ATPase, then the high local current density leads to irreversible denaturation of the protein and formation of a pore in, through, or around the enzyme. There are other sites of pore formation which have not yet been identified.

We have become convinced that understanding of many effects of electric fields on cells and organisms will come down to understanding the interactions with membrane proteins, and perhaps in particular on transport proteins. For example, based on the above observations, we may apply the concept of electroconformational change to the understanding of the electroporation phenomenon. At least two distinct steps must be included. The first step resembles that of the voltage dependent opening/closing of membrane channels ($A \rightleftharpoons B$). The second step presumably involves irreversible thermal denaturation of channel proteins because of the large transmembrane currents induced by the intense applied field ($B \rightarrow C$)

$$A \rightleftharpoons B \longrightarrow C \qquad (9)$$
$$\text{(closed)} \quad \text{(open)} \quad \text{(denatured; open)}$$

Resealing of pores is extremely slow, in the minute to hour time range (Kinosita and Tsong, 1977b, 1978) and is, perhaps, accomplished by repair of the membrane rather than by reversible transition from state B (or C) to state A. In the subsequent sections, we develop some concepts which may be of use in coming to an understanding of the interaction between $\Delta\psi$ and membrane proteins. Also, it may turn out that the dynamic nature of the membrane potential is of the utmost importance in biological signal and free energy transduction, and these possibilities are explored in the latter sections.

We would like to pose the question "does electroporation in pure lipid bilayers occur because the electric field thermodynamically stabilizes some well-defined structural site, or must we envision a more physical 'breakdown' of the membrane structure?". At least in the case of moderate field strengths, we prefer the former explanation. If this is true, electroporation can be viewed as another manifestation of electroconformational change, and the concepts presented here and elsewhere (Tsong and Astumian 1986, 1987, 1988; Westerhoff and Astumian, 1987) will apply. This would suggest that a.c. fields of lower field strengths than commonly used may prove useful in optimizing the biotechnological applications of this phenomenon.

TRANSLOCATION OF SUBSTANCES ACROSS MEMBRANES IN VIVO

The lipid bilayer is relatively impermeable to large polar moieties, such as glucose and lactose, and to ions of any size. Thus, a system of membrane proteins has evolved to accomplish the tranport of such molecules across the membrane. These may be broken up into two groups, channel proteins and carrier proteins.

Channels essentially serve as gated pores, which when open allow for the rapid movement of a substance down its electrochemical gradient. Usually channels are specific for one or a few ions, and the opening and closing is regulated predominately either by the concentration of some ligand, as in the acetylcholine receptor which is a Ca^+ channel, or by the membrane potential, as in the voltage-gated Na^+ channel. The fundamental interactions which allow for gating by an electric field can be well understood in terms of the electroconformational interactions discussed above and illustrated in Fig. 2. However, some problems do arise, since in order to explain the extremely high sensitivity of the channel to changes in $\Delta\psi$, a gating charge (which may loosely be described as the equivalent number of elementary charges which must move down the entire membrane potential) of between 3.5 and 7 must often be assumed. Although certainly not impossible, this "equilibrium" mechanism would seem to require quite a massive conformational change in the molecule. Offner (1980, 1984) has proposed an elegant cascade, or trigger mechanism, which may circumvent this requirement. When a channel opens or closes, the ions in the vicinity must redistribute (a sort of non-discrete "induced fit" mechanism) to reflect the new condition. This redistribution allows for a small change in $\Delta\psi$ to trigger the utilization of the bulk of the energy in the ion electrochemical gradient across the membrane to accomplish a large amplification of the input signal.

Carrier proteins behave like transmembrane enzymes in carrying out their catalytic translocation function. In general, for transport to occur, it is thought that first, a substrate molecule S (which may or may not be charged) must bind from the outside, then a protein conformational transition must occur in which the exposure of the binding site is changed to face inside, and finally the substrate must dissociate. Obviously, such a mechanism could also work from inside to out, although as with all enzymes, the catalytic efficiency may be much better in one direction than in the other, with the asymmetry of function reflecting an asymmetry in the rate constant of the catalytic cycle. Thus, such an enzyme acts as a membrane rectifier, rapidly dissipating a gradient in one direction, but only slowly dissipating an equal gradient in the opposite direction.

ELECTROSTATIC REGULATION OF ENZYME CATALYZED MEMBRANE TRANSPORT BY ELECTRO-CONFORMATIONAL INTERACTIONS

The conformational transitions within the catalytic cycle of a membrane translocator enzyme may well be electrically sensitive based on structural considerations. In this case, the fundamental catalytic constants of the enzyme will depend on the membrane potential. Let us consider the simple, four-state model shown in Fig. 4. Here, for the sake of being explicit, the electrical sensitivity is shown to arise from the intramolecular transport of a "charged arm" from one side of the membrane to the other during

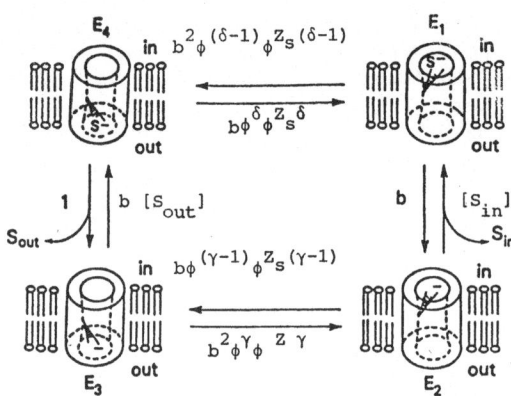

Fig. 4. A schematic mechanism for the enzyme catalyzed translocation of
substance S from one side of the membrane to the other. The only
electrically sensitive transitions are E_4 E_1 and E_3 E_2, and
each involves the intramolecular translocation of a singly negatively charged group ("arm") across the membrane. If additionally
S were charged, the transition E_4 E_1 would have an extra elec-
trically dependent term reflecting this (i.e. $\phi^{\delta z_s}$ and $\phi^{(\delta-1)z_s}$).
The rate coefficients shown on the diagram are based on the assump-
tions that: S bears z_s charge (which may be 0); states E_4 and E_2
have the same zero field intrinsic stability, as do states E_1 and E_1
(i.e., at zero field at equilibrium the cycle is "centrosymmetric")
and the system is arbitrarily normalized such that the kinetic coef
ficient for the transition $E_4 \to E_3$ is unity. In this case, b repre
sents the zero field equilibrium ratio $[E_2]/[E_1]$ $(=[E_4]/[E_3])$, $\phi =$
$\exp(F\Delta\psi/RT)$, and δ and γ are the apportionment constants. For
further details see Astumian et al. (1987) and Westerhoff et al.
(1986).

the catalytic cycle of the enzyme. The rate constants are expressed in
terms of "b", a bias factor which represents the fundamental enzyme asymmetry, and "ϕ", the electric sensitivity and its apportionment coefficients, δ and γ, which represent the partitioning of charge transfer
between the forward and reverse rates, as shown in Fig. 4. z_s is the
charge on the substrate S and $[S_i]$ and $[S_o]$ are its concentrations inside
and outside the cell, respectively. When these concentrations are considere
to be in the steady state (i.e., constant) the thermodynamic driving force
and net cyclic flux can be calculated according to the method of Hill (1977)
Using these equations we can easily plot the flux versus the electric field
which for charged S at equal inside and outside concentrations is the same
as a flow-force plot. A few typical examples are shown in Fig. 5. Of particular interest to note is the non—monotonic behavior of the flow versus
driving force ($\Delta\psi$) in Fig. 5c. Descriptively, at very high fields, the
enzyme is held with its bound charge all on one side of the membrane, thus
preventing catalytic turnover, even though the "driving force" is increased.
Also, note that in the case of the uncharged S, appreciable activity is seen
only within a narrow range of $\Delta\psi$ (Fig. 5b) allowing for effective regulation
of the membrane transporter by modulation of the membrane potential. In
this case, regulation is achieved by shifting a conformational equilibrium
within the normal enzyme catalytic cycle, rather than inducing the molecule
to attain an entirely different "inactive" conformational state outside of
the normal catalytic cycle. Fig. 5A is a "typical" flow force relation for
charge S where the tranlocator transitions are considered to have no intrinsic electrical sensitivity.

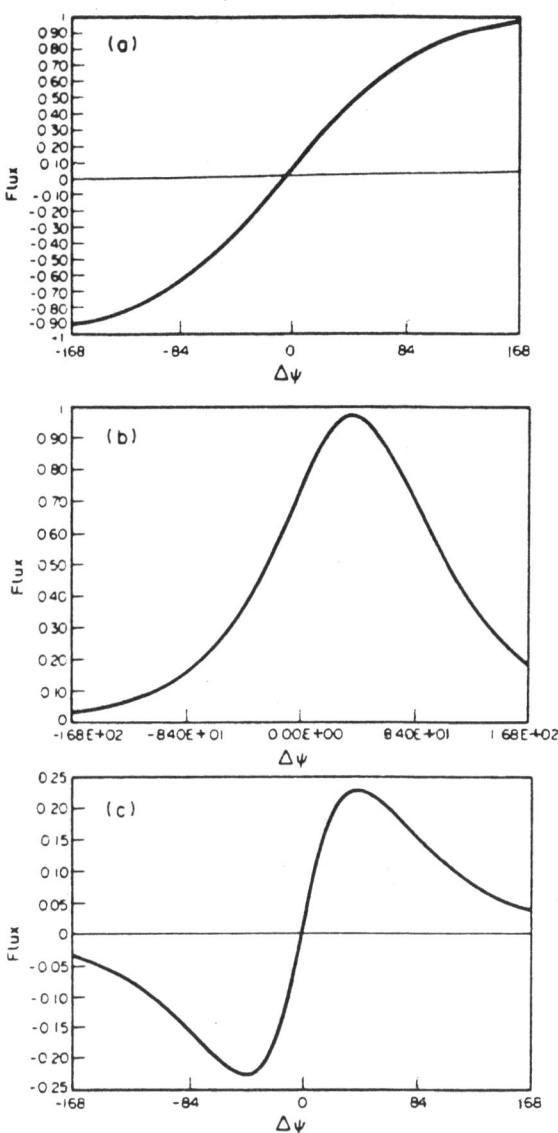

Fig. 5. Calculation of the transport flux of S catalyzed by the enzyme cycle of Fig. 4 for three different cases. For simplicity, all zero field rate constants were taken to be equal to 1, and all the apportionment factors were assigned to be 1/2. (a) S is considered to bear a charge of 1 but there is no intrinsic electric field dependence due to the enzyme transition $E \rightleftharpoons E^*$. In this case, $\Delta\psi$ is part of the thermodynamic driving force for the translocation of S, but plays no role in determining the catalytic constants K_m and k_{cat} for the enzyme. The concentration ratio $(S_{in}]/[S_{out}] = 1$ is maintained constant. (b) S is considered to be uncharged, but the transition $E \rightarrow E^*$ is intrinsically electrically sensitive due to the intramolecular transfer of a negative charge across the membrane potential. In this case, $\Delta\psi$ does not represent a driving force for translocation (hence the constant ratio $[S_{in}]/[S_{out}] = 0.1$) but does govern the catalytic constants of the enzyme. Thus, the protein is catalytically active only within a narrow band of membrane potentials. (c) In this case, S is considered to be charged ($z_S = 1$) and the enzyme transition to be electrically active. Thus, although the magnitude of the driving force is large at high positive or negative potentials, the flux induced drops off since the enzyme loses catalytic activity, essentially being trapped in one state or the other.

THE EFFECT OF EXTERNAL OSCILLATIONS OR FLUCTUATIONS ON THE BEHAVIOR OF CARRIER PROTEINS (AND ENZYMES IN GENERAL)

Spontaneous oscillations of either chemical concentrations or thermo-dynamic parameters have in recent years been shown to occur in many biological contexts. For example, in the glycolytic pathway, the enzyme phosphofructo-kinase (PFK), by virtue of a complicated feedback regulatory scheme, engenders an oscillatory ATP output from a nonoscillatory input of phosphofructose (for review see Hess and Boiteux, 1971). In this case, it is the ATP concentration ([ADP] and [P_i] are considered to be stationary) which undergoes periodic oscillations. In many excitatory (e.g., Chay and Rinzel, 1985) and secretory (e.g., Chay and Keizer, 1983) cells, $\Delta\psi$ undergoes large amplitude, spontaneous oscillations (+ or - 40 mV in some cases), where the trigger for this behavior

is often unknown. Also of recent interest has been the effects of periodic
input signals on enzyme-catalyzed reactions. Serpersu and Tsong (1984) have
shown that oscillating electric fields symmetric about zero potential can
induce the Na,K ATPase to "catalyze" the active transport of K^+ (or Rb^+)
against its electrochemical gradient. Tsong and Astumian (1986), and subsequently Westerhoff et al. (1986), have shown that this can be understood
in terms of the interaction between an oscillating electric field and
protein conformational equilibria. Schell et al. (1987) have theoretically
studied the effects of an oscillatory ATP level (due to PFK for example) on
the efficiency of utilization of the hydrolytic energy of ATP to pump proton
against its electrochemcal gradient.

In order to understand how an enzyme can transduce energy from an
oscillating perturbation, even if that perturbation does not at any time
provide a "thermodynamic driving force" around the catalytic cycle as
calculated from steady-state theory, consider the four-state translocator
shown in Fig. 4. If b>1, states 4 and 2 are more stable than states 1 and
3. In the limit that b>>1, states 1 and 3 may be treated as "steady-state
intermediates". In an a.c. field, the positive phase (+in) favors the righthand states, E1 and E2, while negative phase (-in) favors states E3 and E4,
as shown below.

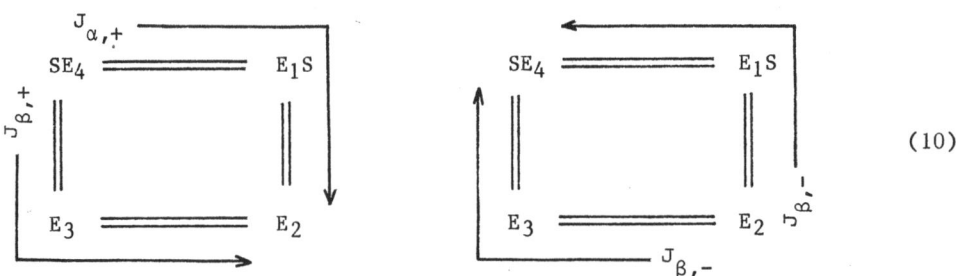

$$(10)$$

During each phase, both clockwise (J+) and counterclockwise (J-) flux
occurs. The instantaneous excess clockwise flux (= J+ - J-) is equal to
$J_\alpha - J_\beta$. This quantity integrated over one field cycle is the net flux
induced per field cycle. If $[S_i]>[S_o]$ and yet net clockwise flux is induced
this proves that work is being done, and the energy for doing this work
must come from the a.c. field (which we know in fact does contain energy as
witnessed by an a.c.-driven motor), even though the field does directly
interact with the output $S_o \rightarrow S_i$ reaction. In Fig. 6, we have plotted the
flux for a simple case as a function of time. These results are in agreemen
with those obtained by numerical integration of the rate equations (Tsong
and Astumian, 1986; Astumian et al., 1986), as well as with the subsequent
calculations done by matrix inversion with $[S_i]$ and $[S_o]$ held constant
(Westerhoff et al., 1986; Tsong and Astumian, 1987). Since the only interaction is between the field and the enzyme in terms of mechanisms for
absorbing energy from the perturbation, this presents a case of energy
transduction in which the enzyme is the sole intermediary between the input
energy source, and the output reaction in which the energy is stored. From
an evolutionary point of view this may be a very important consideration,
since if the structure of the protein evolves in such a fashion that it has
an electrically sensitive conformational transition, energy can be transduce
from an electrical or Coulombic input to whatever reaction the enzyme happen
to catalyze. Thus, instead of asking how the synthesis of ATP is a priori
influenced by protons moving across the membrane, we could inquire as to how
the ATPase is Coulombically influenced by the movement of a positive charge,
e.g., proton, but perhaps other cations as in alkalophiles (Guffanti et al.,
1983), from one place to another. Tsong and Astumian (1986) and Astumian
et al. (1986) have proposed a mechanism for the harvesting of Coulombic ener
from a proton gradient, where the F_o portion of the F_oF_1 ATPase acts as a
gated proton channel. Westerhoff et al. (1987) have shown how Coulombic

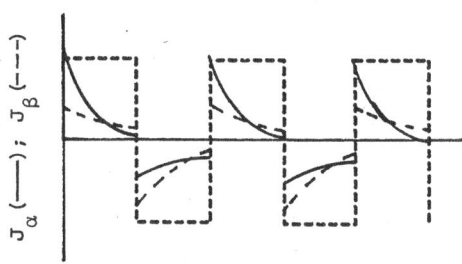

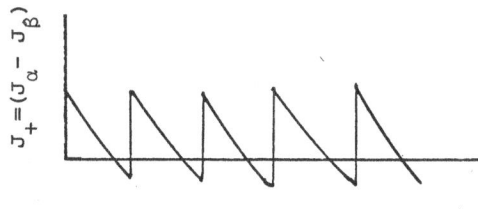

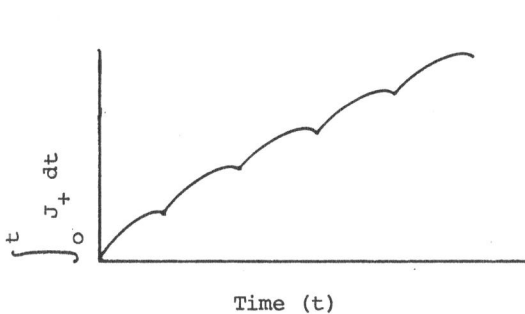

Time (t)

Fig. 6. When a perturbation is applied to an enzyme system such as shown in Fig. 4, the system must adjust to the new condition, and there will be a redistribution between states SE_4 and E_2. This redistribution can take place via two paths, one involving binding and release of S from inside (the α branch) and the other from the outside (the β branch). Because of the kinetic asymmetry (predominately characterizeed by the bias, b, but there are other possibilities), at any moment in time, the α-flux, J_α, may be different than the β-flux, J_β. The instantaneous clockwise flux, $J_+ = (J_\alpha - J_\beta)$, represents the rate of transfer of S from outside to inside at any instant. The amount of S transferred from the beginning of the perturbation up to any time, t, is given by the integral $\int_0^t J_+ dt$. (a) This figure illustrates schematically the α and β fluxes as a function of time for the case $b>1$, $[S_{in}] = [S_{out}]$. Notice that the two curves cross during each period such that $\int_0^\infty J_\beta dt$ (for $z_s = 0$), if a constant perturbation were allowed to persist long enough for equilibrium to be attained. The reversal of the field, however, "catches" the system at a point where net clockwise transport has occurred <u>for both signs (phases) of the perturbation</u>. (b) The clockwise flux, J_+, as a function of time, where $J_+ = J_\alpha - J_\beta$. (c) Demonstration of the <u>net</u> clockwise flux, $\int_0^t J_+ dt$, as a function of time.

energy can be harvested even when $\Delta\mu H^+$ is at all times zero so long as the pH and $\Delta\psi$ portions are fluctuating in compensation of one another. These concepts have been extended to demonstrate the possibility of direct harvesting of energy from the electron transport chain (ETC) without the pumping of a proton if the ETC protein is in the vicinity of the F_oF_1 ATPase when electron transport occurs (Tsong and Astumian, 1987).

Note that the above phenomena are purely kinetic in nature. Energy transduction occurs because the magnitudes of the rate constants may be such that transient net flux accumulates during the initial stages after reversal of the field. For the above case, the averge flux (and in fact the flux for each system individually) for a number of systems at steady state, where some systems are at $+\Delta\psi$ and others are at $-\Delta\psi$, would be identically zero if $[S_i]=[S_o]$. This phenomena is qualitatively different than in those cases

where the oscillating parameter is a part of the driving force of the output
reaction. Here, the induced net flux (with $[S_i]=[S_o]$) can be demonstrated
by an ensemble average. If the oscillation causes the output free energy
(ΔG_o) to oscillate symmetrically about its steady-state value, energy can
be transduced by enzyme rectification (Astumian et al., submitted) additiona
to energy transduction via any electroconformational interactions with the
enzyme. If the ΔG_o oscillation is not symmetric, energy can be transduced
directly from the "excess" oscillatory free energy, as well as by the above
two mechanisms.

Even more common in biological systems are stochastic fluctuations of
concentrations or environmental parameters. Any membrane protein in the
vicinity of an ion channel, for example, is bound to experience a large
amplitude fluctuation in the local electric field strength each time the
channel opens or closes. It should be remembered that such fluctuations
are motivated by the ionic electrochemical gradients across the membrane.
Nonequilibrium fluctuations, unlike those occurring at equilibrium, can
provide a mechanism for triggering the release of stored energy to do
useful work. Horsthemke and Lefevre (1984) have extensively investigated
the effects of "noise" on phase transitions showing that random temporal
variation in a system's environment can cause a shift from one stationary
state to another. They have termed these effects to be "noise-induced"
transitions. As discussed before, Offner (1980, 1984) has suggested that
stochastic fluctuations may trigger a cascade allowing for the utilization
of the energy in an ion gradient for signal amplification. More recently,
Astumian et al. (1987) have shown that random fluctuations can drive an
ordinary enzyme to catalyze the reaction for which it is specific away from
equilibrium. This can be understood in terms of the previous work of Tsong
and Astumian (1986) and Westerhoff et al. (1986) since the noise used in
the above study, as is the case for most noise patterns, could be expressed
as a Fourier series of sinusoidally oscillating components (Astumian et al.,
in preparation).

For the special case of dichotomous Markovian noise, where one consider
only two levels of the noise parameters ($\Delta\psi$ in our case), the behavior of
the system can be investigated analytically, where the transitions from one
noise state to another are treated as "chemical" transitions within a chemic
kinetic diagram (Astumian et al., 1987 a,b,c). For the four-state model of
Fig. 5, this gives an overall eight-state model as below.

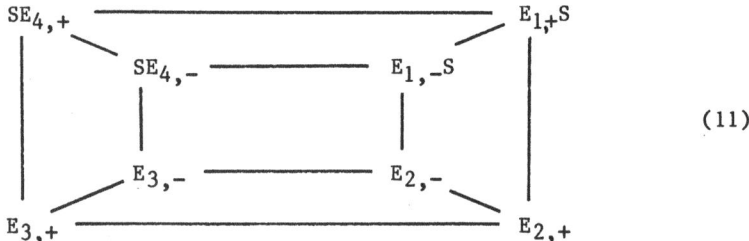

(11)

The outer "box" represents the enzyme operating at $+\Delta\psi$ and the inner "box"
the enzyme working at $-\Delta\psi$. The diagonal transitions are the switches from
+ to - noise states. If the noise is generated externally to the system or
internally by, e.g., the opening and closing of an ion channel across which
there is a large ion electrochemical gradient, the diagonal transition rate
constants may be considered to be independent of the state from which they
orginate. This condition leads to work being done on the system, and has
been referred to as autonomous noise (Astumian et al., 1987a). On the other
hand, noise arising in a system at equilibrium would have transition rate
(and equilibrium) constants which would be dependent on the state (i.e.,
the electrical properties of the state) from which the noise transition

originates, and when these correlations are taken into consideration, no
work results. This equilibrium fluctuation has been termed endogenous
noise. It must be remembered that even if the macroscopic parameters (such
as the "average" potential and the standard deviation of $\Delta\psi$) are identical
between autonomous and endogenous noise, microscopically, the lack of correlation between the noise transition and enzyme state in the former reflects
the occurrence of a free-energy dissipating process (transport of ions down
their electrochemical gradients or input of energy by an external field
generator), whereas the latter, state-correlated noise, does not require
any free energy source. Typical results for external noise-induced free-
energy transduction are shown in Fig. 7.

COUPLED TRANSPORT BY COULOMBIC PROTEIN-PROTEIN INTERACTIONS

Based on the above considerations, we propose a mechanism by which the
energetically downhill transport of one substance can be coupled to the
uphill transport of another, based solely on protein-protein interactions.
Thus, the flows of two uncharged (or one charged and one uncharged, as in
the lac permease system) substrates can be linked. Let us imagine two
enzymes similar to the one depicted in Fig. 4, where one such enzyme cata-
lyzes the equilibration of substance S across the membrane and the other
the equilibration of substance P. The combination of these two four-state
diagrams into a single diagram requires 16 states (Astumian et al., 1987
a,b). If the two enzymes are localized in the membrane close to one another
(we took a radial distance of 10 Å in the calculations we will discuss),
the electric sensitivity parameters for each transition can be calculated
directly from Coulomb's law and the Pythagorean theorem. The fluxes of S
and P, J_S and J_p, can be directly calculated by matrix inversion for the
16-state model given input ΔG_i and output ΔG_o free energies. We assumed
P to be the energy supplier, moving down its gradient, and S to be driven
against its gradient. The question of the preferred direction of transport,
and whether the coupled system acts as a sym- or antiport is answered by
the values of the bias factor for the S and P transporters. If b>1, the
preferred direction of transport will be out to in, and if both b_S and b_P

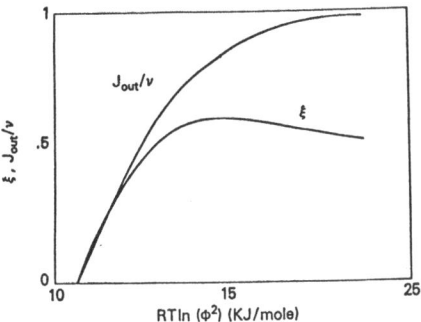

Fig. 7. Results of calculations based on the eight-state diagram arising
        from Fig. 4 subjected to dichotomous electric noise, where the
        noise transition constants were considered to be autonomous of
        influence by the "target" enzyme. Once again, all parameters were
        identical to those shown in Fig. 4, except that the ϕ during the
        "negative" noise pulse was taken as the inverse of that during the
        "positive" noise impulse. ξ, the "effectiveness" is defined
        as $\xi = -J_{out} \cdot \ln(\rho^2/\ln(\phi))$. The parameter values were b = 500,
        $\phi = \exp(-\Delta\mu_S/2\,RT) = .13$, and $\delta = \gamma = 1/2$. What this figure
        shows is that external noise can effectively maintain a significant
        concentration gradient of a substance with perfectly reasonable
        values of the interaction parameter, ϕ, and that the yield
        (cycles/ "noise" pulse) can approach unity.

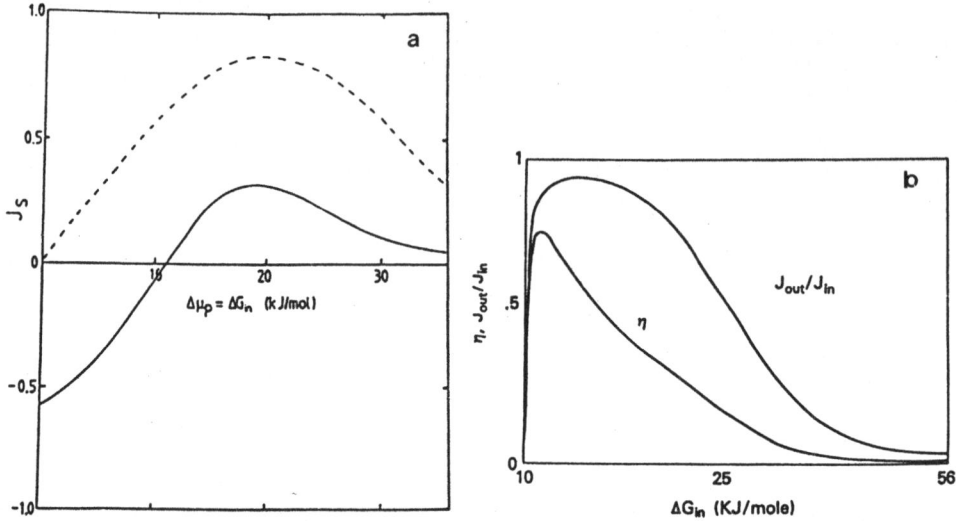

Fig. 8. Results of calculations based on the combination of two four-state
translocators such as shown in Fig. 4, one for substance S and one
for substance P. The resulting diagram contains 16 states, and the
calculations were done by matrix inversion. Both S and P were
uncharged, and all interactions were considered to be fully reciprocal. The enzymes were assumed to be 10 Å apart (radial distance)
and thus all interactions could be calculated on the basis of
Coulomb's law. Parameters were (a) $b = 500$, $\phi = 16$, $\delta = \gamma = 1/2$;
(b) $b = 5000$, $\phi = 1600$, $\delta = \gamma$ $1/2$. (a) We see that although only
electrostatic interactions are considered, the presence of $\Delta\mu_p$ can
sustain a gradient of S with $\rho = .13$, and even cause the S translocator to cycle against its "equilibrium" direction (solid line)
or induce sizable flow in the absence of a gradient of S (dashed
line). Interestingly, we see that the flow of S decreases at very
high input driving force. (b) Here we see that the thermodynamic
efficiency, η, can be as high as .75 and that the flow ratio can
approach unity, both values being consistent with what is observed
experimentally for biological energy transducers. Thus, so long
as we know the simple enzyme characteristics for uncoupled enzymes
and have a detailed knowledge of the molecular mechanisms of the
catalysis by which interaction parameters (ϕ) can be determined,
the energy transduction by the coupled enzymes can be calculated
from first principles and arise solely from the properties of the
simple enzyme catalysis and from the physical proximity of the two
molecules.

are greater than (or less than) unity, the coupled system will act as a
symport. Typical results obtained are shown in Fig. 8, where in 8a the
flow of S against a gradient of 10 kJ/mole at steady state is plotted vs.
the input free energy in the P gradient. Fig. 8b illustrates that the
flow ratios and efficiencies of energy transduction based solely on
enzyme-enzyme interactions can be of similar magnitude to those seen
experimentally. Here, the calculation was based on the assumption of
triply negative charged arms and an effective dielectric constant of 12,
giving $\phi=1600$.

SUMMARY AND CONCLUSION

In this paper, we have discussed how structures responsible for trans-membrane transport interact with electric fields across the bilayer ($\Delta\psi$) and how this interaction modifies their function. Electroporation was treated as an extreme example of this, where an initial electroconformational event is followed by thermal denaturation induced by very rapid field-induced charge transport. Less drastic mechanisms for regulation of membrane permeability were pointed to in terms of influencing protein conformational equilibria by modulation of the static membrane potential.

Dynamic electric fields were implicated in energy transduction and active transport. It was shown that energy-driven fluctuations or oscillations of $\Delta\psi$ can cause an ordinary enzyme, with no special properties other than an electrically sensitive conformational transition, to drive the reaction for which it is specific away from rather than towards equilibrium. This result may be of importance in understanding in vivo energy transduction, since local electric fields across membranes are inherently "noisy" due to the opening and closing of ion channels.

Finally, these concepts were used in formulating a model for coupled transport based solely on Coulombic interactions. It is not so important that the interactions be only Coulombic, but rather we stress that the coupling of two fluxes can be accomplished solely by protein-protein inter-actions which are large or small, depending on the location of the enzymes within their catalytic cycles (i.e, in state space). It was shown that calculations based on physically realistic values can reveal efficiencies well within the ranges observed for physiologically viable cotransporters.

ACKNOWLEDGEMENT

This work was supported in part by NIH and ONR grants to T.Y.T.

REFERENCES

Astumian, R. D., Chock, P. B., and Tsong, T. Y., 1986, J. Electrochem. Soc., 133:124c.

Astumian, R. D., Chock, P. B. Tsong, T. Y., Chen, Y.-d., and Westerhoff, H. V., 1987a, Proc. Natl. Acad Sci. U.S.A., 84:434-438.

Astumian, R. D., Chock, P. B., Westerhoff, H. V., and Tsong, T. Y., 1987b, in: "Enzyme Dynamics and Regulation," P. B. Chock, C. Huang, L. Tsou, and J. H. Wang, eds., Sprinter, New York, in press.

Astumian, R. D., Chock, P. B., Tsong, T. Y., and Westerhoff, H. V., 1987c, J. Chem. Phys., submitted.

Chay, T. R. and Keizer, J., 1983, Biophys. J., 42:181-190.

Chay, T. R. and Rinzel, J., 1985, Biophys. J., 47:357-366.

Hess, B. and Boiteux, A., 1971, Ann. Rev. Biochem., 40:237-258.

Hill, T. L., 1977, "Free Energy Transduction in Biology," Academic Press, New York.

Johnstone, R. M. and Laris, P. C., 1980, Biochim. Biophys. Acta, 599:715-730.

Kinosita, K., Jr. and Tsong, T. Y., 1977a, Biochim. Biophys. Acta, 471: 227-242.

Kinosita, K., Jr. and Tsong, T. Y., 1977b, Nature (London), 268:438-441.

Kinosita, K., Jr. and Tsong, T. Y., 1978, Nature (London), 272:258-260.

Knox, B. E. and Tsong, T. Y., 1984, J. Biol. Chem., 259:4747-4763.

Lo, M. M. S., Tsong, T. Y., Strittmatter, S. M., Hester, L. D., and Snyder, S., 1984, Nature, 310:792-794.

Neumann, E., Schaefer-Rider, M., Wang, Y., and Hofschneider, P. H., 1982, EMBO J., 1:841-845.

Offner, F. F., 1980, J. Phys. Chem., 84:2652-2662.

Offner, F. F., 1984, Biophys. J., 46:447-461.

Schell, M., Kundo, K., and Ross, J., 1987, Proc. Natl. Acad. Sci. U.S.A., 84:424-428.

Serpersu, E. H. and Tsong, T. Y., 1983, J. Membrane Biol., 74:191-201.

Serpersu, E. H. and Tsong, T. Y., 1984, J. Biol. Chem., 259:7155-7162.

Sowers, A. E., 1986, J. Cell Biol., 102:1358-1362.

Teissie, J. and Tsong, T. Y.,1980, J. Membrane Biol., 55:133-140.

Tsong, T. Y., 1983, Bioscience Reports, 3:487-505.

Tsong, T. Y. and Astumian, R. D., 1986, Bioelectrochem. Bioenerg., 15:457-476.

Tsong, T. Y. and Astumian, R. D., 1987, Prog. Biophys. Mol. Biol., in press.

Tsong, T. Y. and Astumian, R. D., 1988, Ann. Rev. Physiol., in press.

Westerhoff, H. V., Tsong, T. Y., Chock, P. B., Chen, Y.-d., and Astumian, R. D., 1986, Proc. Natl. Acad. Sci. U.S.A., 83: 4734-4738.

Westerhoff, H. V. and Astumian, R. D., 1987, in: "Towards a Cellular Enzymology,", A. Klysov, S. Varfolomeev, and G. R. Welch, eds., Plenum Press, New York.

Westerhoff, H. V., Kamp, F., Tsong, T. Y., and Astumian, R. D., 1987, in: "Mechanistic Approaches to Interactions of Electromagnetic Fields With Living Systems," M. Blank and E. Findl, eds., Plenum Press, New York, in press.

Witt, H., Schlodder, E., and Graber, P., 1976, FEBS Lett., 69:272-276.

CELL BEHAVIOUR IN ELECTRIC FIELDS

V. Pastushenko

Frumkin Institute of Electrochemistry of the
USSR Academy of Sciences
Moscow, USSR

The electric field is one of the most important and
widely used instruments in biological studies. A number of
papers were published recently on cell movement in nonhomogeneous alternating electric fields[1-4] and also in rotating
electric fields.[5,6] Apart from their scientific value, these
phenomena may be used in some practical applications, such
as cell separation, cell fusion, etc.[1,7] As it is well recognized, cell fusion is one of the promising approaches in
gene engineering. Well known is the so-called electrically
induced fusion which is based on the phenomena of cell dielectrophoresis and electric breakdown of the cell membrane.[7]
On the other hand, study of the electric breakdown may help
in elucidating the questions of the mechanical and energy
properties of cell membranes.[7-10] Theoretical analysis of the
electric breakdown as the process of the through-pore formation and growth shows that this phenomena has much in common
with traditional electrokinetic effects, because this process
is facilitated by electric forces acting on the pore walls.[11]
    The aim of the present paper is twofold:
1. To demonstrate the intrinsic connection between the phenomena of cell dielectrophoresis and electrorotation /cell rotation in rotating electric field/.[12,13]
2. To present some recent theoretical results from the field
of electric breakdown, namely to demonstrate the possibility
of the appearance of the new type of dissipative structure
by analysing the behaviour of cell/lipid vesicle in strong
electric field.
    It should be mentioned that the theory of cell dielectro-
phoresis has its own history, the essential part of which is
presented by studies of dielectrophoresis in its pure form,
i.e. movement of the dielectric particle in dielectric fluid
in the presence of nonhomogeneous electric field[1], and also
the studies of dipolophoresis in colloidal chemistry /in which
case non-zero electric conductance is accepted for both particle
and external medium/.[2-4]
    The theory of dielectrophoresis of biological particles
was presented by Pohl.[1] Unfortunately, his approach cannot
be considered to be satisfactory because it fails to give

reasonable results in the low-frequency domain. The correct
theory for the simpliest possible model of cell membrane of
the type of Pauly and Schwan[14] was presented in papers of
Sauer[2], and Chizmadzev and Kuzmin[12] The phenomenon of electrorotation was studied in a series of papers[5],[6] Separately from
the phenomenon of dielectrophoresis for practically arbitrary
cell membrane structure. The results of Pastushenko et al.[13]
demonstrate the possibility of a unified approach to both dielectrophoresis and electrorotation. The general outline of
this theory is as follows:

1. Starting from the Maxwellian stress tensor, it may
be demonstrated that in the presence of weakly nonhomogeneous
field the force $\vec{F}$ acting on a particle in a liquid, whether
conducting electric current or not, may be expressed in terms
of its dipole moment $\vec{d}$ and the gradient of the external electrical field $\vec{E}_{ext}$, i.e.

$$\vec{F} = (\vec{d}\, \nabla)\, \vec{E}_{ext}.$$

Essentially, $\vec{d}$ is the coefficient in the expression for the
electric potential φ far enough from the particle:

$$\varphi = -\vec{E}_{ext}\, \vec{r} + \frac{\vec{d}\, \vec{r}}{4\pi\varepsilon r^3}, \quad r >> R.$$

Here $\vec{r}$ is the radius-vector of an arbitrary point and R is
the radius of the particle / for simplicity, spherical geometry
is assumed/. The quantity ε is the real permittivity of the
external medium.

2. The parameter d is conveniently calculated as the
real part of the complex dipole moment D. It may be shown that
the torque acting on the particle is proportional to the
imaginary part of D. Introducing the complex polarizability of
the particle χ defined by the equation

$$D = \chi E_{ext},$$

one may state that the connection between dielectrophoresis
and electrorotation is given by the well-known relations of
Kramers and Kronig.

3. Analytical results were obtained for the nearly arbitrary structure of the spherical cell shell/membrane/ in the form:

$$\tilde{\chi} = 1 - \frac{3}{W+1}; \qquad W = \rho_e \left(\rho_c + \frac{Z}{R}\right)^{-1};$$

$$\rho_e = (\sigma_e + i\omega\varepsilon_e)^{-1}; \quad \rho_c = (\sigma_c + i\omega\varepsilon_c)^{-1}; \quad V \equiv \frac{4}{3}\pi R^3.$$

Here $\tilde{\chi} = D/V E_{oext}$ is the normalized complex polarizability,
E_{oext} is the amplitude of the external field at the point of
the cell centre location, ω is the angular frequency, σ_e, ε_e
and σ_c, ε_c are specific conductivities [S/m] and permitti-
vities [F/m] of the external medium and the inner volume of
the cell respectively, Z is the specific impedance of the cell
shell /membrane/.

4. Characteristic lines of the $\tilde{\chi}$ function were analyzed
on the complex plane W. The definitions of the characteristic
lines are given by equations Re $\tilde{\chi}$ const and Im $\tilde{\chi}$ const,
respectively.

The computation of the electric forces acting on the pore walls during electric breakdown is also based on the use of Maxwellian stress tensor. For the case of two pores present on the cell/vesicle membrane the time-domain dynamics of the pore radii was analysed.[15] It was shown that due to electric interaction between the two pores the forces acting on the pores may not be described as the gradient of a function of the pore radii. These observations suggest the existence of dissipative structure in the statistical ensemble of the cells/ vesicles posessing two pores. The general properties of this structure were analysed. The qualitative changes in the behaviour of the system were demonstrated with increasing external field strength.

REFERENCES

1. Pohl, H., 1973, Dielectrophoresis, Cambridge University Press, Cambridge.
2. Sauer, F.A., 1983, in: Coherent Excitations in Biological Systems, H.Frolich and F.Kremer, eds., Springer-Verlag, Berlin Heidelberg, pp.134-144.
3. Petkanchin, I., Stoilov, St., 1972, in: Poverkhnostnye sily w tonkikh plionkakh i dispersnykh systemakh, Nauka, Moskwa, pp.134-136 / in Russian/.
4. Shilov, W.N., Estrela-Ljopis, W.R. Ibid, p.115-132.
5. Arnold, W.M., Zimmermann, U., 1982, Z.Naturforsch., v.37 c, p.908.
6. Fuhr, G., Glaser, R., Hagedorn, R., 1986, Biophys.J., v.49, No.2, pp.395-402.
7. Zimmermann, U., 1982, Biochim.Biophys.Acta, v.694, pp.227-277.
8. Abidor, I.G., Arakelyan, V.B., Chernomordik, L.V., Chozmadzhev Yu.A., Pastushenko, V.Ph., Tarasevich, M.R., 1979, Bioelectrochem.Bioenerget., v.6, pp.37-52.
9. Pastushenko, V.Ph., Chernomordik, L.V., Chizmadzhev, Yu.A., 1985, Biologicheskie membrany /Russ./, v.2, No.8, pp.313-319 /in Russian/.
10. Pastushenko, V.Ph., Petrov, A.G., 1984, in: IY Winter school on Biophysics of membrane transport. Poland.
11. Pastushenko, V.Ph., Chizmadzhev, Yu.A., 1982, Gen.Physiol. Biophys., v.1, pp.43-52.
12. Chizmadzhev, Yu.A., Kuzjmin, P.I., Pastushenko, V.Ph.,1985, Biologicheskie membrany /Russ/, v.2, No.11, pp.1147-1161 /in Russian/.
13. Pastushenko, V.Ph., Kuzjmin, P.I., Chizmadzhev, Yu.A.1985, Studia Biophysica, v.110, No.1-3, pp.51-57.
14. Pauly, H., Schwan, H.P., 1959, Z.Naturforsch., B.14b, H.2, pp.125-131.

TOPICS OF BIOELECTROCHEMISTRY EXTENDED BY ELECTROMAGNETIC FIELD EFFECTS

Hermann Berg
Academy of Sciences of the GDR
Central Institute of Microbiology and Experimental Therapy, Department of Biophysical
Chemistry, DDR-6900 Jena (GDR)

1. HISTORICAL INTRODUCTION

In the eighteenth century the term galvanism (animal electricity) and, later on, electrophysiology was commonly used /1/. Nowadays, the general denotion "bioelectrochemistry" has been introduced and may be considered as a "daughter" of electrochemistry and a "granddaughter" of chemistry. It is worthwhile to remember briefly the beginning of Luigi Galvani's wellknown discovery (1786) of the convulsion of muscles in contact with different metals. Galvani wrongly believed that many little organic Leyden yars are producing electricity in the muscle. This hypothesis of "animal electricity" was suggested by the actions of the battery like organs of "electric fish".

The experimental approach of L. Galvani /2/ was carefully modified by Alessandro Volta (1745-1827) /3/, who pointed out that this kind of electricity comes from outside to the muscle - from the contact between both metals. The muscle acts as an indicator (electroscope) only, whereas the nerves are working as conductors. Volta's interpretation of galvanism as contact electricity was erroneous, too. The right description of Galvani's and Volta's experiments was given by Johann W. Ritter (1776-1810) /4/ at Jena, who pointed out the relationship between Volta's series of metal potentials and their oxidation states. Furthermore, Ritter discovered chemical reactions at electrode surfaces and showed unambiguously that it is the chemical process which is responsible for the galvanic action of living beings: "Solange wir keine strengeren Beweise haben, daß chemische Kräfte es nicht allein sind, die im lebenden Körper wirken, so lange dürfen wir auch keine neuen Kräfte dafür annehmen." (Fragment 393 (1810)). Because Ritter describes the animal electricity in terms of chemistry he is, as Wilhelm Ostwald pointed out /5/, the founder of scientific electrochemistry. Moreover, Ritter made contributions to describe phenomena like excitation, accommodation, and other sensoric effects depending on the voltage of his piles. Therefore J.W. Ritter founded also bioelectrochemistry and stimulated scientists like

Th. v. Grotthuss (1785-1822) /6/, E. Du Bois-Reymond (1816-1896) /7/, F. Hoppe-Syler (1825-95), H. v. Helmholtz (1821-1894) and others /12/.

2. TOPICS, PROCESSES AND AIMS

The development of electrochemical research in Biology during 200 years reveals that in all biochemical reactions also electrical charges are inseparably involved as in the processes of exchange, transport, interaction, induction, proton transfer etc. connected with internal electromagnetic fields and soliton excitation, too.

Consequently, the primary aim (I) of bioelectrochemistry is the elucidation of such processes in living systems (in vivo) utilizing the current knowledge of electrochemistry. The secondary task (II) of bioelectrochemistry is the analysis of all ingredients in situ or in vitro, e.g. biopolymers and monomers, membranes, effectors, inductors etc. The third proposition (III) of bioelectrochemistry consists in the extension of applications, e.g. in
- basic processes,   - gentechnology,
- biosynthesis,   - neurosciences,
- biotechnology,   - medicine etc.
Of course there are more and more overlapping interrelations with other disciplines as organic electrochemistry, analytical chemistry, biophysics, biophysical chemistry, bioelectricity, electrophysiology, bionics etc.

Nevertheless bioelectrochemistry today contributing to discoveries in life sciences is an independent discipline developed by international societies, too, e.g.: Bioelectrochemical Society (BES), Bioelectromagnetic Society (BEMS), Bioelectric Repair and Growth Society (BRAGS) and by international symposia e.g. :
- Symp. on Electrochem. in Biolog. and Medicine, New York, April 1953 /14/.
- I. Jena Symp.: Die Polarographie in der Chemotherapie, Biochemie und Biologie, 1962 /8/.
- III. Jena Symp.: Elektrochemische Methoden und Prinzipien in der Molekularbiologie, 1963 /9/.
- Internat. Symp. Bioelectrochemistry at 1. Rome 1971 /10/, 2. Pont a Mousson 1973, 3. Jülich 1975, 4. Woods Hole 1977, 5. Weimar 1979, 6. Kiryat Anavim 1981, 7. Stuttgart 1983, 8. Bologna 1985 /11/, 9. Szeged 1987.
- Workshop: Biotechnology of electron transfer processes (models and approaches), November 1977, Philadelphia
- US-Australia joint seminar on bioelectrochemistry, Pasadena, Calif. 1979.
- Gordon Conferences on Bioelectrochemistry : Tilton N.H. 1980, Proctor Acad. 1982, Tilton N.H. 1984, Tilton N.H. 1986.
- Schools on Bioelectrochemistry : Erice (Sicily) 1982, 1984 /13/.
- Symp. Bioenergetics (Structure and Function of Energy Transducing Systems, May 1986 Nagoya.
- IX. Jena Symp.: Photodynamic and Electric Effects on Biopolymers and Membranes 1982 /39/.
- XI. Jena Symp.: Bioelectrochem. in Biotechnol. 1987 /40/.
- XII. Jena Symp.: Trends in Bioelectrochemistry of Biopolymers and Membranes 1988.

Besides conference proceedings a lot of monographs and reviews have been published /12-40/, unfortunately a textbook on bioelectrochemistry is still missed. Characteristic topics of bioelectrochemical research according to the three levels I-III mentioned will be enumerated :

I. Bioelectrochemistry of living systems

Since life processes proceed in colloid solutions of polyelectrolytes, electrolytes and ionic effectors mostly at charged membranes the bioelectrochemical field of research is an universal one for all biosystems. The sensors for measurement are microelectrodes combined with potentiometric, amperometric and conductometric equipments, or noninvasive techniques as dielectric measurements, nuclear magnetic resonance (e.g. for pH-determinations in tissues), electric and magnetic fields. The basic processes for electron exchange, proton transfer, ion transport, electrostatic interaction and current induction are :
- the respiratory chain /13, 23, 28/,
- the photosynthesis /12, 24, 28/,
- the ATP-synthesis /12, 23/,
- phase transitions of membranes and membrane transport /20, 22, 25, 26, 33, 35/,
- vision and photoreceptor potential /31, 32/,
- transfer of action potentials and synoptic transduction /28, 31, 34, 35/,
- cell surfaces /12, 29/,
- anaesthesia /12/.

Recently an electrochemical interpretation of the whole metabolism, that means a hypothesis of the living cell as an electrochemical system, has been postulated by M. Berry /46/. More complex bioelectrochemical processes are involved in : electrocardiography, electroencephalography, electroacupuncture, electroanaesthesy, enzymatic clotting processes etc.

II. Bioelectrochemistry of cellular ingredients

Besides the properties of isolated biological membranes, and artificial bilayers as models /14, 20, 22, 24/ the main interest is focussed on the ionic and (or) polyelectrolyte behaviour /50/, conformational transitions and interactions of
- proteins (enzymes) and aminoacids /17, 47, 48, 49, 53,54/,
- nucleic acids, nucleoproteins, and bases /11, 17, 49, 51, 52, 66, 67, 68/,
- polysaccharides, lipopolysaccharides /55/,
- radicals, ion effectors, coenzymes etc. /12, 16, 35, 55, 56, 57/.

III. Bioelectrochemical applications

The broad fields of bioelectrochemical sensors and techniques for electrophoretic separations is devoted mainly to electroanalytical chemistry /17/. However, this implication may be useful also for applications in bioelectrochemistry:
- basic electrochemistry : accellerating electrode processes (biocatalysis) /17/,
- analytical chemistry : polarographic determinations /15/, ion-selective electrodes /58/, enzyme electrodes /59/, electrophoretic separations /53/,

- electrosynthesis, cofactor reoxidation /54/, electrochemical transitions of steroids /60/,
- biotechnology : biological fuel cells /61, 62/, electrophoretic cell separations /12/, microbiological membrane
  electrodes /59/,
- genetechnology : dielectrophoresis of dividing cells /12/,
- neurosciences : ion-selective electrodes /58/,
- medicine : hormone electrodes /59/, immunselective electrodes /63/, implantable electrodes /13, 17/, stimulation electrodes (pacemaker) /17/.

A fast extension of such examples can be recognized for the
future /35, 40/.

The same holds true for integral theoretical concepts to
elucidate complex problems. For example the electrical network theory representing a new way (bondgraphs) for energy
storage and dissipation in biological hierarchies, inclusive
membrane diffusion (with and without a transducer), proton
pump, coupling of ATP reactions and protein synthesis etc.
/65/. Close connected with these kinds of membrane transport
are a lot of further models especially for pore formation by
the influence of electric field pulses /40/.

Since 1979 (5th Symposium on Bioelectrochemistry of the
Bioelectrochemical Society (BES)) different electromagnetic
fields turn out as an increasing research field in Bioelectrochemistry /42, 35/.

IV. Electromagnetic field effects

Two types of electric fields (electropulsation) have
been used successfully in biosystems :
    A) single high electric field pulses,
    B) weak pulsating electromagnetically currents and
       simple alternating currents.
Under both conditions fast events occur as polarization at
membrane surfaces, dipole induction, e.g. in proteins, ion
displacements followed by
in case A:
- pore formation and extension : electroporation
- conformational changes of
  membrane proteins etc.   : electrotransconformation
- enhancement of permeability,
  connected with release of
  cell ingredients   : electropermeabilization
- incorporation of substances   : electroincorporation
    --- genetransfer   : electrotransfection
    --- celltransformation   : electrotransformation
- fusion of cells and(or) pro-
  toplasts organelles etc.   : electrofusion
in case B (electrostimulation):
- phase transition in membranes : electrotransition
- conformational changes of
  proteins especially in pores : electrotransconformation.

Electroporation has been studied by the groups of E.
Neumann, T. Tsong and U. Zimmermann and others since more
than 10 years. Models for pore formation and permeation /40/
have been developed by many authors: J. Crowley /70/, D.
Dimitrov /71/, Yu. Chismadjew /72/, L. Chernomordik /73/,
V. Pastushenko /74/, I. Weaver /75/, J. Teissie /76/,

H. Berg /77/, I. Sugar /78/, U. Zimmermann /79/, V. Markin /80/, M. Blank /81/, R.W. Glaser /82/.

For practical purposes appropriate constructions for discharge chambers with suitable electrodes have been developed. In the case of electroporation and electroincorporation (electrotransfection) only simple chambers containing two flat electrodes and in between the cell suspension in electrolyte solution is required.

For electrofusion, however, the closest contact between the cell membranes is necessary before pulse treatment, which can be provided in the following ways /11, 32, 36-45/.
  a) mechanical pressure between movable electrodes by means of sedimentation or by the cell growth on limited surfaces,
  b) agglutination of cell membranes by polyethylene glycol (PEG), dextran etc.,
  c) dielectrophoresis in low conducting solutions or magnetophoresis of cells labeled by iron compounds.
According to these demands a)-c) the discharge chambers are constructed, some of them for microscopic observation others for large scale purposes (/40, 42, 45, 83/) e.g. :
types A: thin needle electrodes movable by a micromanipulator; flat electrodes in a small chamber, which can be handed in a centrifuge;
types B: large disc electrodes in a plastic macrochamber; coaxial fixed cylindrical electrodes for discontinuous treatment of cell suspension;
types C: parallel wire electrode (distance < 1 mm) or sputtered metal layers with gaps on slides /40/.

Electrofusion started in 1979 with two kinds of plant protoplasts (Rauwolfia and Hordeum; microscopic evidence) /84/ on the one hand and with auxotrophic yeast protoplast mutants (genetic evidence) /42, 85/ on the other.

Nowadays more than 100 examples of electrofusion are published /40, 83/, however, only in some cases biological tests and observations have been performed afterwards during longer times.

Protoplasts of microorganisms and plants electrofused are mainly : yeast /85, 86/, bacteria /87/, fungi /88/, barley and maize /89/, tobacco /90/.
The problems are :
  - the identification of fused cells (heterocaryons or hybrids) and the cultivation of the new plant from fusion products.
Animal cells can be fused directly : erythrocytes /91/ myeloma and lymphocytes /92, 93, 94, 95/ mouse blastomeres /96, 97/, sea urchin eggs /98/.
The problems are :
  - application of soft field energy in order to warrant viability and breeding of fusion products to get the new animal.

Electrotransformation started for eucaryotic cells with mouse lyoma cells (TK⁻)+virus plasmid /99/, for procaryotic cells with Bac. cereus + plasmid pUB110 /100/.
Nowadays mainly Igk genes are incorporated in mouse cells

/101/ or other plasmids into S.cerevisiae /102/, into tobacco /103/ maize /104/ etc.

V. Electrostimulation

Since 1973/74 the pulsating (pulsatile) electromagnetically induced currents (PEMIC, PEMF, PMF), simple alternating currents (AC) or even direct currents (DC) have been applied to cells tissues and organisms in order to stimulate the membrane permeabilization and some pathways of the metabolism /105/. In principle there may be some similar processes, e.g. polarization, ion displacement, dipole induction leading to conformational changes of pore proteins, enzymes and phase transitions of membranes, however, explanations by models are in the very beginning /115, 116/. Under the general thesis of an "electrochemical information transfer on cell membranes" the membrane proteins are an important kind of targets. Following PEMIC results of inducing active transport of K^+, Rb^+, Ca^{2+} by ATPases and ATP synthesis by ATP-synthetases a model for transduction of field energy to conformational states of different electric moment has been presented. Similar activations are supposed for these enzymes. In contrast the microelectrophoretic model postulates changes in stay time spent by a ion or an effector near membrane receptors, protein channels or binding sites of ionophores, and also membrane binding. The enhancement of permeabilization is described by it from the point of view of substance movement. It seems likely that both groups of models have to be combined in order to understand the phenomena better.

In the biological material a pulsating field strength in the millivolt to volt/cm range is produced by inductive or capacitative triggering. The inductive method uses a pair of Helmholtz coils yielding a rather homogeneous pulsing magnetic field lower than 20 Gauss with a peak- or pulse burst (5 ms duration): Frequency of < 100 Hz, whereas each pulse width is in the order of microseconds. Moreover static and oscillating fields have been combined. Overlooking many types of equipment it seems that not the shape of the field signal but its frequency is more important for electrostimulation /106, 107, 108/.
Typical examples for electrostimulation are :
- cAMP synthesis in embryonic cells /109/,
- DNA synthesis in bone cells /107/, in human fibroplasts /110/,
- protein synthesis in Chinese hamster cells /108/, in rat cells /111/,
- membrane transport into bone cells /112/,
- growth of chicken ganglia /113/.

The comparison between the effects of strong single pulses (> 500 V/cm, SSP) and of weak periodic pulses (< 500 mV/cm, PEMIC) shows similarities regarding polarization and ion displacement. In both cases the membrane permeabilization is enhanced, however, by different mechanisms :
- with SSP by electroporation mainly: electroincorporation of large molecules possible,
- with PEMIC by extension of protein channels: influence on ion exchange occurs.
Under appropriate conditions electrofusion takes place also by an alternating current of ≤ 300 V/cm, i.e. by the field

strength for dielectrophoretic alignment /114/.

From these few examples one can draw the conclusion that this young development of bioelectrochemistry will grow up quickly turning out a valuable tool for gene- and biotechnology.

REFERENCES

1. H. Berg, Histor. roots of bioelectrochem., in: Bioelectrochem. Bioenerg., an Interdisciplinary Survey, Separ. Experientia 36:1247 (1980, G. Milazzo, ed.

2. L. Galvani, Abhandlungen über die Kräfte der Elektrizität bei der Muskelbewegung, (1791), in:"Ostwalds Klassiker d.exakt.Wiss.", Bd.52, V.W. Engelmann, Leipzig (1894)

3. A. Volta, Briefe über thierische Elektrizität, (1792), in:"Ostwalds Klassiker d.exakt.Wiss.", Bd.114, A. v. Oettingen, ed., V.W. Engelmann, Leipzig (1900)

4. J.W. Ritter, "Ostwalds Klassiker d.exakt.Wiss.", Bd.271, H. Berg, K. Richter, eds., Akad. Verlagsges. Leipzig (1986)

5. W. Ostwald, "Elektrochemie, ihre Geschichte und Lehre", Akad. Verlagsges. Leipzig (1896)

6. Th. Grotthuss, Abhandlungen über Elektrizität und Licht, in:"Ostwalds Klassiker d.exakt.Wiss.", Bd.152, R. Luther, A.v.Oettingen, eds., V.W. Engelmann,Lpzg.(1906)

7. E. Du Bois-Reymond, "Untersuchungen über die thierische Elektrizität", Verlagsges. Reimer, Berlin (1848)

8. Proceedings of the I.Jena Symp. (1962): Die Polarographie in der Chemotherapie, Biochemie und Biologie, H. Berg, ed., Akad. V., Berlin (1964)

9. Proceedings of the III.Jena Symp. (1963): Elektrochemische Methoden und Prinzipien in der Molekularbiologie, H. Berg, ed., Akad. V., Berlin (1964)

10. Proceedings: Biological Aspects of Electrochemistry, G. Milazzo, ed., Experientia, Suppl. 18 (1971)

11. Proceedings of Symposia 2.-8., published in: Bioelectrochem. Bioenerg. 1974 (1986)

12. F. Frauenberger, "Illustr. Geschichte der Elektrizität", Aulis V., Köln (1985)

13. Bioelectrochemistry I, Biological Redox Reactions, 1983, G. Milazzo, M. Blank, eds., Plenum Press, N.Y., II, (1987)

14. Electrochemistry in Biology and Medicine, Th. Shedlovsky, ed., John Wiley, N.Y. (1955)

15. M. Brezina and P. Zuman, "Polarographie in der Medizin, Biochemie und Pharmazie", Akad. Verlagsges., Leipzig (1956), english edition: Interscience N.Y. (1958)

16. Ions in Macromolecular and Biological Systems, D.Everett, B.Vincent, eds., Colston Res.Soc., Bristol, V.29 (1978)

17. Topics in Bioelectrochem. Bioenerg., G.Milazzo, ed., John Wiley, N.Y., Vol. 1-5 (1976-83)

18. Bioelectrochem. Bioenerg., an Interdisciplinary Survey, G.Milazzo, ed., Birkhäuser, Basel, 36:1243 (1980)

19. L.I. Boguslavsky, "Bioelectrochem. Phenomena and Phase Boundary" (in Russian) Nauka, Moscow (1978)

20. J. Koryta, "Ions, Electrodes, Membranes", John Wiley, Chichester (1982)

21. Comprehensive Treatise of Electrochemistry, Vol.10, Bioelectrochemistry, S. Srinivasan, Yu. Chismadjew, eds., Plenum Press, N.Y. (1984)

22. E. Heinz, "Electric Potentials in Biological Membrane Transport", Springer V., Heidelberg (1981)
23. H. Morowitz, "Foundations of bioenergetics", Academic Press, N.Y. (1978)
24. P. Mitchell, Chemiosmotic Coupling in oxidative and photosynthetic phosphorylation, Biol.Rev. 41:445 (1966)
25. N. Lakshminarayanaiah, "Equations of Membrane Biophysics", Academic Press, N.Y. (1984)
26. U. Lassen, and B. Rasmussen, in"Membrane Transport in Biology", Vol.I, G. Giebisch et al., eds., Springer V., Heidelberg (1978)
27. H. Stieve, "Transduction of Light Energy to Electrical Signal in Photoreceptor Cells", Cambridge Univ. Press London (1982)
28. "Biophysik", W. Hoppe et al., eds., Springer V. Heidelberg (1982), engl. Edition (1984)
29. "Bioelectrochemistry, Ions, Surfaces, Membranes", M. Blank, ed., Americ. Chem. Soc. (1980)
30. H. A. Pohl, "Dielectrophoresis, the Behavior of Matter in Nonuniform Electric Fields", Cambridge Univ. Press, London (1978)
31. "Coherent Excitation in Biological Systems". H. Fröhlich, and F. Cremer, eds., Springer V., Berlin (1983)
32. "Charge and Field Effects in Biosystems", M.Allen, and P.Usherwood, eds., Abacus Press, Tunbridge Wells (1984)
33. "Electronic Conduction and Mechanoelectrical Transduction in Biological Materials", B. Lipinski, ed., Marcel Dekker, N.Y. (1982)
34. "Nonlinear Electrodynamics in Biological Systems", W.Adey, and A.Lawrence, eds., Plenum Press, N.Y. (1984)
35. "Modern Bioelectrochemistry", F. Gutman, H. Keyzer, eds., Plenum Press, N.Y. (1986)
36. "Interaction between Electromagnetic Fields and Cells", A. Chiabrera, C. Nicolini, and H. Schwan, eds., Plenum Press, N.Y. (1985)
37. "Electroporation and Electrofusion in Cell Biology", E. Neumann, and A.Sowers, eds., Plenum Press, N.Y. (1987)
38. "Handbook of Bioelectricity", A. Marino, ed., Marcel Dekker, N.Y. (1987)
39. Proceedings of the IX. Jena Symp. (1982): Photodynamic and Electric Effects on Biopolymers and Membranes, H. Berg, and H.-E. Jacob, eds., studia biophysica 94, Akademie V., Berlin (1982)
40. Proceedings of the XI. Jena Symp. (1986): Bioelectrochemistry in Biotechnology, H. Berg, and H.-E. Jacob, eds., studia biophysica 119, Akademie V., Bln. (1987)
41. H. Berg, studia biophysica 90:169 (1982)
42. H. Berg, Zellfusion, Transformation und Pharmacoinkorporation durch Elektrostimulation, Sitzungsber. Sächs. Akad.Wiss., Bd.118, H.3, Akademie V., Bln. (1985)
43. H. Berg, in: "Electrochem. in Res. and Developm.", R. Kalvoda, and R.Parsons, eds., Plenum Press,N.Y. (1985)
44. U. Zimmermann, Biochimica Biophys. Acta 694:227 (1982)
45. W. Arnd, and U. Zimmermann, in:"Biological Membranes", 5: 389, D. Chapman, ed., Academic Press, N.Y. (1984)
46. M. Berry, FEBS Letters 134:133 (1981)
47. F. Oosawa, and S. Asakura, "Thermodynamics and polymerization of protein", Academic Press, N.Y. (1975)
48. "The Impact of Protein Chemistry on the Biomedical Sciences", A. Schechter, A. Deen, and R. Goldberger, eds., Academic Press, N.Y. (1984)

49. D. Eisenberg, and D. Crothers, "Physical Chemistry with Applications to the Life Sciences", Benjamin Publ. Co., Menlo Park (1979)
50. H. Berg, Polyelectrolytes and their Interactions, in:"Biophysik", W.Hoppe et al., eds., Springer V.,Bln. (1983)
51. H. Berg, G.Horn, Bioelectrochem. Bioenerg. 8:167 (1981)
52. H. Berg, Der Formenwandel einer DNA-Doppelhelix in Lösung, an der Elektrode und seine molekularbiologische Bedeutung. Abh. Sächs. Akad. Wiss. Bd.54, H.4, Leipzig, Akademie V., Berlin (1981)
53. "Two-dimensional Gel Electrophoresis of Proteins", J.Celis, and R. Bravo, eds., Academic Press, N.Y. (1984)
54. G. Dryhurst, K. Kadish, F. Scheller, and R. Renneberg, Biological Electrochem., Academic Press, N.Y. (1982)
55. "Metabolic Pathways", D. Greenberg, ed., Academic Press, N.Y. (1960)
56. "Current Topics in Cellular Regulation",V.23, B.Horecker, and E.Stadtmaur, eds., Academic Press, N.Y. (1984)
57. "Current Topics in Bioenergetics", V.13, C. Lee, ed., Academic Press, N.Y. (1984)
58. "Ion-selective Microelectrodes and their Use in Excitable Tissues", E. Sykova, P. Hnik, and L. Vyklickly, eds., Plenum Press, N.Y. (1981)
59. J. Schindler, and M. Schindler, "Bioelektrochemische Membranelektroden", W.de Gruyter, Berlin, N.Y. (1983)
60. K. Ponsold, and H. Kasch, Z.f. Chemie 22:157 (1982)
61. "Living Systems as Energy Converters", R.Buvet, M.Allen, J.Massue, eds.,North Holland Publ.Co.,Amsterdam (1977)
62. K. Tanaka, C. Vega, and R. Tamamushi, Bioelectrochem. Bioenerg. 11:289 (1983)
63. G. Rechnitz, Science 214:287 (1981)
64. A. Pilla, Ann. N.Y. Acad. Sci. 238:149 (1974)
65. J. Schnakenberg, "Thermodynamic network analysis of biological systems", Springer V., Berlin (1977)
66. M. Levitt, Cold Spring Harb.Symp.Quant.Biol.47:251 (1983)
67. H. Sobell, Cold Spring Harb.Symp.Quant.Biol.47:293 (1983)
68. H. Berg, G. Horn, J. Flemming, and J. Glezers, Bioelectrochem. Bioenerg. 4:467 (1977)
69. R. Goodman, and A. Henderson, Bioelectrochem. Bioenerg. 15:39 (1986)
70. J. Crowley, Biophys. J. 13:711 (1973)
71. D. Dimitrov, J. Membrane Biol. 78:53 (1984)
72. Yu. Chismadjew, V. Arakelyan, V. Pastushenko, Bioelectrochem. Bioenerg. 6:63 (1979)
73. L. Chernomordik, S. Sukharev, I. Abidor, and Yu. Chismadjew, Biochim. Biophys. Acta 736:203 (1983)
74. V. Pastushenko, and Yu. Chismadjew, Gen. Physiol. Biophys. 1:43 (1982)
75. J. Weaver, R. Mintzer, H. Ling, and S. Sloan, Bioelectrochem. Bioenerg. 15:229 (1986)
76. J. Teissie, and T. Tsong, J. Membrane Biol. 55:133 (1980)
77. H. Berg, studia biophysica 90:169 (1982)
78. I. Shugar, and E. Neumann, Biophys. Chem. 19:221 (1984)
79. U. Zimmermann, Trends in Biotechnology 1:149 (1983)
80. V. Markin, M. Koslov, Biofizika Membran 3:49 (1984)
81. M. Blank, and J. Britton, Bioelectrochem. Bioenerg. 5:528 (1978)
82. R. W. Glaser, A. Wagner, and E. Donath, Bioelectrochem. Bioenerg. 16:455 (1986)
83. G. Hofmann, and G. Evans, IEEE Engin. in Medic. and Biolog. Magazine (1986) 6

84. M. Senda, J. Takeda, Sh. Abe, and T. Nakamura, Plant Cell Physiol. 20:1441 (1979)
85. H. Weber, W. Förster, H. Berg, and H.-E. Jacob, Curr. Genetics 4:165 (1981)
86. H. Halfmann, C. Emeis, and U. Zimmermann, Arch. Microbiol. 134:1 (1983)
87. N. Shivarova, W. Förster, H.-E. Jacob, H. Berg, and R. Grigorova, Bioelectrochem. Bioenerg. 11:181 (1983)
88. W. Künkel, I. Groth, H.-E. Jacob et al., studia bio-physica 119 (1987)
89. G. Bates, Planta 165:217 (1985)
90. H. Kohn, and O. Schieder, Int.Symp. Plant Tissue Cell Cult. Appl. to Crop Improvem., Olomuc (CSSR) 1984
91. P. Scheurich, U. Zimmermann, Naturwiss. 68:45 (1981)
92. R. Bischoff, R. Eisert, I. Schedel, J. Vienken, and U. Zimmermann, FEBS Lett. 147:64 (1982)
93. D. Berg, I. Schumann, and A. Stelzner, studia biophysica 94:101 (1983)
94. M. Lo, T. Tsong, M. Conrad, S. Strittmatter, L. Hester, and S. Snyder, Nature 310:792 (1984)
95. A. Abel, P. Stolley, and G. Dreyer et al., studia bio-physica 119 (1987)
96. H. Berg, Bioelectrochem. Bioenerg. 9:223 (1982)
97. A. Kurischko, and H. Berg, Bioelectrochem. Bioenerg. 15:513 (1986)
98. H.-P. Richter, P. Scheurich, and U. Zimmermann, Develop. Growth and Differ. 23:479 (1981)
99. E. Neumann, M. Schäfer-Ridder, Y. Wang, and P. Hofschneider, Embo J. 1:841 (1982)
100. N. Shivarova, W. Förster, H.-E. Jacob, and R.Grigorova, Z. Allg. Mikrobiologie 23:595 (1983)
101. F. G. Falkner, E. Neumann, and H. G. Zachau, Hoppe-Seyler's Z. Physiol. Chem. 365:1331 (1984)
102. H. Hashimoto, H. Morikawa, Y. Yamada, and A. Kimura, Appl. Microbiol. Biotechnol. 21:336 (1985)
103. R. D. Shillito, M. W. Saul, J. Paszkowski, M. Müller, and I. Potrykus, Bio-Technology 3:1099 (1985)
104. M. Fromm, L. P. Taylor, and V. Walbot, Proc. Natl. Acad. Sci. 82:5824 (1985, USA
105. A. Pilla in: "Bioelectrochemistry", H. Keyzer, and F. Gutman, eds., Plenum Press, N.Y. (1980) 353
106. C. Bassett, M.Caulo, J.Kort, Clin.Orthop. 154:136 (1981)
107. R. Korenstein, D. Somjen, H. Fischler, I. Bindermann, Biochim. Biophys. Acta 803:302 (1984)
108. R. Goodman, A. Abott, A. Krim, and A. Henderson, J. Bio-electricity 4:565 (1985)
109. D. Jones, J. Bioelectricity 3:427 (1984)
110. A. Liboff, T. Williams, D. Strong, and R. Wistar, Science 223:818 (1984)
111. P. Delport, N. Cheng, J. Multer, W. Sansen, and W. De Loecker, Bioelectrochem. Bioenerg. 14:93 (1985)
112. G. Colaccico, and A. Pilla, Bioelectrochem. Bioenerg. 10:119 (1983)
113. B. Sisken, B. Mc Leod, and A. Pilla, J. Bioelectricity 3:81 (1984)
114. A. Kurischko, and H. Berg, Bioelectrochem. Bioenerg. 15:513 (1986)
115. M. Grattarola, and A. Chiabrera, Bioelectrochem. Bio-energ. 16:561 (1986)
116. E. Neumann, Bioelectrochem. Bioenerg. 16:565 (1986)

PULSE IMPEDANCE METHOD AND ITS APPLICATION IN MEDICAL PHYSICS

Fritz Pliquett

Institute of Biophysics
Karl Marx University
Leipzig, GDR

INTRODUCTION

The change in biological tissues, such as blood, muscle or epithelium as a result of pathological processes or the action of physical or chemical influences is frequently detectable by a variation in their passive electrical properties. Especially the change in the cell shape and size, the intra- and extracellular conductivity and the membrane capacity and conductivity can be detected in the frequency region of β-dispersion (Schwan 1957). This is done by measuring the electrical impedance as a function of the frequency.
Biological tissues can be considered as highly concentrated suspensions of cells. In the frequency region between 10 kHz and 20 MHz the electrical equivalent circuit of such an object is a parallel combination of a RC-serial combination with a resistor (Fig.1).

The measuring of the impedance with many different frequencies requires a long time. A more convenient method of determining the passive electrical properties has been developed on the basis of a change in an electrical pulse passing the measuring object. Using this method passive electrical investigations are carried out in a very short time. Several simple devices (UP1, UP2, UP3, UP4), developed by us dirctly display the equivalent diagramm parameters or useful combinations of these. Thus we have a practical method especially for in vivo measurements.

PULSE IMPEDANCE MEASUREMENT

All information about the passive electrical behaviour of an object (see Fig.2) is determinable by the change in an electrical pulse or a pulse wave (Pliquett and Wunderlich 1983, Pliquett, 1979).
The current passing the measuring object, induced by a rectangular pulse of long duration is

$$I = \frac{U_0}{R + R_0} \left(1 + \frac{R_0 \exp(-t/\tau)}{R_0 R_\beta + R_0 R + R_\beta R} \right) \tag{1}$$

with

$$R = R_m + R_i$$

and
$$\tau = \frac{C\,(\,R_o R_\beta + R_o R + R_\beta R\,)}{R + R_o}$$

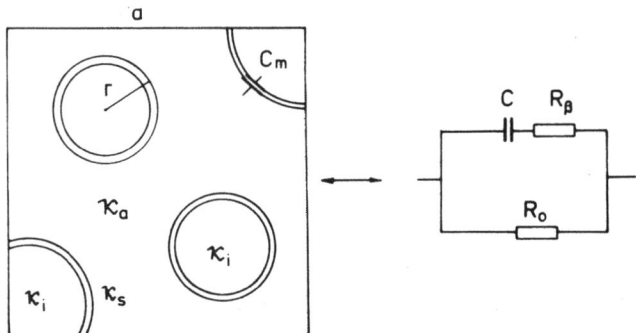

Fig. 1. A model of a biological object: a suspension of shell-shaped cells with radius r and the volume concentration

$$p = \frac{n\ 4\ r^3}{a^3}$$

n is the number of cells, a^3 is the volume of the measuring object, $\varkappa_i$, $\varkappa_a$ and $\varkappa_s$ are the conductivities of the medium inside and outside the cells and of the membrane. C_m is the membrane capacity. C, R_β and R_o are the components of the equivalent circuit.

In the case of a pulse wave with the period 2 T the current wave is given by

$$I = \frac{U_o}{R + R_o}\ \left(\,1 + \frac{2}{1 + \exp\,(-T/\tau\,)}\ \cdot\ \frac{R_o^2\ \exp\,(-t/\tau\,)}{R_o R_\beta + R_o R + R_\beta R}\,\right)\quad (2)$$

From the display of the deformed pulse on the oscilloscope screen (see Fig. 2) we get, if $T > \tau$ at t = 0,

$$h_\beta = \frac{\dfrac{K}{R + R_o}\ \cdot\ 2\,R_o^2}{R_o R_\beta + R_o R + R_\beta R}\quad\quad (3)$$

90

and at the end of the pulse $(t \rightarrow \infty)$,

$$\Delta = \frac{K}{R + R_o} \qquad (4)$$

with $\qquad K = \dfrac{U_o R_m}{A}$, $\qquad U_o$ amplitude of the rectangular pulse

$\qquad\qquad\qquad\qquad\qquad\qquad R_m$ measuring resistance

$\qquad\qquad\qquad\qquad\qquad\qquad A$ deflection factor of the oscilloscope

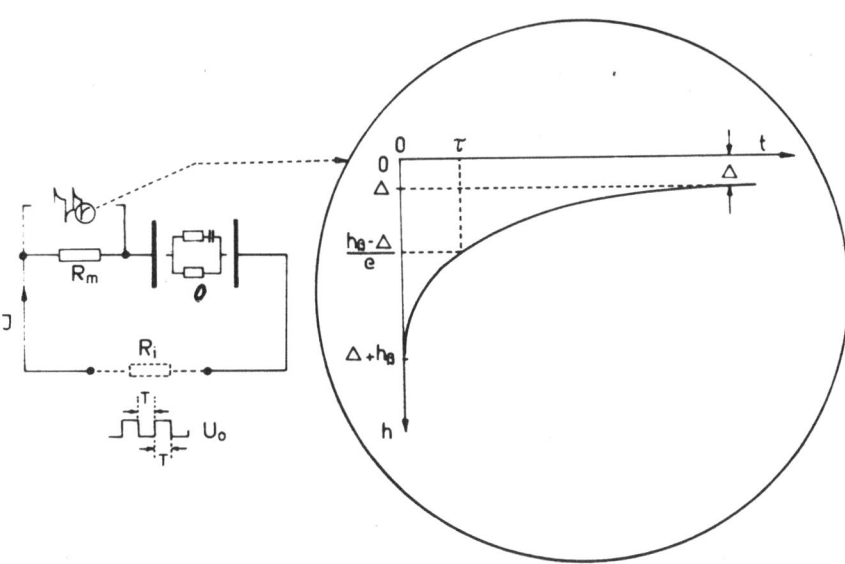

Fig. 2. Wire diagramm of the measuring equipment and deformed pulse. U_o pulse amplitude, R_i interior resistance of the pulse generator, R_m measuring resistance, 0 object Δ , h_β and τ parameters of the deformed pulse, t   time, h deflection on the oscilloscope screen.

The two quantities h_β and Δ are used to characterize the measuring object. If U_o, R_m, A, T, R_i are constant, Δ decreases as R_o increases. Δ depends only on R_o. h_β decreases with the enhancement of R_o.
Moreover h_β slowly increases if R_β/R_o decreases. A noticeable dependence of h_β on τ occurs only if we choose $T \approx \tau$.
Neglecting the electrode polarization R_o equals the resistance of the extracellular medium. R_o increases if the conductivity of the extra-cellular medium decreases and R_o decreases if the effective cell cross section increases, e.g. by reduction of the cell concentration or by diminution of the cell size with constant number of cells.

R_β is the resistance of the medium inside the cells.

R_β in creases if the intracellular conductivity decreases, or if the effectivecross section of the cells decreases, e.g. by diminution of the volume concentration of the cells.

Δ increases und h_β decreases if the effective cross section of the cells or the volume concentration of the cells decreases. h_β increases and remains constant if the intracellular conductivity increases. With an increase of the conductivity of the extracellular medium Δ increases and h_β decreases slowly.

The exact connection between the pulse parameters: h_β, Δ and τ on the oneside and the cell parameters: cell size, cell shape, volume concentration p of cells in a suspension, intra- and extracellular conductivity $\varkappa_i$ and $\varkappa_a$, membrane capacity C_m on the other side is derivable on the basis of a special model of the morphological structure in the tissue. For a suspension of shell-shaped cells surroundet by a insulated membrane and with a low volume concentration we get, on the basis of a model considered by Pauly and Schwan (1959), the following equations to calculate the extra- and intracellular conductivity $\varkappa_a$ an $\varkappa_i$ resp. as well as the membrane capacity C_m:

$$\varkappa_a = \frac{\Delta \cdot z}{K} \; \frac{1+p/2}{1-p} \tag{5}$$

$$\varkappa_i = \frac{\frac{2\,K}{z}\varkappa_a \left(\frac{h_\beta}{2} + \Delta\right) \; \frac{2+p}{1-p}}{\frac{1}{\varkappa_a}\left(\frac{h_\beta}{2} + \Delta\right) - \frac{K}{z}\,\frac{1+2p}{1-p}} \tag{6}$$

$$C_m = \frac{\tau_\beta}{r\left(\frac{1}{\varkappa_i} + \frac{1-p}{2\,\varkappa_a\,(1+p/2)}\right)} \tag{7}$$

with

$$\tau_\beta = R_\beta C = \frac{\tau_a R_b\,(R_a+R_o) - \tau_b R_a\,(R_b+R_o)}{R_o\,(R_b - R_a)} \tag{8}$$

τ_a is the relaxation time determined with a measuring resistance
$R_m = R_a$ and
τ_b with $R_m = R_b$.
R_o is determinable from eq. (4)
z is the constant of the measuring chamber.

Often it is difficult to determine h_β from a picture as shown in Fig. 2. More exactly h_β is calculable by

$$h_\beta = h(t) \cdot \exp(t/\tau) \tag{9}$$

t is to choose as a very short time (about µs).

The derivation of equations connecting the pulse and cell parameters in the case of tissues with a complicated cell structure is not
so easy. In this case we get, from eq. 4,

$$R_o = \frac{K}{\Delta} - R \tag{10}$$

and from (3),

$$R_\beta = \frac{2}{h_\beta} \left(K - 2R\Delta + \frac{R^2 \Delta^2}{K} \right) - R + \frac{\Delta R^2}{K} \tag{11}$$

The capacitor C in the equivalent circuit follows from eq.8

$$C = \frac{\tau_\beta}{R_\beta} \tag{12}$$

Thus the object under investigation is characterized by its equivalent
circuit. R_o is the resistance of the extracellular medium, R_β depends on
the intracellular conductivity and C on the membrane capacity. These three
parameters depend, of course, on the volume concentration of cells in the
tissue discribed only to a first approximation by eqs. 5, 6 and 7.

CHANGES OF PASSIVE ELECTRICAL PROPERTIES DURING CANCEROGENESIS

The literature reports many measurements of the alternating current
resistance (impedance) of cancer tissue in comparison to the normal
tissue. We have to distinguish measurements in vivo and in vitro (post
mortem). The post mortem changes of normal and cancer tissue are different. The rate of p.m. changes of normal tissue is higher than that
of cancer tissue. This is the reason why the results of a comparison of
the impedance of normal and cancer tissue measured in vitro depend
strongly on the time post mortem and on the conditions of the surroundings (temperature, humidity).

In the present paper only in-vivo measurements are taken into consideration.

It is very well known how cancer and normal tissue differ morpho-
logically. We can assume that these differences effect differences in
the passive electrical behaviour.

In experiments with white rats an experimental tumour was induced and
during the development of the tumour the passive electrical properties
were investigated. (Heinitz, 1982, Pliquett and Heinitz, 1987).
While the pulse parameters h_β and Δ of the normal tissue (control) did
not changed these parameters decreased in the case of cancer during the
first five days by more than 50 % of the values at the beginning. (Fig.3)
Till the 9 th day a small increase was observed, and during the next
days a further decrease followed.

The resistance of the extracellular medium increased according to (4). To discuss the change of R_β, we consider eq. 3 for small R

$$\lim_{R \to 0} h_\beta = 2 K \frac{R_0}{R_\beta} \tag{13}$$

Because by (4) R_0 increases and h_β decreases it results that R_β increases more strongly than R_0. The increase of R_0 is explainable by an increase of the extracellular conductivity while the increase of R_β can than be understood only by a decrease of the conductivity in the intracellular medium. This process could be due to water inflow into the cells.

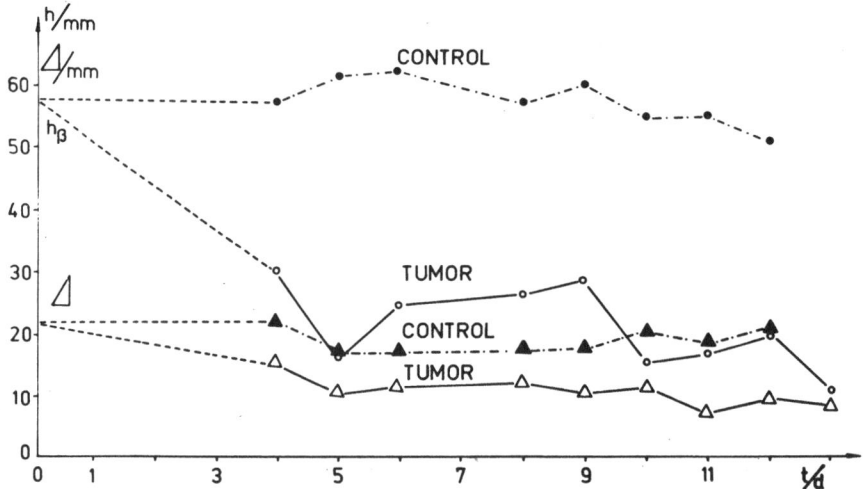

Fig. 3. Pulse parameters h_β and $\varDelta$ of an experimentally induced tumour in rats and normal tissue as control vs. time till the 13 th day after tumour induction. Each Point is an average value of measurements on 10 rats. The standard deviation is less then 10 %

From Fig. 3 it is seen that h_β is a suitable parameter to characterize the condition of cancerogenesis. A appliance for measuring this value quickly and accurately was developed at the Institute for Biophysics in Leipzig.
The h_β-values of portio ephithelium measured in many patients with this equipment are shown in Fig. 4. As in the animal investigations the h_β-value decreases with progressing cancer. As we have shown the pulse parameter h_β is suitable to detect cancer. It provides a quick diagnostic method.

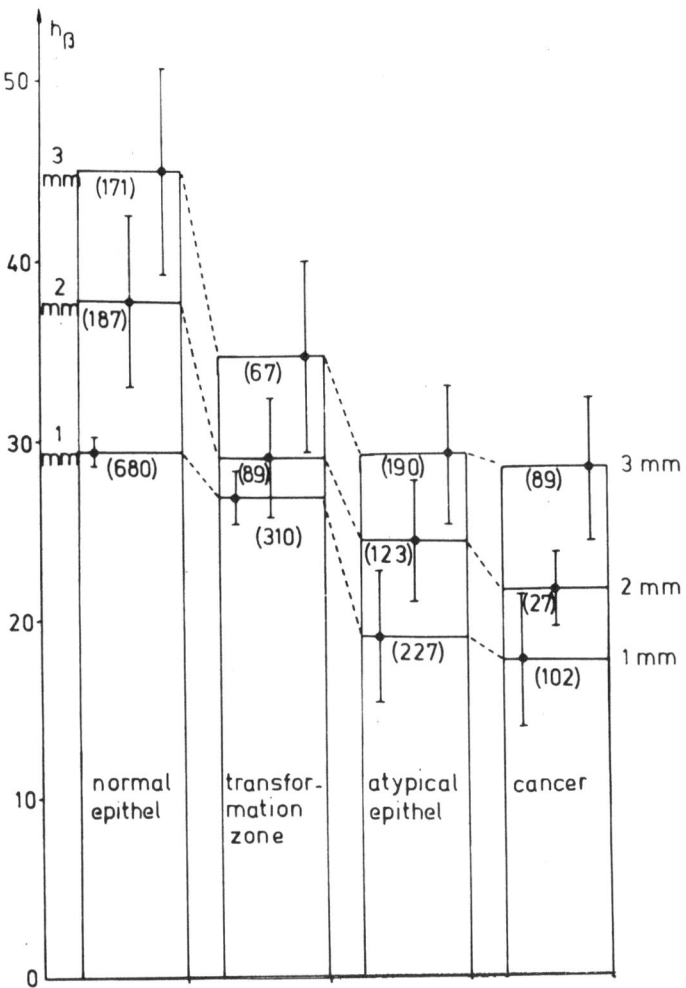

Fig. 4. The pulse parameter h_β of portio-uteri ephithelium
measured at different depts from the surface (1, 2 and
3 mm).
In-vivo measurements of normal and pathologically
changed tissue are presented. The average values and the
standard deviation are marked. The number of patients
is given in brackets.

THE EFFECT OF CRYOTREATMENT ON PASSIVE ELECTRICAL PROPERTIES

Low temperature is used either to conserve or to destroy biological cells or tissues. In both cases it is important to detect the cryo-effects quantitatively. A convenient method is provided by the measurement of passive electrical properties. During the cryodestruction of tissue mainly cell membranes are influenced or destroyed.
In the case of destruction the resistance R_o is decreased and the resistance R_β inside the cells is increased alone in consequence of decreasing the volume concentration p of the intact cells(Pliquett 1987).
Other effects on the passive electrical properties are imaginable and measurable using the impedance method.

Because R_o decreases and R_β increases due to cryodestruction h_β decreases. Such an experiment is shown in Fig. 5. Muscle tissue is exposed several times to a low temperature. After each cryocycle h_β decreases. The greatest effect is observed after the first cycle, but after the next cycles a decrease of h_β is detectable.

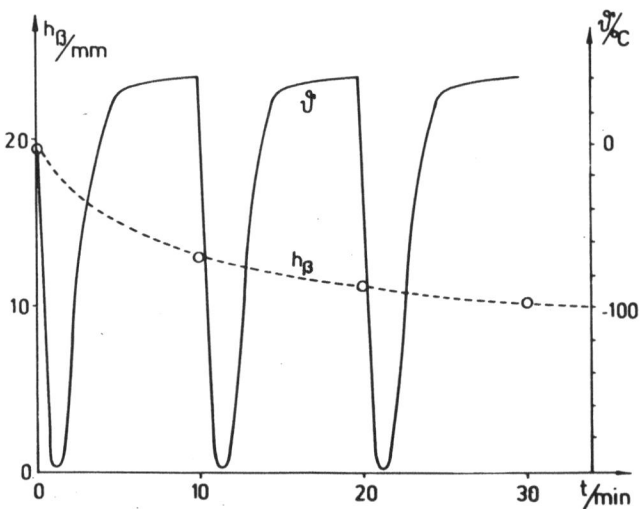

Fig. 5.   The temperature ϑ (solid line) and the pulse parameters h_β (points connected by a dashed line) at 37^oC from M.long.dors. in vitro vs. the time during three cryo-cycles.

This measuring method is adequate to determine the share of destroyed cellsby cryodestruction. A formular for describing the relation between the pulse parameter h_β and the volume concentration p of the yet intact cells can, of course, be given only if the geometrical structure of the tissue is known, but h_β is monotone increasing function of p.

REFERENCES

Heinitz, J.

In vivo Untersuchungen des passiv elektrischen Verhaltens des DS-Karzinosarkoms der Ratte während der Kanzerogense

Med. Diplomarbeit, Leipzig 1982

Pauly, H. and H.P. Schwan

Über die Impedanz einer Suspension von kugelförmigen Teilchen mit einer Schale

z. Naturforsch., 14b (1959) 125-131

Pliquett, F.

Passiv elektrische Untersuchungen biologischer Objekte.

Wiss. Z. Karl-Marx-Universität Leipzig, Math.-Nat.Wiss.R. 28 (1979) 118-127

Pliquett, F.

Nachweis von Kryoeffekten mit passiv elektrischen Methoden.

Dt. Gesundheitswesen 35, 22 (1980) 853-855

Pliquett, F.

Passiv elektrische Verfahren zur Untersuchung von Kryoeffekten in Mathäus (ed.)

Kryotherapie in Ophthalmologie und Dermatologie
Joh.Ambr.Barth Verlag, Leipzig 1987

Pliquett, F. and J. Heinitz

Veränderungen passiv elektrischer Gewebeeigenschaften eines experimentellen Tumors während der Kanzerogenese

(1987) in preparation

Pliquett, F., C. Stafri and K. Kühndel

Passiv elektrische Gewebeparameter von normalem und pathologisch verändertem Epithel

(1987) in preparation

Pliquett, F. and S. Wunderlich

Relationship between cell parameter and pulse deformation due to these cells as well as its change after electrically induced membrane breakdown

Bioelectrochemistry and Bioenergetics 10 (1983) 467-475

Stafri, C.

Passiv elektrische Eigenschaften von normalem und atypischem Epithel der Cerrix Uteri

Med. Diss. Leipzig 1984

LOCUST MUSCLE GLUTAMATE RECEPTORS - SUPPORT FOR AN ALLOSTERIC MODEL

Peter N.R. Usherwood

Department of Zoology
University of Nottingham
Nottingham NG7 2RD, U.K.

INTRODUCTION

Recent comparative studies of receptor-gated membrane channels show that there may exist wide variations in channel gating kinetics between different types of receptors found in the membranes of excitable cells. Studies of nicotinic acetylcholine receptors have led to the conclusion that channel gating follows a sequential scheme (e.g. Colquhoun and Sakmann, 1980) of the type predicted many years ago from macrosysem studies on verte-brate skeletal nerve muscle junctions (Katz and Miledi, 1972), although more complex models have been suggested for this system (e.g. Jackson et al., 1983; Labarca et al., 1985). In this paper I review the literature on a glutamate receptor channel of locust muscle for which a model based upon that for allosteric enzymes has been proposed (Monod et al., 1965).

Locust skeletal muscle is well-endowed with glutamate receptors, which are present in plasma membrane at excitatory nerve-muscle junctions and which also occur extrajunctionally as a mixed population of hyperpolarizing (H-) and depolarizing (D-) receptors (Usherwood and Cull-Candy, 1975). The H-receptors, which gate chloride-specific channels, will not be discussed further in this review. Extrajunctional D-receptors share properties in common with the postjunctional receptors of locust muscle: they gate relat-ively non-specific cationic channels (Anwyl and Usherwood, 1975) and exhibit rapid desensitization in the presence of ligand (e.g. L-glutamic acid) (Usherwood and Machili, 1968; Clark et al., 1979). The ready accessibility of the extrajunctional D-receptors and the ability to block their desens-itization with the lectin concanavalin A (Mathers and Usherwood, 1976; Evans and Usherwood, 1985) makes them excellent subjects for single channel anal-ysis using patch clamp techniques. Channels gated by these receptors also exhibit a high conductance (100-150pS) and can, therefore, be studied using the non-invasive mega-ohm seal technique (Neher and Sakmann, 1976).

EXPERIMENTAL METHODOLOGY

Until recently, it has not been possible routinely to obtain giga-ohm seal recordings from adult locust muscle using patch pipettes, although such recordings have been possible using cultured embryonic locust muscle (Duce and Usherwood, 1986). This review is based entirely on recordings made from extrajunctional D-receptors of extensor tibiae (jumping) muscles of adult locust using mega-ohm seals. Details of the techniques employed in these studies can be found in previous publications from my laboratory (e.g. Patlak et al., 1979; Gration et al., 1982; Kerry et al., 1987a) In all of the experiments the muscles were pretreated with 1-2μM concanavalin A for 30 min to inhibit desensitization of the extrajunctional D-receptors. There is no evidence that this lectin alters the other properties of the glutamate D-receptor channel complex. Attempts to record from D-receptors without concanavalin A pretreatment inevitably results in the appearance of a burst of channels at the beginning of the recording followed by long (often >30 min) periods of inactivity separated by brief bursts of a few channel openings (Gration et al., 1980a,b). These glutamate receptors exhibit rapid desensitization onset kinetics and only slow recovery from this phenomenon when equilibrated with glutamate-containing saline. All of the studies reported herein were undertaken at room temperature (21-23°C) with the muscle fibres clamped (usually at -100mV) using a conventional 2-electrode technique. The signal:noise ratio, with filtering at 3kHz, was usually 3:1. The patch amplifier was similar to that described by Hamill et al. (1981).

DATA REDUCTION AND ANALYSIS

All data were analysed off-line and reduced to a series of dwell times (open times and closed times) that were stored and further analysed on a PDP 11/34 microcomputer. A typical data record, with details of the data reduction scheme, is shown in Fig. 1.

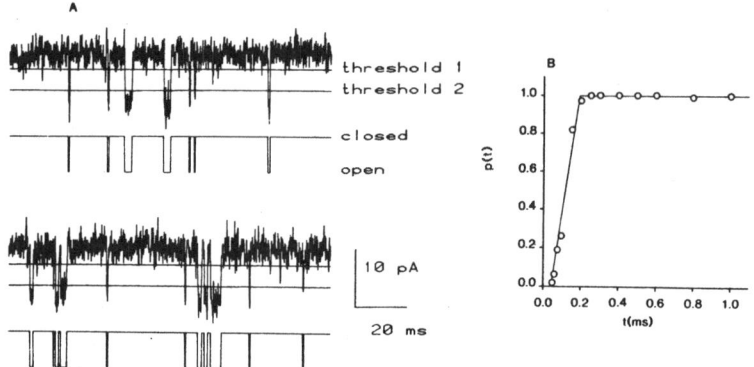

Fig. 1. (A) A segment of a typical data record, obtained in the presence of 10^{-4} M L-glutamate in the patch electrode, with the channel detection thresholds depicted. The beginning of a channel opening (end of a closing) corresponds to the time when the current crosses threshold 2, having previously crossed threshold 1. Similarly, the end of an opening (beginning of a closing) corresponds to the crossing of threshold 1, threshold 2 having previously been crossed. (See also Gration et al., 1982). (B) The probability of detection (p(t)) of brief events is shown as a function of event duration. The points represent the result of the calibration; the solid line is the fitted probability of detection function. (From Kerry et al., 1987a).

An estimate of the number of open states (N_o) and closed states (N_c) of the glutamate receptor channel can be obtained from the analysis of dwell-time distributions in terms of their underlying probability density functions (pdfs). The open ($f_o(t)$) and closed ($f_c(t)$) time pdfs are finite mixtures of exponential distributions with N_o and N_c components respectively viz:

$$f_e(t) = \sum_{j-1}^{N_e} (\alpha_j/\tau_j) \exp(-t/\tau_j)$$

where f_e is open or closed, τ_j is the time constant for either channel opening or channel closing and α_j are the respective mixture preparations. Details of the procedures employed to obtain and curve-fit the open and closed time pdfs are described in Kerry et al. (1987a).

Correlations between successive dwell times have been studied according to procedures outlined in Kerry et al. (1987a). Methods employed to simulate the behaviour of the single glutamate D-receptor channel of locust muscle, which take into account the failure of the recording system accurately to detect channel openings or closings of duration 180µs are also described in Kerry et al. (1987a).

EXPERIMENTAL DATA

For 10^{-4}M L-glutamic Acid

Open and closed time pdfs derived from data sets each containing 8,000 to 16,000 channel events required either four (closed time pdfs) or three (open time pdfs) exponential components for satisfactory fits (Table 1). The first component in each distribution corresponds to very brief closings (0.43ms) and openings (0.42ms).

Weak positive autocorrelation functions (acfs) (see Kerry et al., 1987a) were observed for both open and closed time series, with the correlations persisting for up to 40 openings or closings. The strength and degree of persistence of these autocorrelation functions are determined by, amongst other things, the relative rates of channel isomerizations from open to closed states compared with closed-to-closed and open-to-open transitions. The autocorrelations for locust muscle D-receptor channels suggest that channel isomerizations (open-closed) occur at comparable rates to open-open and closed-closed transitions. The positive autocorrelations exhibited by this system suggest that there are at least three pathways linking the open states with the closed states (Kerry et al., 1987a).

From these studies of the kinetics of the glutamate D-receptor using a single concentration of ligand it was concluded that the receptor channel could occupy a variety of open and closed states and that open and closed states are linked by a number of pathways. A tentative kinetic scheme has been proposed (first alluded to by Gration et al. (1982) and later extended by Kerry et al. (1987a)), which is an allosteric model (Karlin, 1967) based on that for the regulation of enzyme activity (Monod et al., 1965) and described in terms of single channel kinetics by Colquhoun and Hawkes (1977, 1981, 1983). The applicability of this model was tested further by studying the kinetics of the glutamate D-receptor channel over a range of ligand (L-glutamate) concentrations.

For concentrations of L-glutamate between 10^{-6}M and 10^{-2}M

Initial studies of the effects of glutamate concentration (Glu) on single channel kinetics of the extrajunctional D-receptor-channel of locust muscle suggested that this membrane protein probably exhibits complex kinetic properties (Gration et al., 1981a,b). The average probability (P_o) of the channel being in its open state increased to a value close to unity at high [Glu] (c. 10^{-2}M) (see also Fig. 2). This was not unexpected, since the channel opening rate should be dependent upon [Glu] (Fig. 3C). However, what was not expected was the increase in mean channel life-time (m_o) as [Glu] was increased. This suggested that the closing rate of the channel was dependent upon [Glu], a relationship which presumably accounted for the peaking at c. 10^{-3}M glutamate and the decline at higher [Glu] of the plot of open channel frequency versus [Glu].

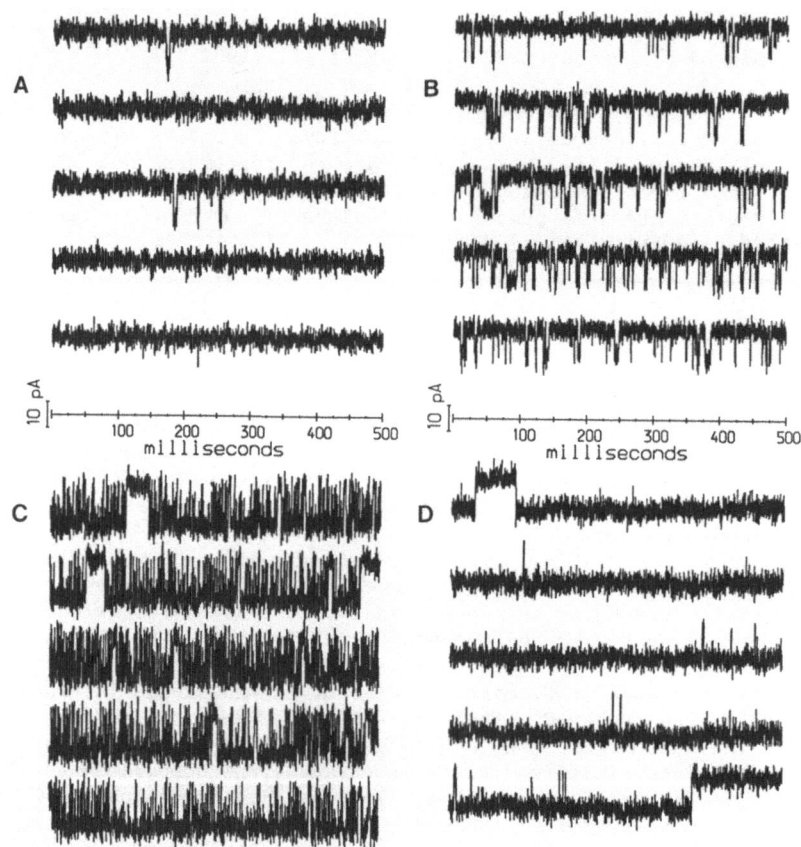

Fig. 2. Representative single channel recordings for 4 glutamate concentrations: (A) 10^{-5}M; (B) 10^{-4}M; (C) 10^{-3}M; (D) 10^{-2}M. For each concentration, 5 consecutive traces are shown, each of 0.512ms duration. The data were filtered at a cut-off frequency of 3KHz on playback, and digitized at 8KHz prior to display. Channel openings are downwards. (From Kerry et al., 1987b).

These initial results by Gration et al. (1981a,b) have now been confirmed and extended by more detailed studies by Kerry et al. (1987b) involving 20 different concentrations of L-glutamate and about 250,000 channel events. From these data a plot of P_o vs. [Glu] (Fig. 3A) reveals a maximum value of P_o 1, which suggests an efficacy of 1. A [Glu] of 8×10^{-4}M is required to produce a P_o of 0.5. Hill plot analysis of the P_o vs. [Glu] curve yields a Hill coefficient of 1.6, a result which implies positive cooperativity in the binding of L-glutamate to the glutamate D-receptor, i.e. more than one Glu binding site per receptor protein.

m_o increases with [Glu] from 0.32ms at 10^{-6}M to c. 100ms at saturating [Glu] (Fig. 3C). This result confirms the earlier findings of Gration et al. (1981a,b), although the improved time resolution in the more recent studies (recording cut-off frequency of 3kHz compared with 1kHz) reveals a lower m_o for [Glu] $\leq 10^{-4}$M. The mean closed time (m_c) decreases from c. 30s at 10^{-6}M glutamate to a minimum of c. 2ms at 6×10^{-4}M and then rises to c.10ms at saturating [Glu] (Fig. 3D).

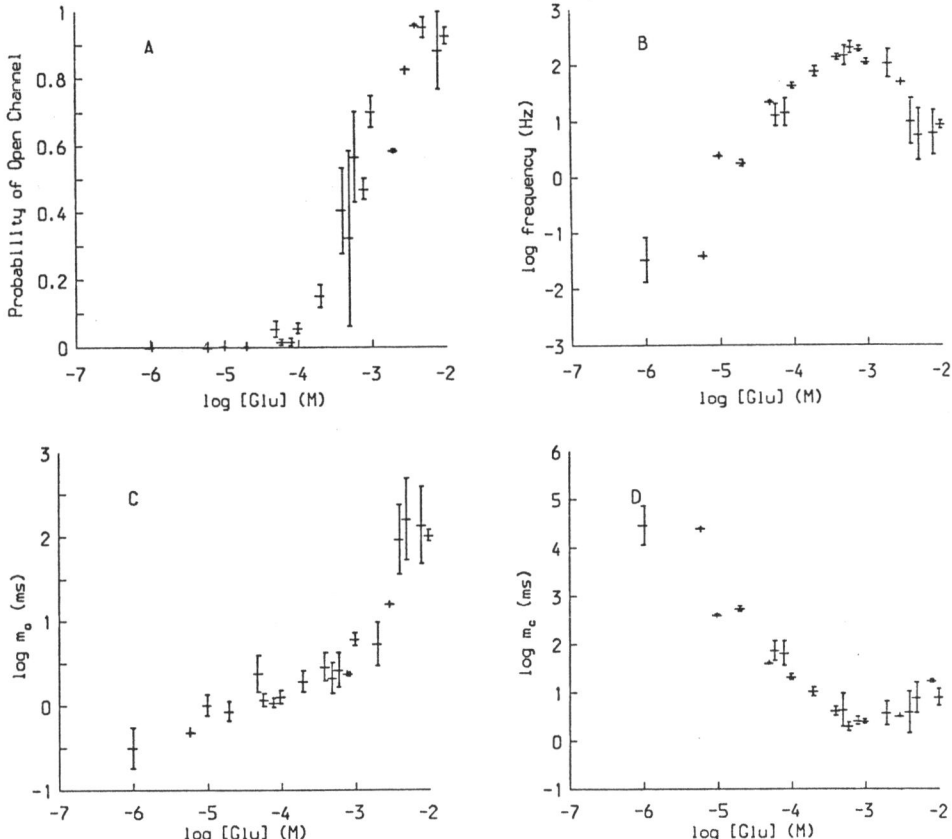

Fig. 3. Overall kinetic parameters for the glutamate D-receptor
single channel database. (A) The probability of the channel
being open plotted as a function of [Glu]. Each point is the
average over a number of recording sites for a given [Glu].
The error bars indicate the standard deviation from the
mean. (B) The channel event frequency (where an event is
defined as an opening plus the following closing) as a
function of [Glu]. Again, the points are averages for each
[Glu]. (C) and (D) are the mean open times and mean closed
times, respectively, as functions of [Glu]. (From Kerry et
al., 1987b).

Data from open and closed time pdfs for four [Glu] extending over three orders of magnitude, are illustrated in Table 1. These show that the complexity in the pdfs seen at 10^{-4}M Glu which were described by Kerry et al. (1987a) are also present over this wide range of [Glu]. Unfortunately it was not possible to obtain enough data at $[Glu] \geq 10^{-5}M$ to construct meaningful pdfs because of the very low channel opening rate obtained when $[Glu]$ falls below $10^{-5}M$. However, since m_o is c.0.3ms at $10^{-6}M$ glutamate it seems likely that only the briefest open state is occupied at this and lower $[Glu]$. At high $[Glu]$ at least four open states are accessible to the glutamate D-receptor channel and the occupancy of the longer life-time open states increases with increasing $[Glu]$. It seems likely that with $[Glu]$ $10^{-4}M$ there are four kinetically distinct closed states and as the glutamate concentration is increased, the relatively contributions of the states shift in favour of briefer closings.

Table 1. Open and closed time pdfs for the glutamate D-receptor channel for four $[Glu]$|

			OPEN			
$[Glu]$ (M)	N_o	i	α_i	CV	τ_i (ms)	CV
10^{-5}	2	1	0.616	0.086	0.310	0.141
		2	0.384	0.138	1.25	0.092
10^{-4}	3	1	0.379	0.366	0.424	0.269
		2	0.532	0.178	1.22	0.229
		3	0.089	0.908	2.94	0.279
10^{-3}	4	1	0.192	0.172	0.512	0.160
		2	0.386	0.143	2.49	0.177
		3	0.371	0.166	7.49	0.126
		4	0.051	0.357	27.7	0.140
10^{-2}	4	1	0.241	0.077	0.516	0.109
		2	0.176	0.117	4.39	0.160
		3	0.291	0.103	50.6	0.135
		4	0.292	0.111	232.	0.082

			CLOSED			
$[Glu]$ (M)	N_c	i	α_i	CV	τ_i (ms)	CV
10^{-5}	2	1	0.103	0.237	0.746	0.135
		2	0.897	0.027	468.	0.047
10^{-4}	4	1	0.206	0.048	0.432	0.070
		2	0.350	0.153	7.52	0.114
		3	0.389	0.131	22.5	0.093
		4	0.055	0.259	122.	0.105
10^{-3}	4	1	0.633	0.087	0.380	0.091
		2	0.303	0.168	1.22	0.129
		3	0.047	0.508	9.74	0.312
		4	0.017	1.483	32.4	0.357
10^{-2}	4	1	0.486	0.088	0.311	0.129
		2	0.264	0.133	1.41	0.216
		3	0.181	0.176	7.76	0.192
		4.	0.069	0.492	34.1	0.160

For each concentration the pdf was derived by combining open times from a number of recording sites into a single normalized histogram. The coefficient of variation (CV) is given for each parameter estimate. (From Kerry et al., 1987b).

Studies of the dwell time acfs suggest that at 10^{-5}M glutamate there are two pathways linking the open and closed states and that at higher [Glu] this number rises to three.

Analysis of the correlations between adjacent open and closed intervals (McManus et al., 1985) indicates that, at least at 10^{-4}M glutamate, there is a strong negative correlation, i.e. briefer openings are adjacent to longer closings and vice versa.

These studies of the effects of glutamate concentration on the glutamate D-receptor channel of locust muscle provide further support for an allosteric model to describe the gating kinetics of this channel. What has emerged is a co-operative scheme containing four identical glutamate binding sites, and a minimum of four open and five closed states for the receptor channel:

$$
\begin{array}{ccccccccc}
C & \underset{}{\overset{4K_B}{\rightleftharpoons}} & CA & \underset{}{\overset{\frac{3}{2}K_B}{\rightleftharpoons}} & CA_2 & \underset{}{\overset{\frac{2}{3}K_B}{\rightleftharpoons}} & CA_3 & \underset{}{\overset{\frac{1}{4}K_B}{\rightleftharpoons}} & CA_4 \\[4pt]
\updownarrow L & & \updownarrow \alpha L & & \updownarrow \alpha^2 L & & \updownarrow \alpha^3 L & & \updownarrow \alpha^4 L \\[4pt]
O & \underset{}{\overset{4\alpha K_B}{\rightleftharpoons}} & OA & \underset{}{\overset{\frac{3}{2}\alpha K_B}{\rightleftharpoons}} & OA_2 & \underset{}{\overset{\frac{2}{3}\alpha K_B}{\rightleftharpoons}} & OA_3 & \underset{}{\overset{\frac{1}{4}\alpha K_B}{\rightleftharpoons}} & OA_4
\end{array}
$$

where C is the closed channel, O the open channel, and A is a molecule of agonist (e.g. L-glutamate). The equilibrium properties of the model are determined by three parameters, i.e. α, K_B and L, where α is the relative affinity of glutamate for the open and closed channel, K_B is the microscopic binding constant for association of glutamate with the receptor channel and L is the equilibrium constant for closed-open transitions. It follows, therefore, that αK_B is the microscopic binding constant for the open receptor channel.

The concentration dependence of P_o is given by

$$
P_o = (1 + [Glu]\,\alpha K_B)^4 / [(1 + [Glu]\,\alpha K_B)^4 + L^{-1}(1 + [Glu]\,K_B)^4]
$$

Thus all three equilibrium parameters may be estimated by fitting the relationship P_o versus [Glu]. This procedure results in a value for K_B of $2.7 \times 10^3 M^{-1}$, a value for α of 14.2 (i.e. the affinity of glutamate for the open receptor channel is 14.2 times greater than that for the closed channel) and a value for L of 1.4×10^{-4}. Kerry et al. (1987b) have extended the quantitative description of the scheme outlined above by estimating channel isomerization rates. Values for closed to open channel transitions range from 530 ($CA_3 \rightarrow OA_3$) to $1.5 \times 10^{-12} s^{-1}$ ($CA_4 \rightarrow OA_4$). In principle, one problem with this scheme is that there is no evidence for channel openings in the absence of ligand, i.e. for the $C \rightarrow O$ transitions. However, in practice if the unliganded open state is explicitly excluded from the model the resultant changes in the equilibrium parameters are not great (Kerry et al., 1987b).

In their early single channel studies of locust muscle extrajunctional D-receptors, Gration et al. (1981b) pointed out that in recordings from a single receptor channel the number of events in successive time intervals

(Δt) did not follow Poisson statistics. In particular, the ratio of the variance (V) to mean (m) number of events in Δt was greater than unity. Ball et al. (1985) subsequently showed that V/m values greater than unity are not necessarily indicative of non-stationary channel kinetics, but could result from non-linear receptor-channel gating schemes. The scheme described above for the locust muscle, glutamate D-receptor channel would generate such modal behaviour, with the duration of the different kinetic modes being determined by the (low) rates of switching between different closed to open isomerization pathways.

CONCLUSIONS

In this brief review I have summarized current ideas on the biophysics and pharmacology of the glutamate D-receptor channel of locust leg muscle. Studies currently in progress in my laboratory are designed to determine the influence of ligand structure on the kinetics of the receptor channel. It has already been shown that the conductance of the channel is not influenced by the nature of the gating ligand, but that the relationships between m_o and ligand concentration and P_o and ligand concentration are agonist structure dependent (Gration et al., 1981a). It will be of interest to determine whether the results of these studies of glutamate receptor agonists confirm currently held views about the allosteric properties of this glutamate receptor channel complex.

ACKNOWLEDGEMENTS

I wish to thank my many working colleagues, past and present, who have contributed towards the information presented in this review.

REFERENCES

Anwyl, R. and Usherwood, P. N. R., 1974, Voltage clamp studies of glutamate synapse, Nature, 252: 291.

Ball, F. G., Sansom, M. S. P. and Usherwood, P. N. R., 1985, Clustering of single glutamate receptor-channel openings recorded from locust (Schistocerca gregaria) muscle, J. Physiol. (Lond.), 360:66P.

Clark, R. B., Gration, K. A. F. and Usherwood, P. N. R., 1979, Desensitization of glutamate receptors on innervated and denervated locust muscle fibres, J.Physiol., 290:551.

Colquhoun, D. and Sakmann, B., 1985, Fast events in single-channel currents activated by acetylcholine and its analogues at the frog muscle endplate, J. Physiol. (Lond.), 369: 501.

Duce, J. A. and Usherwood, P. N. R., 1986, Primary cultures of muscle from embryonic locusts (Locusta migratoria, Schistocerca gregaria): developmental, electrophysiological and patch-clamp studies, J. Exp. Biol., 123:307.

Evans, M. and Usherwood, P. N. R., 1985, The effect of lectins on desensitization of locust muscle glutamate receptors, Brain Res., 358:34.

Gration, K. A. F., Lambert, J. J. and Usherwood, P. N. R., 1980a, A comparison of glutamate single-channel activity at desensitizing and nondesensitizing sites, J. Physiol. (Lond.), 310:49P.

Gration, K. A. F., Lambert, J. J. and Usherwood, P. N. R., 1980b, Glutamate-activated channels in locust muscle, Adv. Physiol. Sci., 20:377.

Gration, K. A. F., Lambert, J. J., Ramsey, R. L., Rand, R. P. and Usherwood, P. N. R., 1981a, Agonist potency determination by patch clamp analysis of single glutamate receptors, Brain Res., 230:400.

Gration, K. A. F., Lambert, J. J., Ramsey, R. L. and Usherwood, P. N. R., 1981b, Non-random opening and concentration-dependent lifetimes of glutamate-gated channels in muscle membrane, Nature (Lond.), 291:423.

Gration, K. A. F., Lambert, J. J., Ramsey, R. L., Rand, R. P. and Usherwood, P. N. R., 1982, Closure of membrane channels gated by glutamate receptors may be a two-step process, Nature (Lond.), 295:599.

Hamill, O. P., Marty, A., Neher, E., Sakmann, B. and Sigworth, F. J., 1981, Improved patch-clamp techniques for high-resolution current recording from cells and cell-free membrane patches, Pfluegers Arch. Eur. J. Physiol., 391:85.

Jackson, M. B., Wong, B. S., Morris, C. E., Lecar, H. and Christian, C. N., 1983, Successive openings of the same acetylcholine receptor-channel are correlated in their open times, Biophys. J., 42:109.

Katz, B. and Miledi, R., 1972, The statistical nature of the acetylcholine potential and its molecular component, J. Physiol. (Lond.), 224:6654.

Karlin, A., 1967, On the application of a plausible model of allosteric proteins to the receptor for acetylcholine, J. Theor. Biol., 16:306.

Kerry, C. J., Kits, K. S., Ramsey, R. L., Sansom, M. S. P. and Usherwood, P. N. R., 1987a, Single channel kinetics of a glutamate receptor, Biophys. J., 51:137.

Kerry, C. J., Ramsey, R. L., Sansom, M. S. P. and Usherwood, P. N. R., 1987b, Glutamate receptor-channel kinetics: the effect of glutamate concentration, submitted to Biophys. J.

Labarca, P., Rice, J. A., Fredkin, D. R. and Montal, M., 1985, Kinetic analysis of channel gating: application to the cholinergic receptor channel and the chloride channel from Torpedo californica, Biophys. J., 47:469.

Mathers, D. A. and Usherwood, P. N. R., 1976, Concanavalin A blocks desensitization of glutamate receptors of locust muscle, Nature (Lond.), 259:409.

McManus, O. B., Blatz, A. L. and Magleby, K. L., 1985, Inverse relationship of the durations of adjacent open and shut intervals for Cl and K channels, Nature (Lond.), 317:625.

Monod, J., Wyman, J. and Changeux, J.-P., 1965, On the nature of allosteric transitions: a plausible model, J. Mol. Biol., 12:88.

Neher, E. and Sakmann, B., 1976, Single-channel currents recorded from membrane of denervated frog muscle fibres, Nature (Lond.), 260:799.

Patlak, J. B., Gration, K. A. F. and Usherwood, P. N. R., 1979, Single glutamate-activated channels in locust muscle, Nature (Lond.), 278:643.

Usherwood, P. N. R. and Cull-Candy, S. G., 1975, Pharmacology of somatic nerve-muscle synapses, in: "Insect Muscle", P. N. R. Usherwood, ed., Academic Press, London, New York, p207.

BASIC PROBLEMS OF ELECTROMAGNETIC BIOLOGY

Yu.A. Kholodov

Institute of Higher Nervous Activity and Neurophysiology, Academy of Sciences of the USSR
USSR, Moskow

Recent interest in the problems of the biological influence
of electromagnetic fields /EMF/ is often connected with the beginning of the Space Era in the early 1960's. The leading countries in space research - the USSR and the USA - are publishing
the predominant part of the papers concerning to be a chapter
of biophysics studying the influence of external natural and
artificial EMF on different biological systems. In this paper
the emphasis will be on Soviet research which represents more
than 60 per cent of magnetobiological studies.

The ecological trend was strongly developed in the first
stages of the development of electromagnetic biology, which
were connected with the problems of the possible orientation
of migrating animals towards a geomagnetic field /GMF/ /Pressman,
1968; Dubrov, 1974/; with the correlation between the oscillations of the GMF value and different important biological processes /Opalinskaja et al., 1984; Sidjakin et al., 1985;Wasilik,
1986/; with the influence of magnetic anomalies on the biological systems /Travkin, 1971/; with the possible role of EMF
generated by biosystems /Brown et al., 1984; Kasnatcheev et al.,
1985/. As can be seen, such investigations are still carried out
today, including not only correlations, but also experimantal
approaches, as well as the biological sigificance of the hypomag-
natic environment /Kapanev and Shakula, 1985/.

The problems of diagnosis and therapy using EMF have been
developing since the's seventies and the number of papers in
this particular area is permanently increasing /Bogoljubov, 1978;
Demetzkii and Alekseev, 1982; Kholodov, 1982/ has to be stressed.
The predominance of the empirical phenomenological approaches,
but the theoretical studies on the problem of electromagneto-
therapy are being developed. With a view to the activation of
these studies the Committee on Magnetobiology and Magnetotherapy
in Medicine the Ministry of Public Health was organized in the
USSR in 1983.

Hygienic standards for different EMF were developed because
significant changes in the natural electromagnetic background
have been observed recently both on Earth and in Space. EMF
become a global factor which changes the conditions of life for
the biosphere in general and in this way influences the biosystems with any level of organization - from membrane to biosphere.

Terms as "electromagnetic pollutuion" and "electromagnetic star-vation" have already been introduced in the literature and some authors /Akoev, 1983/ speculated about the existence of areas of electromagnetic comfort for particular biosystems.

The problems of public hygiene become of great importance. It is known that the Soviet and US standards for the radiofrequency range differ about a thousand times. The limit in the USA is 10 W/cm^2. Limitations for constant magnetic fields /CMF/ exist only in the USSR - the permissible level is 10 mT. The hygienic problems of electromagnetic biology are developed in the book "Hygiene of Labour under the Influence of Electromagnetic Fields" published by Meditzina, Moscow, 1983. The theoretical problems of electromagnetic biology and more specifically the problems of electromagnetic neurobiology will be discusse(

The analysis of the literature /more than 6000 papers/ demonstrates the ability of many living organisms to respond to the changes in natural and artificial /increased or decreased/ EMF. It is considered that every particular biosystem responds to the influence of this global factor.

It has been established that every system of the organism of mammals /above all the nervous vascular and endocrine systems/ can respond to EMF. The data concerning the reactions of different systems of the organism are presented in Table 1. It is seen that every system of the organism responds to applied EMF. EMF as non-ionizing radiation affects every particular living cell, but the most sensitive cell components are estimated to be the membrane, mitochondria and cell nuclei.

The ideas of I.M.Sechenov, N.E.Vvedenskii, I.P.Pavlov are still valid in theoretical biology and in practical medicine. This tendency is very well expressed in the formation of the

Table 1.   Responses of different systems to electromagnetic field stimulation

Systems	Organism level		Isolated system
	total	local	
Nervous	+	+	+
Endocrine	+	+	
Sensory organisms	+	+	+
Blood - vessels	+	+	+
Blood	+	+	+
Muscles	+	+	+
Digestive	+	+	+
Respiratory	+	+	
Secretory	+	+	
Skin	+	+	+
Bone	+	+	+

electromagnetic biology which investigates the responses of different biosystems to natural and artificial EMF, as well as the biological significance of EMF generated by biosystems.

The practical problems of electromagnetic neurobiology are inevitably connected with the theoretical investigations of the physiological mechanisms of the EMF influence on the nervous sytem. Such studies were carried out at Moscow University and were continued later at the Institute of Higher Nervous Activity and Neurophysiology of the Academy of Sciences of the USSR.

It has been established that EMF /from constant electric and magnetic fields to superhigh frequency field/ possess a number of biotropic parameters, such as intensity, vector,gradient, electric or magnetic component, frequency, pulse shape, localization and exposure. The biological efficiency of EMF increases upon variation of the biotropic parameters during stimulation /Kholodov and Shishlo, 1979; Plehanov, 1979; Gandhi, 1980/.

With a view to their participation in the total influence of EMF, the physiological systems can be ordered in the following way: nervous, endocrine, sensory organs, blood - vessels, blood, digestive, muscles, secretory, respiratory, skin, bones. The fact that at local EMF influence responses of all systems have been observed suggests the obligatory participation of the regulatoty sytems of the organism /nervous and endocrine/. However, the existence of reactions when EMF was applied on isolated systems may be considered as evidence of the direct influence of EMF on any tissue of the organism.

Because of the penetration action of EMF one can speculate about a new form of the organism's reaction, which may be named general reorganization reaction. Thus, the general simultaneous character of the electromagnetic treatment will be underlined. This reaction is not yet completely developed and characterized in detail, but some of its pecularities may be discussed. When man is exposed to electromagnetic influence, the reaction often occurs at subcellular level with switching on the slow reaction starting system.

The total reorganization reaction of the organism differs from the usual reaction which occurs at the beginning through the sensory organs, by the involvement in the reaction of several physiological systems. This reaction to such a weak stimulus /as EMF is considered to be/has to provoke adaptational changes in the organism, appearing minutes or hours after the beginning of the influence. We studied the conditioned reflex, the sensory reaction and electroencephalogram. We did not observe summation of the effects if 1-3 min magnetic field influence was repeated after 10-20 min. It is quite probable that the initial reaction is in this time interval. At exposure longer than 20 min, a summation of the effects /evaluated by the conditioned reflex method/ was observed. This event was observed when EMF was applied daily. Generally speaking, the reaction of the organism is manifested by long after-effects. The influence of a 30-min exposure lasts one week, while if MF was applied for 6 hours, the result is observable for nearly one month. The reaction may be termed as adaptational. While the initial reaction mainly includes the nervous system and sensory organs, the adaptational reaction also involves other systems of the organism, mainly the endocrine system.

The therapeutic application of EMF should be discussed. /Bogoljubov et al., 1978/. It has been proved that MF with different inductions and a large frequency range can provoke adaptational reactions in the organism which lead to increase of the resistivity to different infections, to temperature influences, to ionizing radiation, etc.

These reactions are considered to take place mainly at

the organism level, while at system or cellular level they are not so well demonstrated.

Garkavi et al./1979/ enlarge Selye's stress theory with additional stages of general response of the organism to a stimulus with increasing intensity. They consider that the triad consisting of the reactions of training, activation and stress may appear several times during increased intensity or prolonged duration of the stimulation and is realized at different "stages".

The electric resistance of the skin, the white blood picture and the character of the autoflora are the principal informational indices for a given stage of the reaction. These demonstrations of the reaction are accompanied by considerablechanges in neurohormonal regulation, including immunological processes.

The reaction of training is supported by a weak stimulus, the reaction of activation - by a stimulus with medium amplitude, the stress reaction - by an intensive stimulus. The stable reaction of activation is typical of the healthy organism.

When the duration of EMF increases, the adaptation reactions may turn into pathologival processes connected with destructive changes in the cells. The estimation of this physical factor is a problem of social hygiene. The morphological changes observed in tissues and organs after EMF influence are non-specific and reversible.

Further two forms of the initial reaction of the nervous sytem to the applied magnetic fields and microwaves /when the duration of the influence is less than 1 min/ will be dicussed. Some parameters of slow and quick reaction are presented in Table 2 on the basis of literature data and our own results.

From the initial reaction of the organism we studied in greater detail the sensory reaction of man, as well as the conditioned reflex and the electroencephalographic reaction of rabbit.

Changes of the motor activity and correction of the reaction of the organism to the EMF influence simultaneously with other stimili /light, sound, electirc current, etc./ occured several minutes after the beginning of the EMF influence. These reactions will not be considered now.

There was no summation of the effects when 1 min action of EMF was applied again after 5-10 min. Probably, the starting electroencephalographic reaction of the rabbit occurred in this time interval. The registration of sensory indication of EMF involved switching off the generator during simple motor response of the patient. The duration of the influence decreased and consequently the intervals between EMF applications dropped to 1 min. In this way we succeeded in estimating the duration of the reaction.

The reactions to EMF have longer /about 20 s/ latent period compared with the initial human sensory reactions to typical stimuli /light, sound/ which are two orders of magnitude longer. Probably for this reason some authors have not established any effect. The variations of the biotropic parameters of EMF /induction, frequency, pulse shape, localization/, even if they change the duration of the latent period, do not provoke a quantitative jump. In all cases a peculiar slow system of the initial reaction /with duration of about 10 s/ is functioning.

The detailed characteristic of the slow system of the starting reaction should include not only the basic reaction observable during EMF action, but also the reaction of stopping which appears about 15 s after switching off the field /this reaction lasts for about 10 s/.

112

Other systems of slow adaptational reaction are manifested in the reaction of the organism to EMF when the exposure time increases. Moreover, the slow system of the initial reaction is not directly connected with the pecularities of the studied stimulus. For instance, alternating magnetic fields can provoke the involvement of the system of quick reaction, when EMF is addressed to the eye. The so-called magnetophosphen is observed, i.e. a sensation for lightning under the action of alternating MF with amplitude higher than a definite limit of induction to the human head. On an analogy with this term /connected with the quick sytem of initial reaction/ the sensory reaction connected with the slow system of initial reaction may be named "magneto-touch". It is relevant to remark that for microwaves there also exist slow /demonstrated in our experiments/ and quick systems of the initial sensory reaction.

The modality of the slow initial system of the sensor reaction at action of alternating magnetic fields or microwaves is similar and this may be considered as an indication for the nonspecificity of the reaction. The same EMF provoke electroencephalographic synchronization reaction in rabbits, which characterizes the slow system of the initial reaction, while short EMF stimulation or usual stimuli provoke quick electroencephalogarphic reaction of desynchronization.

Thus, the existence of slow and quick systems of initial reaction was demonstrated by analysis of the reaction of the nervous system to the factors of electromagnetic origin. It can be assumed that the slow system can be observed under the action of non-electromagnetic stimuli.

It is possible to develop electrodefensive conditioned reflex in rabbits, but it will be less stable compared with the reflexes to normal stimuli /light, sound/. The latent period of the conditioned reflex to sound is often equal to 1 s, to the MF - 12 s. The changes of the electrical activity of the brain upon EMF stimulation also differ from the electroencephalographic reactions to normal stimuli or to the influence of impulse EMF. Normally on the electroencephalogram is observed a quick desynchronization reaction appearing after several milliseconds: lowering of the amplitude and increase of the frequency of biopotentials, while at EMF stimulation the synchronization reaction appears several seconds after the beginning of the stimulation, accompanied by an increase of the amplitude and decrease of the frequency of the biopotentials.

The experiments with isolated parts of the animal brain show that EMF can influence the brain not only by reflex pathways, but also directly, as EMF penetrate through the skull. When preparations of isolated parts of cortex are studied, the reaction to applied EMF was better pronounced than in intact brain.

Therefore, to the traditional reflex pathway of the action of any stimuli should be added the direct influence of EMF on the structures of the central nervous regulation /this distinguishes the reaction of the whole organism to EMF from the reactions to traditional stimuli.

In short, the slow system of initial reaction functions adequately at organism and system level. This concerns the qualitative and quantitative parameters of the reaction, as well as their identity in different experimental objects.

One can assume that neurons are secondarily involved in the system of slow reactions to EMF action and a significant role in these reactions belongs to other structures of the nervous tissue, neuroglia and blood - vessels.

Table 2. Parameters of the reaction, realized on different level of organization of the biological object with participation of quick and slow systems of the starting reaction at influence with MF and microwaves

Level of organization	Object	Method	EMF	System of starting reaction	Latent period s	Length of reaction s	Character of reaction
Organism	Man	Sensor indication	MF	quick	0.2	2	Light
				slow	20	10	Touch
			Microwaves	quick	0.2	2	Light
				slow	20	10	Touch
Central nervous system	Rabbit	Recording of slow electric activity of brain	MF	quick	0.1	2	Desinchronization
				slow	20	10	Sinchronization
			Microwaves	quick	0.1	2	Desinchronization
							Sinchronization
Neuron	Skate	Registration of impulsa electric activity of neurous cell	MF	quick	1	1	Increased frequency
	Rabbit			slow	10	10	Increased and decreased frequency
			Microwaves	quick	1	1	Increased frequency
				slow	10	10	Increased and decreased frequency

The histological investigations on animal brain exposed to EMF action prove the primary hypoxy reaction of the neuroglia /Alexandrovskaja et al., 1966; Nachilnitzkaja et al., 1978/, as well as its participation in the physiological response of the brain to MF, which is another manifestation of the difference of EMF stimulation from the action of other stimuli.

An important role in the processes of training and memory is attributed to the neuroglia. It was demonstrated that the repeated influence of EMF mainly distrurbs both processes: training and memory. Once formed, it is difficult to change the conditioned reflex under EMF influence. It is considered that EMF influence on glia and blood - vessels can be explained by their influence on the hematoecephalic barrier.

In this way one can explain also the non-specific electro-encephalographic reaction of synchronization, which appears at influence of different EMF on the animal head. The intensity of switching on the slow system of initial reaction of different parts of rabbit brain decreases as follows: hypothalamus, sensomotor cortex, visual cortex, specific nuclei of the thalamus, non-specific nuclei of the thalamus, hypocampus, reticular formation of the midbrain.

Generally speaking, EMF provoke two types of reactions of the different structures of the nervous system. Some reactions are distingished by the existence of a slow system of initial reaction, reaction of switching, long after-effect, as well as participation of all structural elements of the nevrous system in reactions which are specific for EMF. The second type of reactions is connected with excitation of the specific receptors and is explained by the indication of electromotive force under alternating field influence, being similar to the reaction under the influence of traditional stimuli.

One can discuss several primary mechanism and one should mention not only electromagnetic induction under influence of alternating MF or when the object is moving in CMF, but also the significance of ferromagnetic particles as magnetite in the biological objects; the chemical polarization of nuclei and electrons, etc.

The biological effects do not always increase with the increase of the MF intensity. More correct seems to be the discussion of the existence of an amplitude-frequency window in which the biological effects are better pronounced.

The problem emerges of the functional importance of an artificial magnetic field, which does not differ very much from the natural ones.The validity of this hypothesis is very well demonstrated by several species of electric fishes which use their own EMF for orientation and communication.

It can be considered that in addition to the synaptic connection between different parts of the brain, electromagnetic bonding exists as well. The effects of EMF increase upon varying one or more of the biotropic parameters, which should be taken into account when hygienic standards and physiotherapeutic devices are developed.

The final biological effect of EMF depends also on such pecularities of the object itself as age /the reactions of children and old people are stronger/, sex /men are more sensitive than women/, initial physiological conditions /the working organ has a stronger reaction/, as well as individual capabilities. It is quite probable that some of these peculiarities are connected with the function of biological membranes.

The listed factors indicate the necessity of biological analysis of the observed physiological reactions of the organism

to EMF influence. The methods of membranology may help in
studing the magnetobiological effects.

ELF-PULSATING MAGNETIC FIELD (PEMF)-INDUCED ACOUSTIC EFFECTS IN VESSEL WALLS - AN ADEQUATE STIMULATION OF BARORECEPTORS ?

Ulrich Warnke

FB 15,4 Biology
University of Saarland
W. Germany

INTRODUCTION

In the year 1979 we published the results of our preliminary experiments.[1] PEMFs with impulse package frequencies in the ELF-range modify the infrared radiation (3-10 μm) of local areas of the body surface in humans. This occurs in single cases already at magnetic induction values of $dB/dt < 500$ mT/sec when the field is applied to the head-thorax-area (Fig. 1).

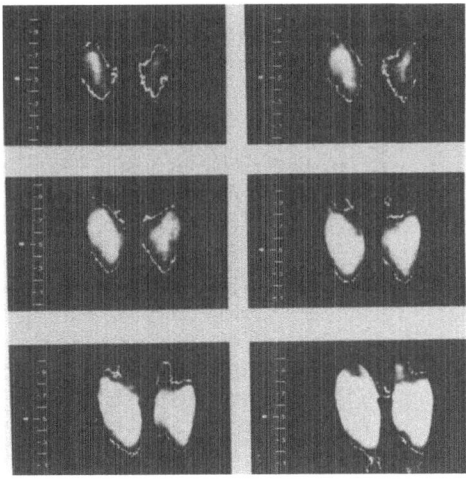

Fig. 1. Modification in the IR-radiation (wavelength 5μm) of the human legs during application of PEMF at the neck. Characteristics of the field are described in the text. The time between each picture is 2 minutes.

The reaction appears mostly within two minutes in the arms, hands and legs. The thermograms and the method of IR-plethysmography clearly indicate that the continuously increasing energy irradiation of the body originates at the blood vessels. The reason for this is definitely a dilation in the larger vessels.

The direct consequence of a modified blood circulation in tissue caused by certain pulsating magnetic fields is an alteration of the oxygen partial pressure in tissue. PEMF with special characteristics can rapidly increase the pO_2 measured transcutaneously.[2]

In new experiments we got further results: PEMFs with different special impulse forms, impulse repititions and amplitudes increase or decrease the respiration cycle, the heart rate, the blood pressure and vessel perfusion (Fig. 2).

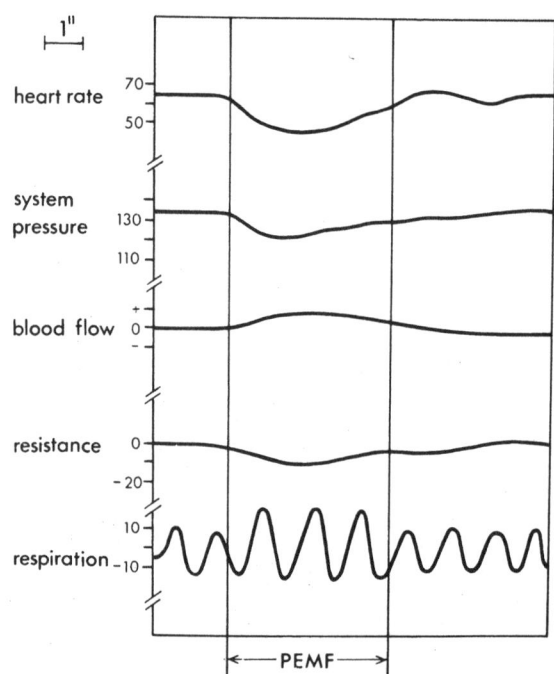

Fig. 2. Central effects of PEMFs with specific characteristics on the circulation system of the human body. The PEMF is applicated at the neck.

The magnitudes induced by the PEMF accordingly affect the information memory and information conveyance of the automatic nervous system.

The registered effects are subjected to great fluctuation and poor reproducibility. They can be explained by a central reduction and alternatively by a local increase in the activity of the sympathicus.

It was obvious that the results from our experiments so far done were extremely similar to the results which were known from the mechanical or electrical stimulation of baro-receptors in artery walls.

STIMULATION CHARACTERISTICS OF MAGNETICALLY INDUCED ELECTRICAL MAGNITUDES (AMOUNT OF ELECTRICITY)

The amount of electricity applied to the organism represents a standard value for the stimulation character.

Stimulation parameters can be modified independently from the current intensity by electric incitation. But, in the case of magnetically induced stimulation, the induced amount of electricity is determined by the magnitude and the time variation of the induced voltage surge. This surge is completely independent of the duration of constant stimulation coil parameters, e.g. of the MF impulse duration.

The duration of the induction is determined alone by the time constant of the current ascent in the stimulation coil-

$$\tau_c = \frac{L}{R} \frac{[H]}{[\Omega]}$$

For the consideration of the amplitude of the induced voltage impulse and its duration an adequate magnitude is available: the voltage time integral:

$$\int_{t=0}^{t=t_i} U_{ind} \, dt$$

This is the voltage-time-area.

The voltage-time-area is conditioned by the induction changes of the stimulating magnetic field: it depends on the timing of the current flow in the stimulation coil. Therefore, the voltage surge and its timing over the areal resistance of the observed organism determine the amount of induced electricity and thus the characteristic of the stimulation.

The induced voltage widely differs from the curve form of the voltage impulse in the stimulation coil with a decreasing time constant. Thus, a small time constant, e.g. a square pulse as coil current or as magnetic field, is reproduced to a great extent in rectangular form. Compared to this, the induced curve is completely distorted.

ELECTRICAL POLARIZATION OF CAROTID SINUS

The carotid sinus together with parts of the aorta forms a continous loop with the homogenous medium blood feeling the induction of the PEMF applicated in the neck. The field of the coil (Ø 20 cm, 40 wdg.) should be situated perpendicularly to the observed area of the organism (Fig. 3).

The loop is closed by the conductor fluid (electrolyte) outside the blood vessels with parts of the blood vessel wall as interfaces.

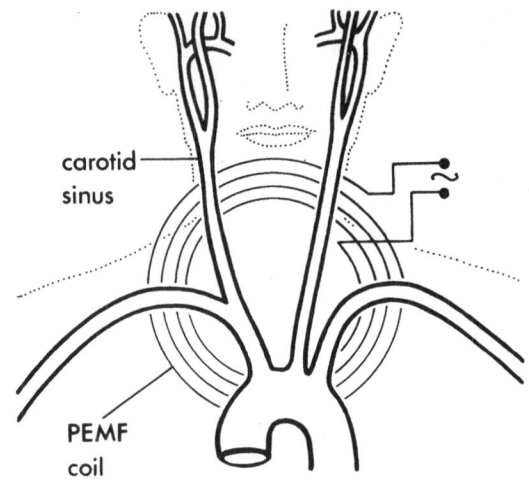

Fig. 3. PEMF-induced eddy currents
into the aorta-carotid sinus
loop with polarization of
vessel wall in baroreceptor
regions.

The induced electromotive force (emf) around the loop is
proportional to the rate of change of magnetic field:

$$emf = -10^{-4} dB/dt \cdot \pi r^2$$

with r radius of the loop.

Emf decreases proportionally to the distance between
stimulation coil and the body. The magnetic field intensity H
decreases in inverse ratio to the cube power of the distance,
but the spherical areas through which the magnetic induction
acts as a magnetic flow in the organism, augment by the square
power of this distance.

The resulting electric field E for a homogenous part of the
loop is given by

$$E = emf/\sigma AR$$

where σ is the conductivity of the fluid, A the current-carrying
cross section plane and R the total resistance of the observed
part of the fluid in the loop.[3]

The ions of the fluid will be separated through field-
force F by electrophoresis with $F = q \times E$; q as the net charge
of an ion. The charges shift on the field lines producing thus
a displacement current in blood, electrolyt and vessel wall
tissue. At the boundaries blood/ vessel wall/ surrounding fluid
change in electrical conductivity caused charges build up,
until continuity of current is possible. The accumulation of
charges at boundaries develop a back emf (emf_b) cancelling the
induced field within a specific time constant. In this way
the vessel wall with components of finite resistivities and
larger time constants becomes polarized. In the vessel wall
the induced emf and the back emf will be added to the same
vector direction; that is a macroscopically reinforcement of
the polarising field ocurrs in the wall (Fig. 4).

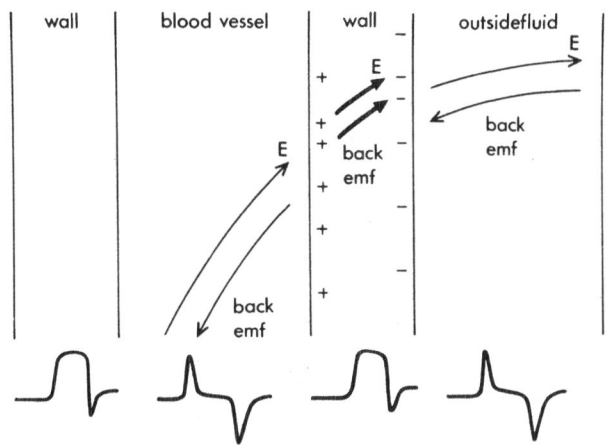

Fig. 4. In the vessel wall the induced emf
is reinforced by the "back emf"
while in the fluid the induced emf
is cancelled by the "back emf"

The eddy current is continous if

$$(\dot{E}-\text{emf}_b)/r_f = (\dot{E}+\dot{\text{emf}}_b)/r_w$$

with r_f as the resistance of the fluid and r_w the total
resistance of the vessel wall.

The electric field in the vessel wall is modulated by some
important parameters, which are able to change shape and
magnitude of the field:

- the field component produced by generation of electrical
  charges in the moving blood fluid (frictional electricity).
  At discontinuities free charges may be liberated from the
  double layer to form a moving space charge;

- the field component produced by inhomogenities of the
  wall-material as build up by microscopic dipols by time
  dependent free charge seperation in spots with low
  resistivity. The seperated charges will accumulate at
  the boundaries to higher resistivity, thus decreasing or
  increasing local amplitudes. The local amplitude also
  depends on the time rate of change of the polarization
  field and is correlated to the RC of the local region;

- the field component induced by the electric displacement
  vector in regions with high and low dielectriccoefficients.
  If a region with low dielectric constant ε_L is surrounded
  by larger regions with high dielectric constant ε_H (macroscopically the vessel wall surrounded by electrolyt fluids)
  the polarization field will be reinforced up to a factor ε_H;

- the field component built up by a sharp radius of tissue
  curvatures separating structures with different dielectric
  constants, e.g. tunica media and adventitia including
  electrolyte pockets with high dielectric constants similar
  to water. These various forms of inhomogenities give rise

to dispersion phenomena and relaxation processes with different relaxation times.

FORCES IN THE VESSEL WALL

All field components mentioned above are correlated with force-components on charges of membranes and tissues in the vessel wall. Thus electrostriction and Coulomb forces acting inside the wall on regions of different resistivity and capacitance.

At the boundary between two dielectric materials in an electric field there exists a free charge. The charge-density q_0 is given by the difference of the charge-densities q_1 and q_2 of the two materials

$$q_0 = q_1 - q_2$$

In the case that the permittivity ε_1 of one medium is greater than ε_2 of the other medium, then q_0 is positiv. On this free charge q_0 a field force is effective with a direction from ε_1 to ε_2 (Fig. 5).

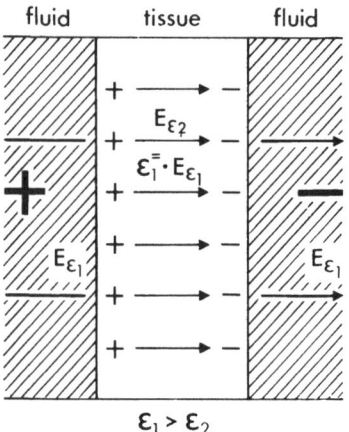

Fig. 5. Dielectric field-reinforcement in the vessel wall. This electrostriction force works independently of field polarity through different dielectric constants.

In an AC field the direction of force is not periodically changed, because the force-effect is proportional to the square of the fieldstrength. Thus electrostriction forces work independent of field polarity (as do Coulomb-forces).

With the boundary condition $\varepsilon_1 E_1 = \varepsilon_2 E_2$ the force per unit area (field component perpendicular to the boundary) is

$$F_\varepsilon = \frac{1}{2} (\varepsilon_1 - \varepsilon_2) E_1 \cdot E_2 \cdot K$$

K=unit vector along the axis. With regard to the boundary of blood and vessel wall, where medium one is a conductor compared to medium two with $E_1 \ll E_2$, the electrostriction force become

$$F_E \sim \frac{1}{2} \, \varepsilon_2 \, E_2^2 K = \frac{1}{2} \, (\varepsilon_1^2 / \varepsilon_2) E_1^2 \cdot K$$

ELECTROSTRICTION-EXPERIMENTS

Artificial Multilayers

It is impossible to control all field components and inhomogenities in vessel walls. For first electromechanical tests we therefore build up an artificial phospholipid membrane as a multilayer. Comparable to the condition of the organism an oscillating magnetic field is used as an electric field generator for polarizing the multilayer with a loop of electrolyte (Ringer solution) as a secondary winding (Fig. 6).

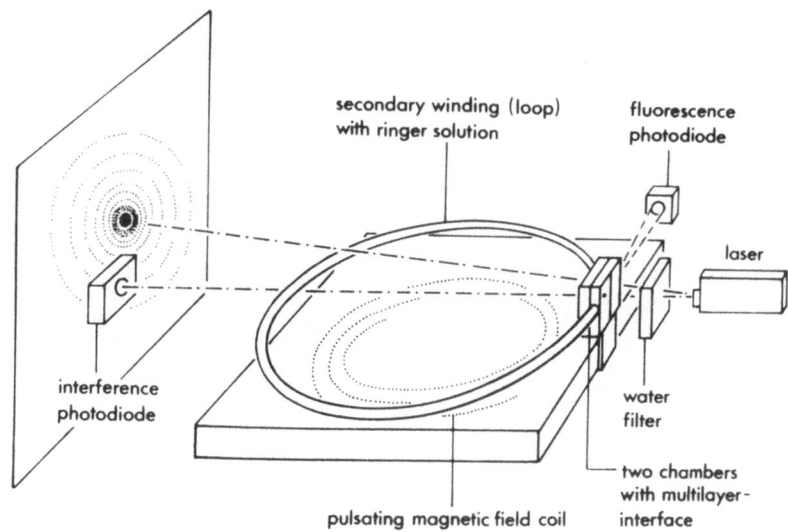

Fig. 6. Arrangement for PEMF induced electrical polarization of membranes and tissue disks. The potential difference can be measured by fluorescence of incooperated dye; the effects of electrostriction and of Coulomb force are observed and measured by interference.

The advantage of this arrangement is that in the model no metal electrodes are in galvanic contact neither with the electrolyt nor with the multilayer preventing artefacts. The measurement of polarization potentials is done by incorporated dye merocyanin as a fluorescence voltage-probe.

The whole sample was treated as a plate condensor with two conductors and the multilayer with the thickness d between the plates of area Ap.

The charge q of the plates is connected with the polarization voltage V by q = CV.

The multilayer capacitance is $C = \varepsilon_0 \varepsilon \, A_p^2 / V$; the dielectric constant ε is depending on the temperature T, the specific volume V m and area a m per molecule.

The magnetically induced potential difference V gives rise to a electrostriction force, which is
a) compressing the multilayer
b) changing the tension in the plane: $\Delta x = -CV^2/2$
The all sided stretching or compression in the plane is

$$\Delta F = \Delta x/h = -CV^2/2h$$

with h as a part of the thickness of the multilayer. The change of its area and thickness and the essential configuration increases the capacity C and the resistance R_e and has an influence too on the dielectric constant ε as mentioned above.

To make these changes visible a very simple method is effective. Polarized monochromatic light irradiation was focused on the multilayer and built up an interference picture. While the pulsating magnetic field is switched on, the induced emf polarizes the multilayer and forces begin to work as it is seen on configuration changes of interference patterns.

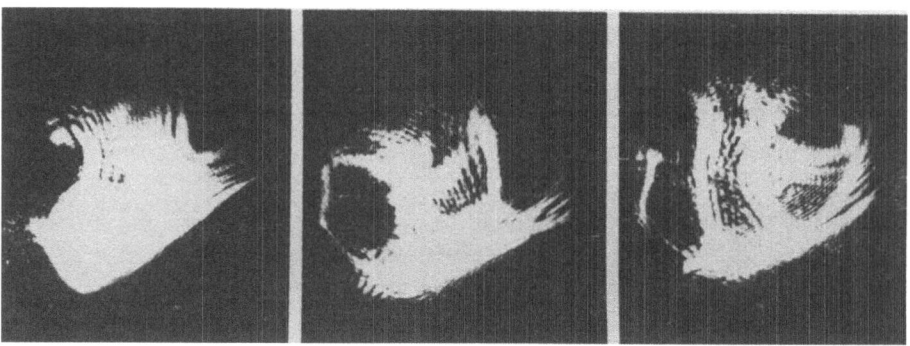

Fig. 7. Changes in interference pattern of a beam of light going through the sample during PEMF induced electrical polarization and consequently working forces.

A photodiode probe makes the influence measureable in a semi-quantitative way (Fig. 8), on conditions that the entire system is placed on a balance table with air cushions preventing the influence of vibration from the roomfloor.

The light picture of the multilayer shows that the membrane is not homogeneous as a result of impurities. Thus small light or dark spots corresponding to different thickness in the membrane can be observed. Interesting enough with the application of the polarizing field these spots change their interference pattern as well as their place as a consequence of different dielectric qualities, which can be described as

$$C_v = C(1+\alpha V^2)$$

where α is a parameter which is correlated to the dielectric constant to the thickness and to Young's modulus.

124

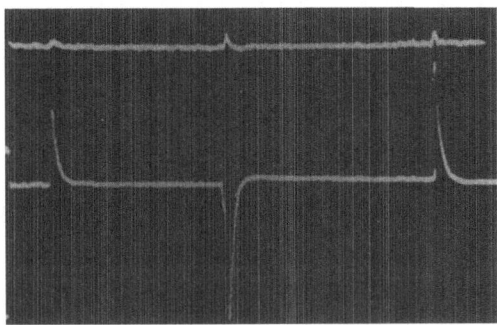

Fig. 8. Effect of electrostriction force
in multilayer (upper curve; relative
peak amplitude 5 mV) triggered by
different polarity (lower curve;
peak amplitude 130 mV). The force
works independent of polarity.
Potential-impuls basis-time about
100 µsec.

Films of Vessel Wall

Vibration measurement are done by changing the artificial
multilayer with layers from the vessel wall, that means films
of the intima, media or adventitia.

Induced acoustic effects are measureable by pulsed
Doppler Ultrasound equipment (max. sensitivity 30 µm vibrating
amplitude) and with a Laser Doppler Vibrometer System (max.
sensitivity: 10 nm). The measurements are based on the detection
of the Doppler frequency shift of the ultrasound as well as the
laser light backscattered from the moving film. The detector
output current contains the Doppler frequency information
which is a linear function of the moving target velocity.

As a result of these running experiments an electromechanical vibration is to be measured in calculated field
amplitude range of about 400 Vp-p/cm. Resonance range can be
shifted by stretching the vessel wall film, that means by
changing the elasticity.

Extremely interesting for examination are parts of arterial
wall, where the wall especially the media is extremely thin
and represents the point at which the nerves of the baroreceptor fan out around it in an annular belt. The thinnest
point is about 200 µm, due to the fact that the smooth muscle
is reduced and replaced by an increase in elastic tissue
existing in the form of fenestrated membranes.

The adventitia at this point is characterised by the fact
that it contains most of the terminations of the afferent nerves
passing by. Secondly the adventitia is noticeably more vascular
than elsewhere. The rich vascularisation of the wall and the
wall thinning has consequences in electrical behaviour in vivo:

-long relaxation time; high hysterisis

-high field strength in relatively low polarisation-potentials.

Although the electrostriction of this points is not exactly proved until mow, we do expect here the strategic points of the working mechanism.

MECHANICAL STIMULATION OF BARORECEPTORS

Arterial baroreceptors are distortion receptors which respond to deformation of the arterial wall in any direction on the basis of the three-dimensional structure of their sensory endings. In their normal physiological function baroreceptors respond to the increasing phase of blood pressure pulses as well as to amplitude and to speed of increasing amplitude.

The experiments with mechanical stimulation of baroreceptors give surprisingly clear results concerning frequency- and amplitude parameters. These results may explain the above mentioned frequency and amplitude windows in whole body induction tests.

Under consideration of all tests so far done, we formed an effective signal of coilcurrent of pulsating magnetic field application.

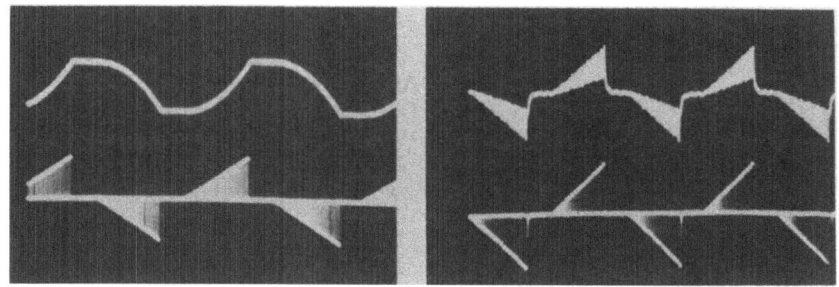

Fig. 9. left: example of the characteristic of a optimized signal
for stimulation of baroreceptor. Upper curve:
coil current; lower curve: induction emf of PEMF
right: Polarization voltage of vessel wall tissue disks
(upper curve) triggered by the induction emf
(lower curve; impulse train frequency 20 Hz).

The induced single rectangular impulse frequency is due to the adding up constant of the baroreceptor-generator-potential by mechanical vibration with amplitudes of 50 µm per fiber (200 Hz). The impulse train frequency is due to adequate stimulation of receptors while double frequency through electrostriction is included (20 Hz).

With such a signal the effects described above on the circulation system in human beings are more reproducible.

126

REFERENCES

1. U.Warnke, The Infrared-Radiation of Human Body as a Physiological Indicator of the Low Frequency Pulsed Magnetic Field, Z. f. Phys. Med. 3,8:166-174 (1979)

2. U. Warnke, Therapy with Pulsating Magnetic Fields: Fundamental Research and Physiological Mechanisms of Action, in: 1. Symposium of the International Society of Bioelectricity, Boston (1983)

3. W.T. Kaune, M.E. Frazier, A.J. King, J.E. Samuel, F.P. Hungate, and S.C. Causey, System for the Exposure of Cell Suspensions to Power-Frequency Electric Fields, Bioelectromagnetics 5:117-129 (1984)

MAGNETOTHERAPY WITH LOW-FREQUENCY ELECTROMAGNETIC FIELD

Nencho G. Todorov

Institute of Physiotherapy, Balneology and Rehabilitation
Medical Academy, Sofia

Magnetotherapy has gained increasing popularity in medical practice over the past decades. Devices for magnetotherapy have been designed in various countries over different periods, but the predominant range of such devices was designed in the Soviet Union. The first Bulgarian devices for magnetotherapy in Bulgaria was designed in 1974 and was modified in 1980 as "Magnit H-80" /Fig.1/.

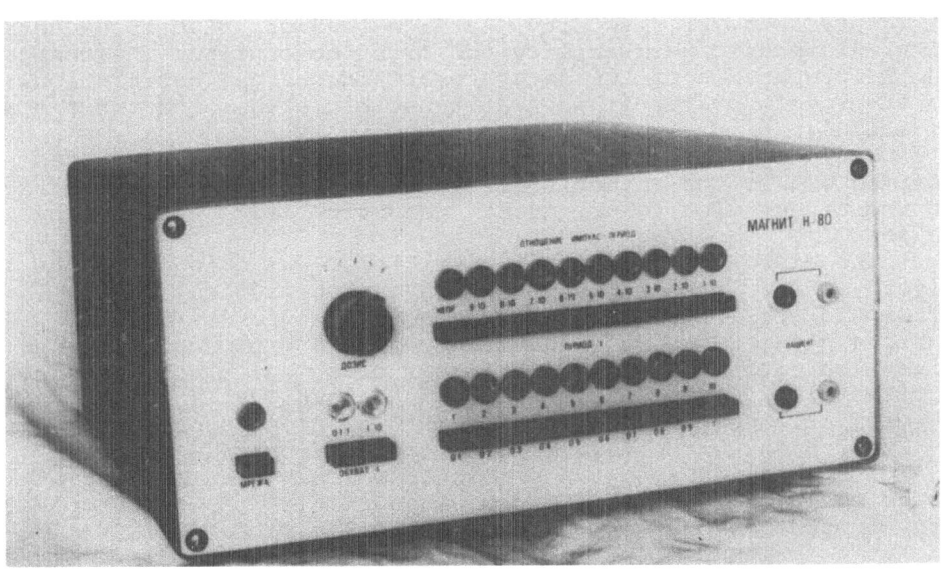

Fig. 1 Apparatus for low-frequency magnetic field with therapeutic application "Magnit H-80"

The device consists of a generator of a low-frequency magnetic field with a sinusoidal shape, which is modulated by rectangular impulses with a frequency between 0.1 and 10 Hz. The ratio between the impulse and the pause may vary from 1:10 to 9:10. Two types of inductors are used; in the larger inductors the current changes from 0.3 to 3 A, whereas the intensity of the generated magnetic field varies from 50 to 500 Oe, accordingly. After prolonged clinical tests of the device, it has been introduced and registered officially in Bulgaria, and it is produced on a mass scale.

The aim of the present paper is to present the most important experimental and clinical results obtained using Magnit H-80.

The action of magnetic fields on the human organism evokes no doubts in medicine and biology today and the magnetic field is becoming an important therapeutic factor.

Changes taking place in the cells at molecular level, as well changes in the functioning of a number of organs, systems and of entire organisms, have been established. Nevertheless, it can be claimed that the mechanisms of action of the magnetic field are still not fully clear. There exist a number of hypotheses which explain this phenomenon. According to Aristarkhov et al./1978/, the principal hypotheses can be summarized in the following order:

1. Interaction of the free radicals with the magnetic field applied;
2. Changes in the velocity and mechanisms of substance transport through membranes;
3. Semiconductor effects occuring in nucleic acids and proteins upon application of a magnetic field;
4. Changes in the polarizability of the molecules;
5. Changes in the valency angle of the molecules.

In our opinion, the magnetic field affects the biosystems by inducing a change in their electrical properties. The participation of water is very important for the realization of the effect of the magnetic field on the cell. The magnetic field can influence directly some physico-chemical properties of water, such as: viscosity, surface tension, electroconductivity, dielectric permeability and light absorption. The changes in the properties of the water result in changed correlations: water-protein, water-lipid, water-nucleic acids, and their combinations.

Many experimental and clinical investigations have been carried out in the author's clinic over the past twenty years on the effect of a low-frequency magnetic field generated by devices from the "Magnit-H" series.

The histological and enzymo-histological changes in the liver, cardiac tissue, skin, spleen and the adrenal cortex after the application of a low-frequency magnetic field with various parameters, were investigated in a group of 20 guinea pigs. The differences obtained after treatment of the adrenal cortex were most significant from a histological point of view. Hyperplastic responses of the cells from glomerulose and fasciculated zones were observed after application of a magnetic field. The azines from the plomerulose zone were combined with large cells indentified through vacuolization of the cytoplasm. Many lympho-histocytes were seen among the fibres building up the azines. The cell bonds in the fasciculated zones are also formed of large vacuoles with rich lipid content. The capillary network in this zone is also wide and fully supplied with blood.

The suprarenal activity was investigated in a group of 58 guinea pigs by determining the content of ascorbic acid using the method of Roc Coutcher. The ascorbic acids was found to decrease upon increasing the frequency and the intensity of the magnetic field applied.

The hormonal activity of 48 guinea pigs was observed with a view to determining the mechanisms of the effects of the magnetic field /with respect to heparin, histamine, ceruloplasmin, etc./. A strong increase in their values in the blood was observed when magnetic field of 180 Oe was applied: the histamine values rise to 0.8 mg %, with a parallel increase in the heparin values, though not to such an extent.

Another series of experiments in a group of 66 guinea pigs were carried out to test the effect of the magnetic field on the prothrombin time, on the amount of fibrinogen and the zeta-potential as a function of the parameters of the magnetic effect.

The prothrombin time increased after the application of weak magnetic fields. Upon increasing the intensity of the field applied, the amount of fibrinogen grew as well. Generally, the values of the zeta-potential decreased for all parameters studied.

The results obtained for ionic permeability of human skin demonstrate a 150-200 per cent increase of the ionic permeability for the ions subjected to ionophoresis under the conditions of a low-frequency magnetic field.

A considerable number of clinical tests have been carried out to study the effect of a low-frequency magnetic field in the treatment of many diseases.

A stimulation effect after the application of a low-frequency magnetic field in bronchial asthma patients was observed in 40 cases under sanatorial conditions and in 20 patients treated at our clinic. After the treatment of the first group of patients, they reported a subjective improvement of their state, which was also accompanied by positive dynamics of some objective parameters, such as: pulmonary volume and gases in the arterial blood. Changes in the partial pressure of O_2 and CO_2 were also observed: CO_2 decrease. There are grounds for believing these changes to be due to the improved function of the adrenal cortex.

The application of the same method in the clinic resulted in improved state of 20 patients with bronchial asthma. Treatment with magnetic field resulted in positive changes both in the patient's subjective assessment of his state, and in a change in the objective parameters: pulmonary volume, gases in the arterial blood, etc. Cortizol values in the blood, as well as the content of aldosterone and ACTH were also evaluated. The principal dynamics of these values shows that magnetic therapy leads to positive changes in the patients' etiopathogenesis.

Patients with cardiac diseases were also subjected to such therapy. Low-frequency magnetic field was applied over the synocarotid zone of a group of 27 patients with hypertension - 1st and 2nd degree. The treatment resulted in a reduction of the subjective complaints: headache, insomnia, stenocardiac pains, dizziness, etc. Positive changes were also found in the blood pressure: the systolic pressure of 10 patients decreased by 10 mm Hg, of 7 patients - by 15 mm Hg, of 4 patients - by 20 mm Hg and of 6 patients - by more than 20 mm Hg. No drugs were administered in the course of the magnetotherapy.

Of particular interest is the application of a low-frequency magnetic field in patients with myocardial infarction.

The following parameters of the electrocardiogram were investigated in dynamics: ST-segment, T-wave, QT-interval, Q-tooth, as well as some laboratory tests: LDK, KFK and thrombo-time of the blood. Magnetotherapy was followed by a rapid decrease of the pain in 12 patients even on the second day after the first treatment. The T-interval was normalized in eleven patients.

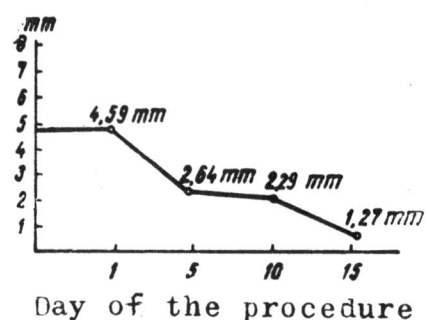

Day of the procedure

Fig. 2   Dynamics of changes of ST-segment

Magnetotherapy was applied on 71 patients with obliterating arthritis, using Poisenille's law. The skin temperature and the rheography were monitored, and some laboratory tests were performed. The following results were obtained: fast and effective improvement in 54,9 per cent of the patients, relatively good effect in 39 per cent, and no changes in 6.1 per cent of the patients treated. Figure 6 illustrates these findings: patient R.L.T. with 4th degree proliferating arthritis before /left/ and after /right/ magnetic therapy.

Magnetotherapy had a very marked positive effect in 51 patients with varicose ulcers of the lower extremities: 81 ulcers with an area between 1 and 150 cm^2. After the treatmnet, 71 of the ulcers were completely closed, while the rest tended to close.

Magnetotherapy was applied on 13 patients with duodenal ulcer. Perforation was found radiographically in six patients. Magnetotherapy produced good results in ten patients. Magnetotherapy was applied on 120 patients with periarthritis, using the "Magnit H-30" device. The following parameters were tested: skin temperature, pain, electroexcitability. Laboratory and radiographic studies were also performed. Thirty patients were subjected to combined therapy with magnetic field and with interference current.

The treatment of 73 patients with arthroses was also successful. Excellent results were obtained for 31 per cent of the patients, good results for 31 per cent of the patients, and 42 per cent of the patients were assessed to be clinically healthy. The application of a low-frequency magnetic field with rheuma-

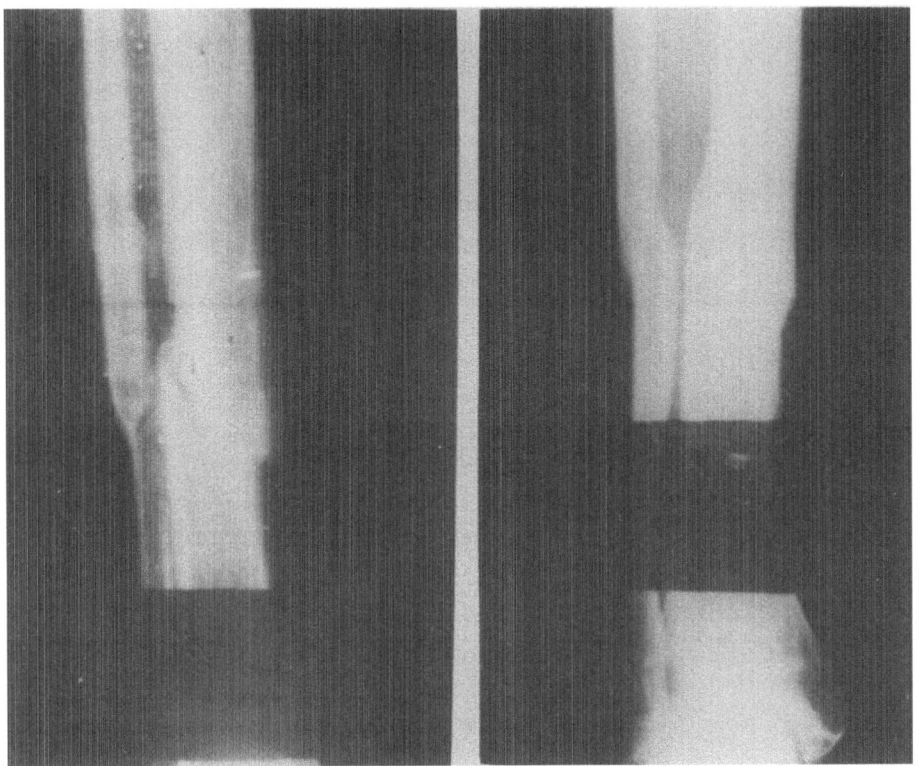

Before treatment   After treatment

Fig. 3   Influence of magnetic field on the consolidation of
         fracture

toid arthritis was also successful; the magnetic field was
applied over the adrenal cortex. The pain subsided considerably,
with a parallel decrease in the oedema and of the disturbed mobility in the morning.

The application of a low-frequency magnetic field to 28
patients with fracture of the lower extremities produced
rather characteristics results. Consolidation of the bone was
observed radiographically after the treatment and the patients
reported a decrease in the pain. Figure 3 shows the state of
patient I.S. before and after the magnetic therapy.

The list of other nosological units in which the magnetic
field induces positive results includes: 40 patients with masti-
tis, 17 patients with endometritis, 29 patients with chronic
gynaecological diseases, 95 patients with urticaria, etc.

In conclusion, we have grounds to claim that our experimental and clinical material is convincing evidence for the
indisputably positive effect of the low-frequency magnetic
field generated by the "Magnit H-80" device in medical practice.

ELECTROMAGNETIC FIELDS - A NEW ECOLOGICAL FACTOR

Marko S. Markov

Department of Biophysics, Biological Faculty,Sofia
University
Bul.Dragan Tzankov 8, Sofia 1000, Bulgaria

The contemporary conditions of life put man in dependence of the complex of physical influences of the environment and in the first place - of the electromagnetic fields /EMF/. The rapid development of science and technology has resulted in the introduction of many new technologies and devices in industry, agriculture and everyday life. On the other hand, during their phylogenetic and ontogenetic development the living organisms are continuously exposed to the influence of different biotic and abiotic factors. The physical factors are included in the first group and the entire evolution of life is connected with an adaptation to the action of these factors.

The available data on the influence of the electric charges, of the natural and artificial electric, magnetic and electromagnetic fields of biological systems testify to the increasing importance of the ecological aspect of the influence of these physical factors on different living organisms. It can be affirmed that the natural EMF are characterized by their continuous and comprehensive action throughout the entire living activity of the organism. Usually the geomagnetic field /GMF/ is included in the class of EMF. GMF may be regarded as a basic ecological factor which reflects the oscillations of GMF, as well as the influence of other sources of natural EMF: the Sun, the Stars, Space.

Until recently the question concerning the possibility of an influence of strong magnetic fields on biological systems was controversial, but now there exist many facts indicating that even natural magnetic fields have biological significance /7/. This is the reason why the natural magnetic fields are considered to be an ecological factor.

The problem of the eclogical importance of EMF is mainly estimated in relation with their role for the emergence of different pathologies, in the prognosis of human health. However it is necessary to estimate also the action of EMF on animals and plants, on the biosphere in general.

Sometimes artifital EMF are called "factors of new ecology" because they sharply although unnoticeably change the usual electromagnetic situation. This class includes radio-background created by radio- and TV-emitters, EMF of industrial origin, electric wiring, as well as therapeutic devices. The

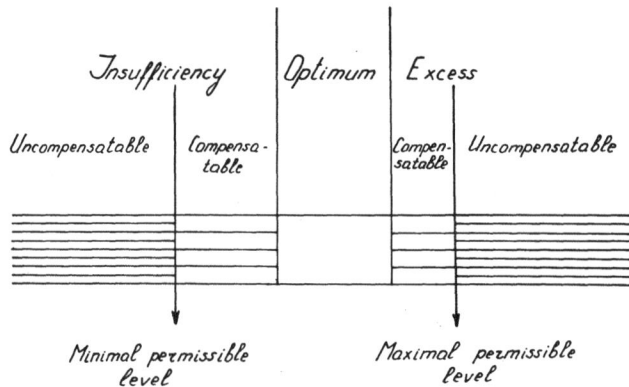

Fig. 1 Possibilities of living systems to accept electromagnetic fields

value of the electromagnetic pollution of the environment has already reached a significant level. Even a rough estimation gives us grounds to affirm that the majority of the population lives in an increased EMF intensity compared with the background.

There are reasons to affirm that the species composition of the biosphere results from the long evolution and from the influence of relatively constant atmospheric, temperature, electromagnetic, etc. conditions. The disturbance equilibrium between environment and organism can lead to ecological discharmonies, to disturbance of the compensation-adaptation machanisma. The estimation of the ecological significance of EMF needs a permanent investigation of the character, degree and basic regularities of their influence on biological systems with a different level of organization. It is also connected with the evaluation of the compensation-adaptation reactivity of different organisms.

The ecological significance of EMF is closely connected with their role in the course of evolution : it is established that all ranges of the electromagnetic spectra have accompanied life on Earth. However, the solution of the fundamental problems is seriously hindered by the absence of enough data on the evolutionary influence of EMF of different type, intensity, frequency, direction, etc. Simultaneously, the absence of objective information on the distribution of natural and artifitial EMF in the biosphere, as well as the lack of methods and devices for measuring and controlling, have an adverse effect. A question of primary importance becomes the elaboration of new norms and the specifying of the statistical norms of EMF under production conditions and in everyday life. Now these parameters are estimated in a different way in the different countries. The neglecting of low values of the electromagnetic energy and the elementary extrapolation for the human organism of data obtained by the experiments with animals may result in incorrect estimation of the man-EMF relationship.

The accumulation of experimental and clinical data gives some authors reason to define a new nozological unit - "radio-wave disease". The manifestation of this disease strongly depends on the intensity of the EMF, on the duration of its influence, on the combination of definite levels of the field intensity and exposure time, on the age and functional state of the organism.

The biological influence of EMF has been observed in many functions of the living organisms. Very often the ecological significance of EMF is discussed when the results obtained in the investigation of men and animals are analysed. It is considered to be proved that EMF basically influence the organism through direct action on the central nervous system and through the reflexes. Such an approach seems incorrectly grounded as it attaches absolute importance to only one side of the biological effects of EMF. It seems to be more correct to discuss the course of physical-chemical processes arising at tissue, cellular and subcellular level. The effects of EMF factors of the environment on the living systems occur on the basis of energy interactions. However, in many cases the biosystems react to the electromagnetic fields with intensity lower than the theoretically calculated limit for producing any energy effects.

The most detailed investigations are carried out currently on the unfavourable influence on the health of low-intensity EMFunder production conditions: disturbances at the level of the system, organ and tissue, as well as different functional

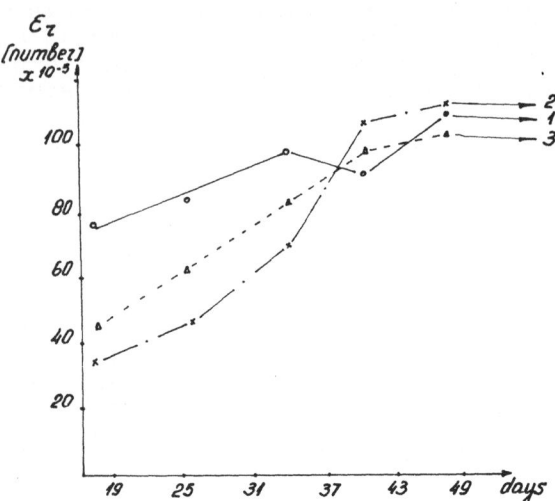

Fig.2 Changes of erythrovyte number in blood of albio rats
under CMF influence
1 - control group
2 - group of direct influence
3 - group of indirect influence

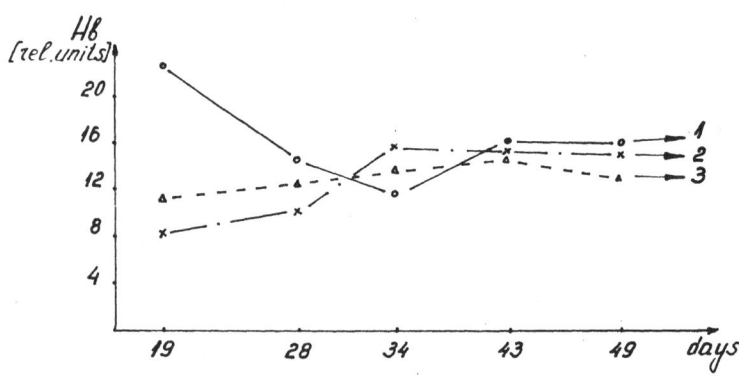

Fig.3 Changes of the content of hemoglobin in rats under CMF
influence
1 - control group
2 - group of direct influence
3 - group of direct influence

changes in the blood-vessel and endocrine systems are observed
/1,3,5,6,8,10/.

Very important is the estimation of the significance of
the changes in the physiological reactivity of the organs and
systems, adapting the organism to the conditions of the eviron-
ment.

The evaluation of the influence of EMF on organisms on the
basis of individual data should necessarily be supplemented
with the data about changes of the same parameters in a large
group of people. In this way only, on the background of the
physiological lability of the individual it is possible to
estimate the reliability of the responce of the group, genus
and species to EMF.

Shandala/4/ defines the hygienic and physiological norms
and notes that the normalizing of the artificial factors of the
environment should be based on a determination of the minimum,
optimum and maximum permissible level of the intensity of the
electromagnetic factor /Fig.1/.

Under the influence of EMF with low and middle intensity,
changes usually occur which are within the limits of the physiological reactivity of the organs and systems responsible for
the adaptation. In this way the investigation of the adaptation
of the organism to the different frequency and intensity of EMF
ranges becomes a question of primary importance. The serious
study of the adaptation mechanism to the EMF influence can open
new practical possibilities for elimination of the unfavourable
results of the electromagnetic influence. An essential importance in the realization of the magnetobiological effects has
the cumulative action of the field, as well as the total time
of the influence and the rhythmof the electromagnetic action.

Biological system were found to possess unexpectedly wide
adaptation possibilities: the same morphological systems and
structures can realize many functions and adaptation processes.
Such an adaptability is very important because it predetermines
a considerable range of adaptation to the changes of the parameters of the environment.

Shandala /4/ using the idea of stage adaptation which takes
into account the estimation of the response of the organism to
gradually increasing intensity of EMF, has established that
when the intensity of an extremely low-frequency field increases,
adaptation and compensation reactions which are manifested by
change of the redox processes and of the oxygen regime in
tissues, take place in the organism.

In experiments carried out at our laboratory it has been
shown that a repeated /48 times/ everyday influence of constant
magnetic field /CMF/ with induction B=45 mT and exposure time
t=30 min on newborn rats leads to adaptational responses. We
have observed changes in the motphological composition of the
blood, in the erythrocytes sedimentation reaction and in the
quantity of hemoglobin.

Figures 2-3 show summarized results of the direct and
indirect influence of CMF on these parameters. As indirect
action of CMF we indicate the influence by including water
treated in CMF with the above-mentioned parameters in the daily
ration of the animals.

The adaptability of the organisms is not unlimited. The

adaptation may occur only to a definite limit above which irreversible changes sometimes arise. All disturbance in the organism originating under EMF influence may be regarded as a result of the influence of the electromagnetic processes connected with the regulation of the physiological functions. Pressman /2/ concludes that in every living organism there exists a system for defence against the inner natural and artificial EMF, which is connected with the phase dependance of the biological effects under EMF influence.

Starting from the viewpoint that the organism is usually adapted to one or an other factor of the environment, Subbota /7/ assumed that some mechanisms of deadaptation have to be sought on the basis of some manifestations of disbalance. It is also necessary to take into account that after stopping the influence a disturbance of the contour of selfregulation created at the new level occurs once more that the organism has to adapt itself all over again.

The evaluation of the ecological influence of EMF on the biosphere has to be connected with the analysis of the possibility of mutagenic effects. At low intensities /which usually exist in the biosphere or are used as artificial sourses/ EMF cannot provoke significant changes in the state of the organism in general. However, they may be accumulated and an unfavourable effect may manifest itself after some generations.

When the biological influence of EMF is estimated, it is necessary to take into account their simultaneous action together with many other physical factors, such as vibrations, noise, temperature changes, etc. What is important, the biological effect of each of these factors not only adds to the effect of the others, they are also subordinated to very complicated laws.

REFERENCES

1. Hauf, R., Wiesinger, J., 1973, Int.J.Biometeotol., v.17, N.3, 213-215.
2. Pressman, A.S., 1968, Electromagnetic fields and living nature, M.Science /in Russian/.
3. Serduk, A.M., 1977, Interaction of the organism with electromagnetic fields as factor of the environment, Kiev, Science.
4. Schandala, M.G., 1975, Hygiene of towns, N.14, 58-68 /in Russian/.
5. Tyler, P.E., 1975, Biological effects of nonionizing radiation, Ann.N.Y.Acad.Sci., v.247, 6+14.
6. Stefanov, B., Solakova, S., 1973, Hygiene of work and occupational diseases, N.7, 44-45.
7. Subbota, A.G., 1968, Hygiene of work and biological influence of electromagnetic field with radio-frequency, M. Science, 148.
8. Tjagin, N.V., 1971, Clinical aspects of radiation with extremely high frequencies, Leningrad, Medicine /in Russian/.
9. Travkin, M.P., 1978, Reactions of biological systems to magnetic field influence, M.Science, 178-198 / in Russian/.
10. Slberti, S., Maggi, G., Salimei, E., 1971, Sociritas, v.56, N.5, 375-398.

ERYTHROCYTE SHAPE IS INFLUENCED BY FREE ELECTRIC AND CHEMICAL ENERGY

R. Grebe[1], M. Zuckermann[2] and H. Schmid-Schönbein[1]

[1]Abt. Physiologie, RWTH Klinikum, Pauwelsstr, D-51 Aachen, FRG
[2]Dep. of Physics, McGill University, Montreal, Canada

INTRODUCTION

The erythrocyte membrane is built up of the phospholipid-cholesterol bilayer, sandwiched by charge carrying proteins, the glycocalix, coupled to the bilayer via the hydrophobic coil of glycophorin A, and the bulk of the extrinsic proteins, coupled via Ankyrin and Band 3. Both the extrinsic proteins and some of the phospholipids of the interior face of the membrane are also carrying charges.

All elements are considered completely mobile in the plane of the membrane, although to a different degree. In addition there is an interbilayer mobility of cholesterol, shown to flip-flop with characteristic times in the order of some seconds. Consequently our concept implies that every local change or perturbation of the membrane will immediately influence the surrounding membrane parts and will trigger equilibration processes over the whole membrane:

LOCAL CHANGE MEANS GLOBAL CHANGE

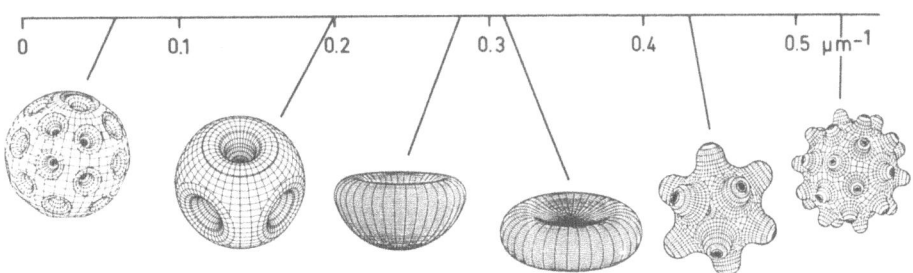

Fig. 1. Computer simulation of the reversible transition of discocytes to echinocytes and stomatocytes using constant surface area(135μm²) and volume(98μm³). The mean mean curvature increases from the sphero-stomatocyte to the discocyte to the sphero-echinocyte.

THEORY

Differential Geometry of Erythrocyte Shapes

Geometrical data of erythrocytes were taken from microphotographs, electron microscopy and data from the literature(1). This data was used to calculate computer models (2) and analyse the curvature changes associated with the well known reversible transition of the discocyte to echinocytes and stomatocytes. Typical results of our computer simulation using constant surface area of 135 μm^2 and a volume of 98 μm^3 are shown in Fig.1. The mean mean curvature increases from the sphero-stomatocyte to the discocyte and to the sphero-echinocyte.

Free Electrostatic Energy

The exterior and interior surface of the erythrocyte membrane can be thought of as equipotential surfaces (3). Under this assumption simple spherical shells provide a basis for calculating the electrostatic properties of the erythrocyte membrane. As shown above every erythrocyte shape is characterised by a typical mean mean curvature K, which gives the radius to calculate the free electrostatic energy of the system erythrocyte membrane and surroundings.

The free electrostatic energy density has to be computed for the external and internal surface by

$$f_e = 1/2 \ (\ V_1 \cdot \sigma_1 + V_0 \cdot \sigma_0) \ .$$

The surface charges include the charged phospholipids as well as the charged extrinsic membrane proteins. It is introduced in the calculation as the mean surface charge density σ_x. The potential V_x is given as the solution of the POISSON-BOLTZMANN equation:

$$\Delta V = - \ \rho \ / \ \varepsilon \ .$$

It depends on the ionic strength, the surface charges, the charges of the cytoplasmic polymer solutes, of temperature and pH.

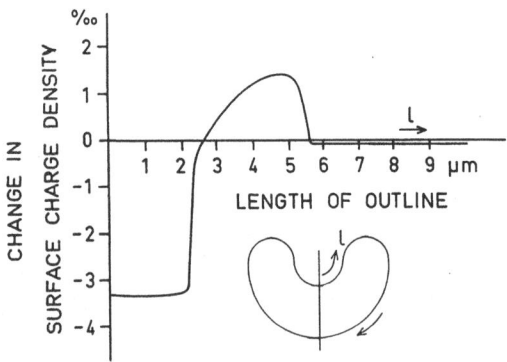

Fig. 2. Change of local surface charge density (charge carrying molecules) along the outline of a stomatocytic shape. The locus of low mean curvature (dimple l=0μm) shows a diminution of charges the locus of high mean curvature (l=5μm) an increase.

Free Chemical Energy

With increasing deviation from the biconcave shape there is an increase in variation of local mean curvature. In equipotential surfaces such a local change in curvature induces a corresponding change in local charge density. Provided the charge carriers are mobile, a chemical potential is formed tending to reestablish an equal charge distribution. In other words: Change in curvature produces free chemical energy, which has to be taken into account when computing the total free energy for any erythrocyte shape.

To compute the free chemical energy one has to know the local densities of surface charges. A method used to optimize electrodes in high voltage electronics (4) has been adapted to biophysical problems. By this method idealized point-, line- and surface charges are used to build up an equipotential surface. Knowing the location and magnitude of these charges the electrostatic characteristics of equipotential surfaces and their exterior space can be calculated (5).

This method was used to imitate the electrostatic properties of the erythrocyte shapes both inside and outside of the cells. At the present stage the method is limited to surfaces with rotational symmetry. As an example the relative change in surface charge density for the outline of a stomatocyte is shown in Fig. 2. An increase in charge density in convex and a decrease in concave surface parts is to be seen.

The free chemical energy density in every point of the interior and exterior membrane surface is given by (6)

$$f_c = kT \cdot \{n_e \cdot \ln(n_e/n_{em}) + (n_o-n_e) \cdot \ln[(n_o-n_e)/(n_o-n_{em})]\}$$

with n_e as local surface density of charged molecules, n_{em} as mean surface density of charged molecules and n_o as surface density of molecules.

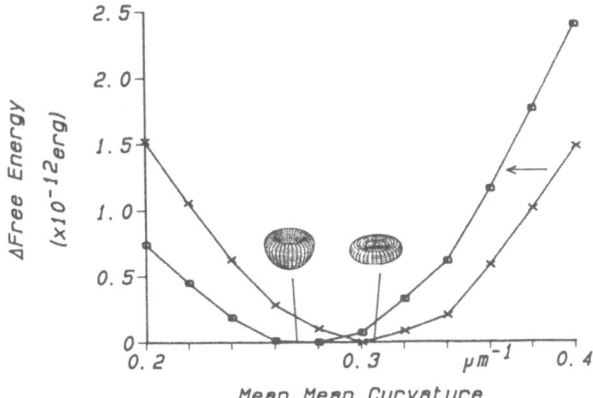

Fig. 3. The graph x gives the free energy for the computer simulated shapes in relation to the mean mean curvature. Reduction of outer surface charge gives a shift of the free energy, graph ■. The shapes related to the free energy minimum are shown.

RESULTS

The total free energy density is given by

$$f = f_e + f_c \ .$$

Using this and the above given equations as well as the results of our calculations on the geometry of the erythrocyte shapes, we are able to calculate the free energy of model cells with different mean mean curvature related to the electrostatic properties of the membrane and the surroundings. The values used for this calculations have been: concentration of ions inside/outside cell 0.3mol/l (debye length 0.792nm), area per surface charge inside 10nm² / outside 9.95nm²(9.94), membrane thickness 5nm, rel. permittivity 76.7 and temperature 37°C.

As a result in Fig. 3 the graph x gives the sum of the two seperately calculated free energy densities integrated over the whole cell. An energy minimum is to be found which predicts an equilibrium shape for a cell with a mean mean curvature of about 0.3 μm.

In addition the outcome of a change in one boundary condition is been shown. The graph marked by ■ shows the mean mean curvature depending free energy after slight reduction of the outer surface charge density of the previous model. This diminution of outer surface charge leads to a more stomatocytic equilibrium shape.

CONCLUSIONS

There is a close relationship between mean mean curvature and erythrocyte shapes. This gives a basis to calculate free electrostatic and chemical energy stored in different shaped membranes. In every case there exists an energy minimum for a distinct mean mean curvature, which gives a special shape. The influence of various electrostatic magnitudes can be examined in these models of red blood cells.

Decrease in outer surface charges shifts the free energy minimum towards more stomatocytic, increase towards more echinocytic RBC shapes, which is in qualitative agreement with our experimental results (7).

It has been shown that in this holistic model free energy, depending on global properties of the membrane and of the inner and outer environment only, predict one special shape out of the given group. The only presumptions in this model are conservation of total inner and outer surface charge, high lateral mobility of all membrane molecules and the restriction of any movement of surface charges from one side of the membrane to the other.

REFERENCES

1 E. Evans and Y. C. Fung, Microvasc. Res. 4:335 (1982).
2 S. Svetina, J. theor. Biol. 94:13 (1982).
3 H. S. Lew, J. Biomechanics 3:569 (1970).
4 D. Metz, ETZ-A 97:121 (1976).
5 R. Grebe, G. Peterhänsel and H. Schmid-Schönbein, submitted.
6 M. J. Zuckermann, A. Georgallas and D. A. Pink, Can.J.Phys. 63:1228 (1985).
7 R. Grebe, H. Wolff, M. Schönfeld and H. Schmid-Schönbein, Phlüg. Arch. 405(2):R23 (1985)P

THE ELECTROCHEMILUMINESCENT CHARACTERISTICS CORRELATE TO THE AGE-DEPENDENT CHANGES IN THE COMPOSITION OF PLASMA MEMBRANES

I. Zlatanov and S. Bachev

Central Laboratory of Biophysics
Bulgarian Academy of Sciences
1113 Sofia, Bulgaria

INTRODUCTION

Chemiluminescence of many biological systems, including plasma membranes, is due to exothermal reactions of oxidizing processes of the lipids and other cell components /Greenwood, 1982, Halliwell, 1985 /. When biological preparations are placed in an electric field the rate of peroxide processes increases several times and the intensity of luminescence increases strongly. Electrochemiluminescence /ECL/ was used in organic chemistry studies /Akins, 1979/. Successful attempts have been made for application of this method in the investigation of peroxide and free-radical processes in biology and medicine /Bard, 1972, Orel, 1976/.

The cell metabolism is accompanied by free-radical processes to accumulation of singlet molecular oxygen in the cell components /Harman, 1968/. This very reactive oxygen can peroxidate lipids /Tappel, 1973/ and causes crosslinkings in the proteins /Zs.-Nagy, 1985/. Age-dependent accumulation of free radicals /for review see Rothstein, 1982/ produces changes in the membrane structure and functions.

This article presents our investigation on the relationship between some developmental alterations in the lipid composition of the rat teste plasma membrane and its electrochemiluminescent /ECL/ characteristics /in suspension/.

MATERIAL AND METHODS

ANIMALS – Male Wistar rats, 1, 4, 7, 10, 13 and 17 months old, were used.

PURIFIED PLASMA MEMBRANES from total testes were isolated and tested enzymatically as described previously by Zlatanov and Smilenov /1986/. All the chemicals and reagents used in the experimental procedures were "analytical grade" and solutions were prepared with bidistilled water.

MEASUREMENTS OF ECL INTENSITY were performed using a specially constructed sample chambre with a central filiform cathode and two symmetrically mounted parallel filiform anodes. Each platinum electrode was 2 cm long and the distances between them were 1 cm. The concentration of the membrane suspension was about Ø.4 mg/ml at a final volume of 3 ml. A constant electric field of 4 V/cm was applied for generation of ECL in the suspension. ECL signals were monitored using a photomultiplier, model "FEU 78" /USSR/. The area /Q/ under the kinetic curves /Fig. 1 / was proportional to the quantum yield /Q/ of ECL and respectively to the concentration of free radicals in the membrane suspension.

PHOSPHOLIPID, CHOLESTEROL and PROTEIN contents of the membranes were determined by the standart methods of Folch /1957/, Kahovkova /1969/ , Sperry /1957/ and Lowry /1951/.

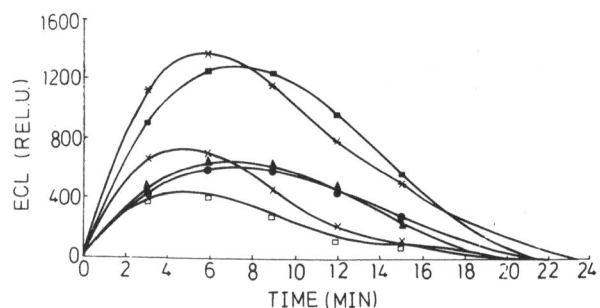

Fig. 1. Kinetic curves of electrochemiluminescence /ECL/ in rat testes plasma membranes. / ◇ / 1 month, / ◐ / 4, / △ / 7, / ✗ / 10, / ◻ / 13 and / ✳ / 17 months old rats

RESULTS AND DISCUSSION

Figure 1 shows that the quantum yield Q of ECL increases with age. As one can see in Table 1, it was augmented more than 3-fold during the investigated period. Like other investigated cells and membranes /Rothstein, 1982/, testes membranes are also subjected to peroxide damages with ageing.

The lipid /L/ and protein /Pr/composition of membranes is decisive for their structure and functions /Barenholz, 1982/. As many authors reported /for review see Rothstein, 1982/, the L/Pr ratio changes specifically for different tissues and cells during development and ageing. Table 1 shows an increase in this ratio for testes membranes in the initial developmental stages from the 1st to the 17th month, when the testes matured. After this period the relative part of membrane protein was enhanced, which might be due to a compensatory effect of the decreased activity of many membrane enzymes and proteins with ageing /Zlatanov, 1986/. Particularly, the lowered protein activities could be due to age-induced increase of the intensity of free radical processes leading to intra- and intermolecular crosslinkings of their chains /Zs.-Nagy, 1985/.

Cholesterol /CH/ is another important component of the

membrane /Yeagle, 1985/. In many tissues its concentration rises with ageing /Rothstein, 1982/. The ratios CH/L and CH/Pr in the membranes under investigation were elevated in the course of ageing / Table 1/. These alterations correlate very well to the age-dependent changes in Q / Table 1/.

Table 1. Influence of age on the ratios cholosterol/lipid /CH/L/, cholosterol/protein /CH/Pr, phosphatidyl-choline/sphingomyelin /PC/SM/, lipid/protein /L/Pr/ and the quantum yield /Q/ of the electrochemilumine-scence /ECL/ in rat testes plasma membranes

Age /months/	CH/L /mol/mol/	CH/Pr /μmol/mg/	PC/SM /mol/mol/	L/Pr /μmol/mg/	Q relative units
1	0.30±0.07	0.35±0.04	3.78±0.03	1.16±0.10	0.28±0.02
4	0.34±0.05	0.46±0.02	4.40±0.05	1.37±0.05	0.46±0.03
7	0.26±0.13	0.51±0.06	5.88±0.07	1.90±0.12	0.44±0.01
10	0.34±0.11	0.54±0.05	6.44±0.05	1.60±0.11	0.43±0.01
13	0.44±0.07	0.55±0.02	6.06±0.01	1.24±0.04	0.90±0.02
17	0.50±0.07	0.65±0.03	6.00±0.05	1.30±0.05	1.00±0.01

It has been established following correlation between ECL parameter Q, proportional to free-radical content and testes plasma membrane compositional parameters: CH/L=0.93; CH/Pr=0.85; PC/SM=0.59; L/Pr=0.39.

Phosphatidylcholine /PC/ fluidizes the membrane, while sphingomyelin /SM/ has the opposite effect. The PC/SM ratio characterizes the physical properties of the membrane. Its value is very important for the functioning of some membrane-bound enzymes and receptors. The PC/SM ratio also correlates with Q, but the correlation coefficient is not very high.

In conclusion, the general parameters correlating to the age-related changes in ECL characteristics of rat testes plasma membranes are the ratios CH/L and CH/Pr, whereas the PC/SM and L/Pr ratios show lower correlation coefficients. This fact indicates the importance of the CH content in respect to the free-radical processes during development and ageing. The PL composition, particularly the content of PC and SM, influences these processes during ageing too.

REFERENCES

Akins, D.G. and Barke, R.G., 1979, Energy Transfer in Reaction of Electrogenerated Aromatic Anions and Benzoyl Per-oxide Chemiluminescence and its Mechanism, Chem.Phys.Lett., 29/3/:428-435.
Bard, A.I., Keszthelyi, C.P. and Jachikawa, H., 1972, On theEfficiency of Electrogenerated Chemiluminescence, in:Chemi-luminescence and Bioluminescense, Proc.Int.Symp.Athens,1972, N.Y.-L.
Barenholz, Y. and Gatt, S., 1982, in: Phoapholipids, Elsevier, North Holland
Folch, J., Lees, M. and Sloane-Stanley, G.H., 1957, A Simple Method for the Isolation and Purification of Total

Lipids from Animal Tissues, J.Biol.Chem., 266:497-509

Greenwood, C., Hill, H.A.O., 1982, in: Oxygen in Biological Systems, Chemy Br., London

Halliwell, B., Cutteridge, J.M.C., 1985, in: Free Radicals in Biology and Medicine, Oxford University Press, Oxford

Harman, D., 1968, Free Radical Theory of Ageing.Effects of Free Radical Reaction Inhibitors on the Mortality role of Male LAF mice, J.Gerontology, 23:476-482

Kahovkova, J. and Odavich, R., 1969, A Simple Method for Analysis of Phospholipids Separated on Thin-layer Chromatography, J.Chromatogr., 40:90-95

Lowry, O.H., Rosebrough, N.J., Farr, A.L. and Randall,R.J., 1951, Protein Measurement with the Folin Phenol Reagent, J.Biol. Chem., 193:265-275

Orel, V.E., 1976, Registration Method of Blood Plasm Electrochemiluminescence- in: Utilisation of Radioelectronics in the Field of Medical and Biological Studies, Naukova Dumka, Kiev /In Russian/ 135-156

Rothstein, M., 1982, in: Biochemical Approaches to Aging, Acad.Press, New York

Sperry, W. and Web, M., 1957, A Revision of Shoenheimer-Sperry Method for Determination, J.Biol.Chem., 187:97-106

Tappel, A.L., 1973, Lipid Peroxidation Damage to Cell Components, Fed.Proc. 32/8/:1870-1874

Yeagle,P.L., 1985, Cholesterol and the Cell Membrane, Biochim.Biophys.Acta, 822:267-287

Zlatanov, I.V. and Smilenov, L.B., 1986, The Effect of Age on the Activities of Membrane-Bound 5-nucleotidase, Guanylate Cyclase and Adenylate Cyclase in Rat Testes, Compt. Rend.Acad. Bulgare Sci., 39:101-104

Zs.-Nagy, I., 1985, Aging of the Cellular Membrane: Basic Principles and Pharmacological Interventions, Geriatrika, 1/8/:102-111

CHANGES IN CONFORMATIONAL STATE OF GLYCERALDEHYDE-3-PHOSPHATE DEHYDROGENASE ADSORBED ON PHOSPHATIDYLINOSITOL LIPOSOMES

Jan Gutowicz, Krystyna Michalak, and Teresa Modrzycka

Department of Biophysics, Academy of Medicine
Chałubińskiego 10, 50-368 Wrocław, Poland

INTRODUCTION

Glyceraldehyde-3-phosphate dehydrogenase /G3PDH/ belongs to the glycolytic enzymes which are found to have ability of association with membranes /1,2,3,4,5/. The association with membranes is controlled by electrostatic interactions. Hence, electrically charged surface of lipid bilayer was chosen as a model system for investigation of the association. Electrostatic binding of the enzyme on monolayers and liposomes modifies the enzyme specific activity /6,7,8/. The modification of specific activity may be a result of several reasons. Conformational modification of the enzyme molecules induced by the adsorption may be one of them. The aim of this work was attempt to test the possibility. In addition changes in conformational state of the enzyme upon the association with membranes may have important implications for other properties of the enzyme. We used the fluorescence probe bound to the enzyme molecule which was formed during the reaction of o-phthaldialdehyde with amino groups of the protein. This fluorophore is an acceptor of the excitation energy transfered from tryptophanyl residues of the protein. Resonance energy transfer efficiency strongly depends on the distance between donor and acceptor of the excitation energy. Any changes in the distance due to conformational rearrangement in protein molecule must effect the energy transfer.

MATERIALS AND METHODS

Glyceraldehyde-3-phosphate dehydrogenase was prepared from rabbit muscle according to Kochman and Rutter /9/. Phospholipid and other chemicals were purchased /Koch-Light Laboratories, Sigma, Fluka/.

The enzyme concentration and specific activity were determined spectrophotometrically by measurements of extinction at 280 nm and change of extinction at 340 nm as function of time, respectively /10,11/. Amino groups of G3PDH were labelled with o-phthaldialdehyde /OPA/ according to the procedure for phosphatidylserine amino group described elsewhere /12/. Phospholipid vesicles were prepared by shaking of lipid suspension in 50 mM triethanolamine buffer, pH = 8.6 for 30 min.

The fluorescence was always measured /Perkin Elmer MPF3L/ after the incubation of the labelled enzyme with phospholipid vesicles for 30 min. at room temperature. Tryptophan and the probe fluorescence were excited at 290 nm and 340 nm, respectively.

RESULTS AND DISCUSSION

Upon binding of the probe to the enzyme molecule its specific activity gradually decreased with increase of the degree of labelling /R/. Hence, the degree of labelling in our experiments did not exceed 6 molecules of the probe per tetrameric molecule of the enzyme. At such degree of labelling the activity was not decreased more than about 12%. The covalent binding of the probe with G3PDH results in a new fluorescence band with emission maximum at 445 nm /Fig. 1/. The labelled enzyme fluorescence excited at 290 nm is quenched in tryptophan emission band and enhanced in the probe emission band /445 nm/. It can be explained by reso-nance energy transfer between tryptophanyl residues and the isoindole fluorophore. To test whether conformational changes in the enzyme may effect isoindole fluorescence and Trp to probe energy transfer efficiency we investigated the effect of urea as well known modifier of conformation on emission spectra of the labelled protein. Addition of 6 M urea to G3PDH solution caused an increase of tryptophanyl fluores-cence intensity of the unlabelled enzyme. Urea added to the labelled protein caused greater enhancement of the Trp residues fluorescence intensity and also quantum yield /not shown/ than for the unlabelled protein and a simultanous decrease of isoindole fluorescence as it is shown in Fig. 2.

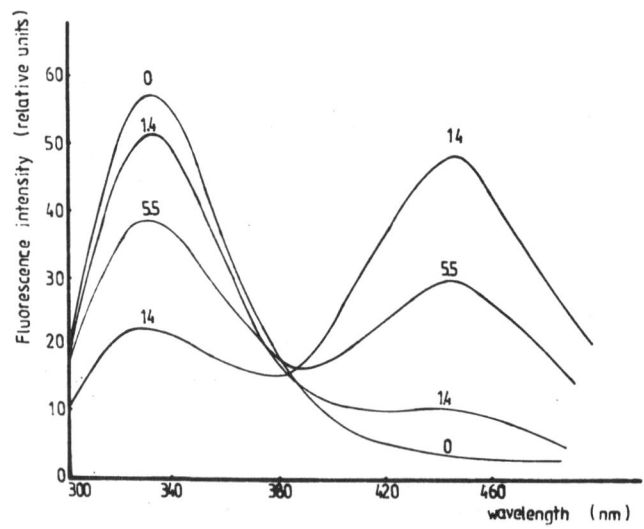

Fig. 1 . Fluorescence emission spectra of glyceraldehyde-3-phosphate dehydro-genase labelled with o-phthaldialdehyde. Excitation wavelength - 290 nm, molar ratio of OPA to G3PDH: 0, 1.4, 5.5 and 14 as indicated in the figure.

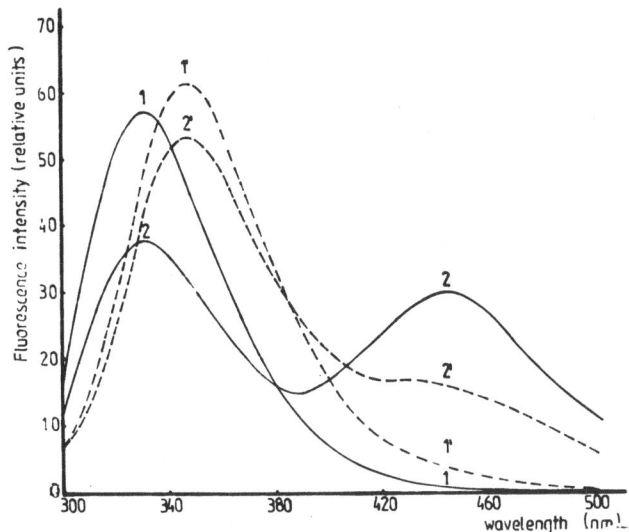

Fig. 2. Fluorescence emission spectra of
the labelled and unlabelled enzyme
in the presence of 6 M urea.
1,2 - spectra of the unlabelled
and labelled enzyme, respectively,
1', 2' - the same samles in the
presence of 6 M urea, excitation
wavelength - 290 nm.

Changes in spectrum of the labelled protein shown in Fig. 2
can be explained as a result of the reduction of Trp to
probe energy transfer efficiency due to increase of donor -
acceptor distance during unfolding of the protein molecule.
The experiment with urea proved that incorporation of the isoindole fluorophore to the enzyme molecules allowed to monitor
of conformational changes. Addition of the liposomes to the
labelled enzyme solution resulted in quenching of the probe
fluorescence when the probe is directly excited $/\lambda_{ex}$ =340nm/
as well as when the donor is excited $/\lambda_{ex}$ = 290nm/ - Fig. 3.
Presumably two mechanism are simultaneously responsible for
the quenching of the probe fluorescence: 1/ quenching by
direct interaction of phospholipid molecules with the isoindole chromophore, 2/ reduction of the energy transfer
efficiency between the two types of fluorophores in the
protein as a result of the change in the distance between
them which means conformational rearrangements in the protein
molecules. It is noteworthy that the quenching of isoindole
fluorescence by lipid vesicles was more efficient when the
fluorescence was excited at 290 nm than that excited at
340 nm /Fig. 3/. The ratio of the isoindole quantum yield
at the excitation wavelength of 340 nm to quantum yield at
that of 290 nm increased with increase of lipid to protein
molar ratio - not shown. This result may be due to occurence
of the second mechanism. Previously observed increase of the
quenching of the enzyme tryptophanyl fluorescence by dynamic
quencher may be interpreted as a dynamic exposure of the
fluorophores /13/. Data presented here suggest rather changes
in a static conformational state.
    Capability of membranes for modification of the enzyme
conformation may have important implications for the enzyme
properties and function in vivo.

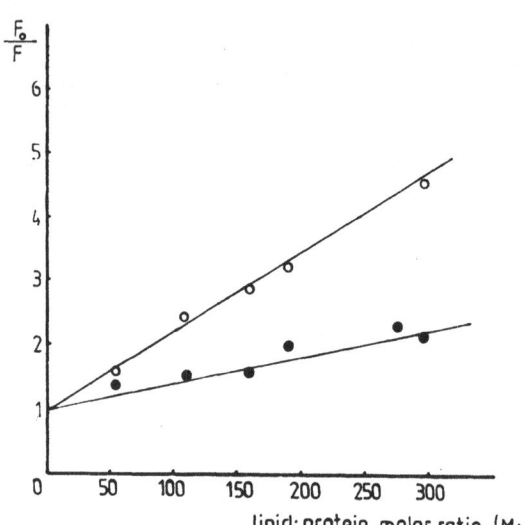

Fig. 3. Stern-Volmer plots of the
quenching of the probe
fluorescence in G3PDH as
function of PI/protein
molar ratio. Excitation
wavelength - 290 nm /—○—/
and 340 nm /—●—/, emission
wavelength - 445 nm.

REFERENCES

1. Nilson O. and Ronquist G. /1969/Biochim. Biophys. Acta
   183, 1-9
2. Duchon G. and Collier H.B. /1971/ J. Membr. Biol.
   138-157
3. Kant J.A. and Steck T.L. /1973/ J. Biol. Chem. 248,
   8457-8464
4. McDaniel D.F. and Kirtley M.E. /1975/ Biochim. Biophys.
   Res. Commun. 65, 1196-1200
5. Kliman H. J. and Steck T.L. /1980/ J. Biol. Chem. 255,
   6314-6321
6. Wooster M. S. and Wrigglesworth J.M. /1976/ Biochem. J.
   159, 627-631
7. Wooster M.S. and Wrigglesworth J.M. /1976/ Biochem. J.
   153, 93-100
8. Gutowicz J. and Modrzycka T. /1978/ Biochim. Biophys.
   Acta 512, 105-110
9. Kochman M. and Rutter W.J. /1968/ Biochemistry 7, 1671-
   1677
10. Dandliker W.B. and Fox J.B. /1955/ J. Biol. Chem. 214,
    275-283
11. Velick S.F. /1955/ Methods Enzymol. 1, 401-411
12. Sidorowicz A. and Michalak K. /1984/ Studia Biophysica
    102, 181-187
13. Gutowicz J. and Modrzycka T. /1986/ Gen. Physiol. Biophys.
    5, 297-306

ACKNOWLEDGEMENT

This work was supported by Research Grant C.P.B.P.04.01.

BIOLOGICAL ACTIVITY OF MACROPHAGES INTERACTING WITH SMALL LIPOSOMES AND MICELLAR-LIKE MEMBRANE STRUCTURES

R.P. Pancheva, M. Naplatarova, and P. Antonov

Department of Physics and Biophysics, Scientific
Medicobiological Institute, Medical Academy
1431 Sofia

Liposomes are potentially useful as a carrier system to deliver pharmacologically active agents to the cell, especially if they can be directed to specific cell types in vivo. At present, the therapeutic applications of liposomes as carriers are limited by their rapid uptake by the liver and spleen. A series of experiments with a wide variety of cell systems were designed to study factors in an attempt to overcome this barrier. One possibility for this purpose is to elabote a macrophage-mediated carrier system.

Recent results demonstrated that the sensitivity of macrophages to the vesicle surface charge and size can be used either to increase or to reduce liposome uptake by this cell type[1,2,3] and that it depends also on the composition and phase state of the phospholipids.[4,5] However, there still remain areas where considerable work is needed in order to elucidate precisely the processes involved, in particular, of practical importance is the relatively low extent of this uptake.[6]

In this communication we report considerably higher biological activity of mouse peritoneal macrophages interacting with micellar-like membrane structures, compared with small liposomes.

EXPERIMENTAL

Lyophilized egg phosphatidylcholine from Avanti Polar Lipids, Birmingam, AL and freshly isolated ^{125}I-labelled lecithin[7] were used.

To obtain small liposomes /SL/ the lipid suspension /20 mM/ was sonicated using the probe-type sonicator for 30 min at 4°C. Multilamellar vesicles and metal impurities from the sonicator tip were removed by ultracentrifugation at 100,000 g for 60 min.

Micellar-like membrane structures /MM/ were prepared from stock solution of lecithin in chloroform-methanol, which was then mixed with recrystallized sodium deoxycholate /Sigma/ in methanol at various V/V ratios. The organic solvents were evaporated and the residue was then suspended in saline to a lipid concentration of 20 mM. The suspensions were subjected to ultrasonic vibration for 20 s. MM structures were characterized by a fluorescence method[8] and NMR techniques.[9]

Macrophages were obtained by peritoneal lavage of stimulated mice. The cell concentration was adjusted to 4×10^6 cells/ml and their viability was determined by Tripan-blue exclusion.

MM and SL were incubated with macrophage for 30 min at $37^\circ C$. In all experiments the lipid concentration in the samples was 0.4 mM/10^6 cells. Quantitatively the interaction was evaluated after five-fold washing of the cells and gamma-counting measurements.

RESULTS AND DISCUSSION

Extensive physico-chemical studied have recently been carried out on the ternary systems lecithin-bile salt-water /LBW/.[9,10] The interest in these systems stems mainly from the fact that they are able to from model micellar-like lipoprotein complexes and to solubilize cholesterol.

According to the literature data /NMR-spectroscopy,[9] X-ray small-angle scattering analysis,[11] quasi-elastic light scattering,[12] fluorescence,[8] Uv-absorption measurements and differential scanning calorimetry[13]/ at physiological lecithin-bile salt ratios /L/B/ in LBW forms micellar-like structures. Flat bilayer fragments and/or metastable liposomes occur above L/B molar ratios 4.0.

Because of the dependence of MM structures on many experimental conditions /pH, ionic strenght, bile salt type, lipid concentration, mode of sample preparation, temperature, etc, the samples were characterized by NMR-techniques confirming corresponding results reported elswhere.[9]

Three series of experiments were carried out: /a/ incubation of macrophages with MM at L/B 1.1, 1.7 and 2.9 ; /b/ incubation at L/B 4.0, 4.6, 5.1 and 6.3, and /c/ with SL. /The values for L/B were selected on the basis of our previous experiments.[8] The data are presented in Table 1.

TABLE 1

Interaction of mouse peritoneal macrophages
with three types of model membrane strucrures

L/B molar ratio	Numer of the samples	Lysis of the cells %	Vitality %	^{125}I incorporation %
/a/ 1.1-2.9	16	84	37	1.5 ± 0.6
/b/ 4.0-6.3	18	0	75.5	16.0 ± 3.5
/c/ SL	9	0	72	0.5 ± 0.1

The results could be summarized as follows: /i/ The macrophages incorporate ^{125}I MM-series /b/ 15 times more compared to SL and MM at low L/B-series /a//Table 1/; /ii/ Microscopic observations show high percentage lysis and nonvital cells at low L/B at highest L/B value-series /b/ ; the cells appear swollen ; /iii/ Incorporation of the radioactivity increases slowly with time after 30-min incubation /the data are not given/.

A few reasons could be suggested for the explanation of these findings: first of all, membrane permeability perturbation in the presence of the detergent, as shown by Riehm.[14] The effect of the membrane curvature and the size of the interacting particles could also play a role. However, additional experiments /mainly physiological/ are required for evaluating the prospects offered by this macrophage-mediated system as a carrier for proteins and/or lipophyllic drugs in vitro.

REFERENCES

1. Schwendner, R.A., Lagicki, P.A. and Rahman, Y.E., 1984, The effect of charge and size on the interaction of uni-lamallar liposomes with macrophages, Biochim.Biophys.Acta, 772:93.
2. Pratten, M.K. and Loyd, J.B., 1986, Pinocytosis and phago-cytosis: The effect of size of particulate substrate on its mode of capture by rat peritoneal macrophages culture in vitro, Biochim.Biophys.Acta, 881:307.
3. Hsu, I.M. and Juliano, R.L., 1982, Interaction of liposomes with the reticuloendo thelial sytem. II Nonspecific and receptor mediated uptake of liposomes by mouse peritoneal macrophages, Biochim.Biophys.Acta, 720:411.
4. Margolis, L.B., 1984, Lipid-cell interaction. Liposome adsorption and cell-to-liposome lipid transfer are mediated by the same cell-surface sites, Biochim.Biophys.Acta, 804:23.
5. Juliano, R.L., Hsu, I.M. and Roden, S.L., 1985, Interaction of the polymerized phospholipid vesicles with cells. Uptake, processing and toxicity in macrophages, Biochim.Biophys. Acta, 812:42.
6. Szoka, F., Jacobson, K. and Papahadjiopoulis, D., 1979, The use of aqueous space markers to determine the mechanism of interaction between phospholipid vesicles and cells, Biochim.Biophys.Acta, 551:295.
7. Antonov, P.A., Pancheva, R.P. and Ivanov, I.G., 1985, Radio-iodination of naturally occuring phospholipids, Biochim. Biophys.Acta, 835:408.
8. Antonov, P.A., Pancheva, R.P., Naplatarova, M. and Stoylov, S.P., 1985, A fluorescence study of the ternary system lecithin-sodium deoxycholate-water showing structural transformations, FEBS Lett., 187:327.
9. Small, D.M., Penkett, S.A. and Chapman,D. 1969, Studies on simple and mixed bile salt micelles by NMR spectroscopy, Biochim.Biophys.Acta, 176:178.
10. Lichtenberg, D., Robson, R.J. and Dennis, E.A., 1983, Solu-bilization of phospholipids by detergents. Structural and kinetic aspects, Biochim.Biophys.Acta, 737:285.
11. Mueller, K., 1981, Structural dimorphism of bile salt/leci-thin mixed micelles. A posible regulatory mechanism for cholesterol solubility in bile? X-ray structure analysis, Biochemistry, 20:404.
12. Mazer, N.A., Benedek, G.B. and Carey, M.C., 1980, Quasi-elastic light-scattering studies of aqueous biliary lipid sytems. Mixed micelles formation in bile salt-lecithin so-lutions, Biochemistry, 19:601.
13. Claffey, W.J. and Holzbach, R.T., 1981, Dimorphism in bile salt-lecithin mixed micelles, Biochemistry, 20:415.
14. Riehm, H. and Biedler, J.M., 1972, Poteation of drug effect

by Tween-80 in Chinese hamster cells resistant to actinomy-
cin D and Daunomycin, Cancer Res., 32:1195.

CONCERNING SOME VARIATIONS OF THE CHARACTERISTICS OF ELECTRO-CHEMICALLY ACTIVE MEMBRANES AFTER PULSE MAGNETIC AND ELECTRO-MAGNETIC FIELD ACTION

N. Kentchev,[x] N. Popdimitrova,[xx] N. Todorov,[xx] and
N. Anachkova
Institute for Foreign Students,[x] Medico-Biological
Institute, Sofia,[xx] Institute of Health Resorts,
Physiotherapy and Rehabilitation, Sofia, Bulgaria

INTRODUCTION

The electrochemically active membranes have been the object of intensive investigations lately [1] , because of the complex nature of action of some external factors and their complex structure [2] . Ultrahigh frequency magnetic field have been found to influence the human nervous system [3] , as well as a number of biochemical processes [4] . According to the latest achievements in science, the magnetic field affects a number of chemical reactions in the organism with the participation of radicals [5] . Of interest is the magnetic and electromagnetic field action on skin permeability and drug transport through the skin, which is the topic of the present paper.

According to the quantitative theory [6] and its supplements [7] , the specific pharmacological drug electrophoresis effect is determined by the amount of substance introduced:

$$P = K.d.B^{-x}(1 - e^{-b \cdot c}). \; I_{eff} \cdot \eta \cdot t, \qquad /1/$$

where $\eta = P_1/P_2$ is a correction factor for the active electrode surface (P_1 is the drug amount introduced for a patient at fixed electrode surface, P_2 - at unit area).

The ionophoretic permeability characterizes both the skin and the procedure peculiarities, it is highly informative and in the course of our work we determined its changes under different conditions of external field action:

$$\varkappa = P/Q_{eff} \cdot s \qquad /2/ \qquad \text{and} \qquad Q_{eff} = I_{eff} \cdot t_{eff} \qquad /3/$$

where Q_{eff} depends on the optimal procedure conditions.

MATERIAL AND METHODS

Drug electrophoresis was applied on the patient's forearms. Sources of pulsating and diadynamic currents of different pulse and pause duration were used. The effective current was determined on the basis of the total quantity of electricity and the procedure duration which are different for various re-

gimes /Fig.1 a, b, c/. To determine the amount of drug introduced and ionophoretic permeability, a computer programme 7 ,
elaborated for the purpose, was used 6 .

Drug electrophoresis was applied twice: before and after
the action of the changeable pulse magnetic or electromagnetic
field. For both cases the experiments were performed on 20 patients, the novocain concentration being 0.25 per cent and adrenaline concentration - 10^{-4} per cent. The regime of the applied
changeable pulse.magnetic field was: 160 er, 1 Hz, 0.2 s, 10 W.
The regime of the pulse ultrahigh electromagnetic field was:
1 Hz, 0.8 s, 10 /or 50/ W.

RESULTS AND DISCUSSION

The effective current of the procedure was determined using
the quantity of electricity for each pulse or a serie of pulses
Q_i, the pulse duration t_1 and the pause duration t_2 /Fig.1 b/.
At alternating current in the general case

$$Q_i = \int_{t_{i-1}}^{t_i} I/t/dt \qquad /4/ \quad \text{or} \quad Q_i = Q_1 = \int_0^{t_1} I/t/dt \qquad /5/$$

as similar current pulses as those applied in physiotherapy,
and the total quantity of electrocity during the procedure

$$Q = \sum_{i=1}^{n} Q_i = n.Q_1 \qquad\qquad /6/$$

Hence the effective current is:

$$I_{ef1} = n.Q_1 / \left[n.t_1 - (n-1) \right] .t_2 \quad \text{at } t_1 \approx t_2$$

or $I_{eff2} = n.Q_1/n.t_1 \approx Q_1/t_1$ if $t_1 \gg t_2$

(according to /4/, /5/, /6/ and in practice determined for each
equipment regime).

The experimental results are pressented in Table 1.

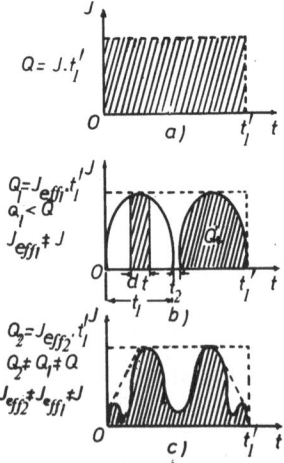

Fig. 1   Current/time dependence

Table 1   Ionophoretic permeability χ $(mg.C^{-I}.cm^{-2})$ after pulse magnetic and electromagnetic field action for 10 min

Regime of action	Control	χ
160 er-1 Hz-0.2 s	0.62	1.08
	0.83	0.95
	0.98	1.10
1 Hz-0.8 s-10 W	0.58	0.15
	0.50	0.40
	0.68	0.18
1 Hz-0.02 s-10 W	0.35	0.88
	0.81	0.48
	0.48	0.81
1 Hz-0.8 s-50 W	0.38	0.33
	0.48	0.40
	0.66	0.58

A specific effect on the skin membrane was established for the two types of field action. When changeable pulse magnetic field was used, the skin permeability increased, as well as the quantity of drug substance introduced /Table 1/. Thus, the combination of drug electrophoresis with changeable pulse magnetic field provokes higher effect of the method.

Experimentally it was found that ultrahigh pulse magnetic field provokes miltidirectional changes in the ionophoretic skin permeability, depending on the regime and the irradiation time /Table 1/. A marked tendency for ionophoretic permeability to decrease was established as well. Therefore the magnetic and electromagnetic field action in different regimes and irradiation intensity enable the regulation of drug electrophoresis velocity depending on the type of drug applied and the pathology. In this wa-, some combined methods could be introduced to optimize the drug accumulation.

REFERENCES

1. Popdimitrova, N., Kentchev, N., Anachkova, N., 1981, Physico-
-Chemical paremeters and quantitative dependence at the

drug electrophoresis, Savremena Meditsina, 12,567.

2. Popdimitrova, N., Kentchev, N., Todorov, N., 1987, Some observations and theoretical considerations on ion and water transporting epithelia /in press Compt. rend. bulgar.Acad. Sci./.

3. Cholodov, Ju.A., 1975, Reakcia nervnoj sistemy na magnitnom pole, M.Nauka.

4. Nachilnitskaja, Z.N., 1974, Nekotorye danie k characteristike PMP kak biologitsheski aktivnogo faktora, Magn. pole m medicina, Frunze.

5. Nagakura, S., Chajasi, Ch., 1986, Magnetic field and chemical reactions, Scientific American, 6, 80.

6. Ulastic, V.S., 1976, Teoria i praktika lekarstvenogo elektroforeza, Minsk.

7. Todorov, N., Kentchev, N., Popdimitrova, N., Stefanov, S., Anachkova, N., Ivanova, M., 1986, Quantitative aspects and computer programme for optmizing the parameters and the procedure of drug electrophoresis /in press, First National Workshop on Bioelectrokinetics, Guiletzsitza/.

EFFECTS OF LOW FREQUENCY MAGNETIC FIELDS ON CHICK EMBRYOS

AND EMBRYONAL CHICK TIBIA

Keijo Saali, Jukka Juutilainen and Tapani Lahtinen[+]

University of Kuopio, SF-70211 Kuopio, Finland and
[+]University Central Hospital, SF-70210 Kuopio,
Finland

INTRODUCTION

It is thought that the stimulating effect of pulsed magnetic fields (PMFs) on bone growth is through the induction of electric currents in the conducting part of the tissue. In our previous studies, exposure to low-frequency magnetic fields (LFMFs) increased the percentage of abnormalities in chick embryos (Juutilainen, 1986). The effect seemed to be independent on the rate of change of the magnetic field (dB/dt) and pulsed fields were found to be as effective as sinusoidal waveforms (Juutilainen et al., 1986; Juutilainen and Saali, 1986). Thus, the results do not support the theory of induced currents as the only possible interaction mechanism between magnetic fields and living organisms.

The aim of this study was to compare the effectiveness of sinusoidal and pulsed magnetic fields on the calsium uptake of embryonal chick tibia. The PMF was similar to that used in bone healing. Measurements of Ca-uptake were made according to the procedures presented by Assailly et al. (1981) and Colacicco and Pilla (1983). Because PMFs have been reported to stimulate the growth of chick embryos (Rooze and Hinsenkamp, 1985), an attempt was made to reproduce this effect with sinusoidally oscillating magnetic fields.

MATERIAL AND METHODS

Exposure system

The exposure system used in studying the Ca-uptake of embryonal chick tibia consisted of a function generator and a power amplifier connected to a 38-turn rectangular coil with dimensions 2.5x50x9.5cm. The coil used for sham-exposing the control tibiae was identical to the exposure system but it was not connected to the signal source. Both coils were on the same table at a distance 70 cm from each other so that the magnetic field at the location of the controls was less than 1/3000 of the exposure field. The systems were open vertically so that the air could flow through the coils to establish the same temperature within $0.5^{\circ}C$ for the control and exposed tibiae. The most efficient type of signal was chosen on the basis of the experiments of Assailly et al. (1981). Each pulse had a duration of 200 µs for the main and 20 µs for the opposite polarity. The burst duration for the pulse train was 5 ms within each 67 ms interval. The rate of change of the magnetic field was 7.5 µT/µs for the main polarity and 75 µT/µs for the opposite polarity.

The exposure system used in studying the effect on the growth of chick embryos consisted of modified rectangular Helmholtz coils. Four exposure systems with different number of turns were used to expose the eggs simultaneously to four different field strenghts. The dimensions of the coils were 11.5x100x2 cm and the distance between the coils in the Helmholtz arrangement was 11 cm. The rate of change of the PMF was 0.7 µT/µs for the main polarity and 7 µT/µs for the opposite polarity. Sinusoidal signals with the frequencies 50 Hz and 1 kHz and amplitudes from 0.1 A/m to 100 A/m were used.

Embryonal chick tibia

The basic medium containing 120 mM glucose, 90 mM NaCl, 6 mM KCl, 1 mM NaH_2PO_4, 1.2 mM $CaCl_2$ and 1.2 mM $MgSO_4$ was made weekly so that it was never more than three days old. For every experiment, the final incubation medium (I-medium) was made daily by addition of 0.1 M NaH_2PO_4 and 1 M $NaHCO_3$. The final concentrations were 4 mM phosphate and 10 mM bicarbonate. One half of the I-medium was reserved for preincubation and the rest for incubation with exposure. In the latter case, $^{45}CaCl_2$ was added to the I-medium in order to get the exposure medium (E-medium) with 0.4-0.5 µCi $^{45}CaCl_2$ per tibia. The initial pH of the I-medium was about 7.3 and it changed gradually to a maximum 8.2 in 120 minutes, which was the total incubation time (preincubation + exposure).

Fertilized eggs of the breed Mäkelä 16 (a variety of White Leghorn) were incubated in a room with humidity and temperature control and air circulation for 9 days at $37.5^{\circ}C$ and 55% relative humidity (RH). After that, the tibiae of each embryo were isolated and placed in closed glass tubes with 2 ml of I-medium. After 60 minutes of preincubation at $37^{\circ}C$, each tibia was transferred into an open test tube containing 1.4 ml of the E-medium. One of the tubes (the first tibia) was placed into the exposure system at $37^{\circ}C$ and 55% RH and another tube (the second tibia from the same embryo) was placed into the control system.

After 60 minutes of exposure, the tibiae were removed, rinsed four times with 2 ml of ice-cold basic medium, transferred to clean test tubes, containing 1 ml of 0.1 M HCl. The closed tubes were then placed for 22 h at $37^{\circ}C$. After that (the extraction of Ca-45 was approximately complete), four 200 µl samples were taken from the supernatant liquid, the scintillation fluid (10 ml ACS) was added and Ca-45 was counted in a scintillation counter (LKB 1210 Ultrobeta) after refrigerating at $+4^{\circ}C$.

Chick embryos

Fertilized eggs were incubated for 4 days. After that the eggs were candled and the living embryos divided into control

Table 1. The effect of the PMF on the Ca-uptake
of embryonal chick tibia

number of tibia pairs	exposed/control (mean±2SEM)
12	1.11 ± 0.12
12	1.01 ± 0.08
9	1.03 ± 0.07
8	1.00 ± 0.07
10	1.03 ± 0.05
9	0.97 ± 0.07
9	1.16 ± 0.04
10	1.11 ± 0.04
10	1.03 ± 0.06
10	1.00 ± 0.09
8	1.05 ± 0.11
10	0.97 ± 0.04
10	0.99 ± 0.06
10	0.99 ± 0.03
Σ137	1.03 ± 0.02

and exposed groups. The eggs were sham-exposed (control) or exposed to PMFs for 6 days, examined macroscopically and living and normal embryos were weighed.

RESULTS AND DISCUSSION

In spite of the large number of tibiae used (n=137), no consistent effect of the PMF on Ca-uptake was found (Table 1). The mean value of the ratio exposed/control was 1.03 ± 0.02 (mean±2SEM). The aim of this study was to compare the effects of pulsed and sinusoidal fields. However, because we could not reproduce the results of the other research groups (Assailly et al., 1981; Cocacicco and Pilla, 1983) with PMF, no experiments with sinusoidal fields were made. The reason to the inconsistency with the other studies is not known.

Exposure to PMF increased the weight of chick embryos. The relative weight of the 30 exposed emryos was 1.078 ± 0.031. This is significantly ($p < 0.001$) more than the weight of the 25 control embryos (1.000 ± 0.027). Also sinusoidal fields increased the weight of the embryos (Table 2). There is no linear dependence of the effect on the field strength. This result resembles our earlier observations of the abnormal development of chick embryos (Juutilainen et al., 1986; Juutilainen and Saali, 1986; Juutilainen, 1986). However, small differences (within 0.5°C) in the air temperature were observed between the egg groups in short-term measurements, which may have caused small differences in growth.

The results of this study are consistent with the general hypothesis that LFMFs stimulate growth. The results do not support the earlier observations that PMFs stimulate the Ca-

Table 2. The effect of sinusoidal magnetic fields on the weight of 10 days old chick embryos

frequency	field strength (A/m)	number of embryos	relative weight ± 2SEM
50 Hz	control	27	1.000 ± 0.021
	0.6	27	1.015 ± 0.019
	1.4	28	1.015 ± 0.027
	10	28	1.050 ± 0.021
	100	28	1.089 ± 0.022
1 kHz	control	28	1.000 ± 0.019
	0.1	14	1.118 ± 0.026
	0.2	14	1.130 ± 0.046
	0.4	12	1.157 ± 0.042
	0.6	14	1.107 ± 0.025
	0.8	14	1.122 ± 0.032
	1.4	14	1.115 ± 0.032
	10	14	1.106 ± 0.023
	100	14	1.112 ± 0.044

uptake of embryonal chick tibiae. No difference between the effects of pulsed and sinusoidal fields could be shown.

In order to exclude thermal effects in further studies, the temperature of the exposed and control objects should be exactly the same.

REFERENCES

Assailly, J., Monet, J.D., Goureau, T., Christel, P., and Pilla, A.A., 1981, Effect of weak inductively coupled pulsating currents on calcium uptake in embryonic chick tibia explants, Bioelectrochem. Bioenerg., 8:515-521.
Colacicco, G., and Pilla, A.A., 1983, Chemical, physical and biological correlations in the Ca-uptake by embryonal chick tibia in vitro, Bioelectrochem. Bioenerg., 10:119-131.
Juutilainen, J., Harri, M., Saali, K., and Lahtinen, T., 1986, Effects of 100-Hz magnetic fields with various waveforms on the development of chick embryos, Radiat. Environ. Biophys., 25:65-74.
Juutilainen, J., and Saali, K., 1986, Development of chick embryos in 1 Hz to 100 kHz magnetic fields, Radiat. Environ. Biophys., 25:135-140.
Juutilainen, J.P., 1986, Effects of low frequency magnetic fields on chick embryos. Dependence on incubation temperature and storage of the eggs, Z. Naturforsch., 41C:1111-5.
Rooze, M., and Hinsenkamp, M., 1985, In vivo modifications induced by electromagnetic stimulation of chicken embryos, Reconst. Surg. Traumat., 19:87-92.
Saali, K., Juutilainen, J., and Lahtinen, T., 1986, A system for exposing biological objects to variable combinations of electric and magnetic fields, J. Bioelectricity, 5:171-186.

THE INFLUENCE OF THE MAGNETIC FIELD UPON THE HATCHING

CAPACITY OF POULTRY AND UPON PLANT SEEDS

Angela Ottová and Štefan Neuschl

Institute of Biotechnology, Slovak Technical
University, Mlynská dolina-D, 812 19 Bratislava,
Czechoslovakia

INTRODUCTION

In our contribution the influence of the stationary
magnetic field upon the embryonal development of poultry as
well as upon the germination of plant seeds is considered
in detail. It is evident that the stationary magnetic field
exerts its influence upon various biological objects. It may
be assumed that biological adaptivity plays a role in the
process.

MATERIAL AND METHODS

The experiments were carried out within the induction
range (B) 0.05-0.5T, exposition time (t) in the magnetic
field taking 5-120 minutes. It came out unambiguously from
our experiments that with the products Bxt= 1.5; 3.0; 4.5;
6.0; ...; 18.0 (T.min), both the hatching capacity of poultry and the germination of plant seeds were increased. This
effect turned out to be particularly conspicuous and significant with eggs and plant seeds, respectively, which were
stored for a longer space of time.

RESULTS AND DISCUSSION

On average the increase amounted to 5-20 per cent related to control, that is to eggs or seeds treated under the
same conditions but not exposed to the magnetic field [2].
The explanation of this effect rests upon molecular-biologi-
cal mechanisms of the orientation of proteins in the biological membranes under the influence of the outer magnetic
field. Within this process the transport channels, through
which ion transport in the cells is implemented, may be opened or closed. Target controlled transport subsequently
leads to an increased activity within the cell which sets in
motion all the functions involved (for example growth, cell
division etc.).

In some cases, in experiments with biological objects,
the cross-correlation analysis could be applied. The authors

suggest therefore to carry out the application of the magnetic field upon biological objects within the rhythm of the so called quasi telegraphic signal, that is the behaviour of the effect of the magnetic field in time is defined by the 50 per cent probability of the appearance or non-appearance, respectively, of the magnetic field in the given time interval. This allows to calculate various dynamic characteristics /for example the transition characteristics/ of biological objects as well as to study the observation of functional changes.

It is assumed in this connection that the adaptivity of the biological objects can be excluded by this.

In this contribution the algorithm of the calculation of dynamic characteristics is discussed. Concerned is the examined biological object. Apart from this the linear properties of biological objects are presumed.

Algorithm for the calculation for the answer of the system in which noises or unexpected factors are present is as the following:

It is supposed that the system responds both to the experimental input quantity $s(t)$ and to the noises $x(t)$ which overlaps these experimental signals. In general the following expression is valid:

$$y(t) = \int_0^\infty x^\circ(t-\varkappa) \cdot f(\varkappa)d\varkappa + \int_0^\infty s^\circ(t-\varkappa) \cdot f(\varkappa)d\varkappa + C$$

where the function f characterizes the property of the signal transfer.

It can be shown that the following expressions are valid:
$$f(0) = 0,$$
$$f(a) = -2R_{ys}(0)$$
$$f(2a) = -2R_{ys}(a) + 2f(a)$$

$$\vdots$$

$$f(na) = -2R_{ys}[(n-1)a] + 2f[(n-1)a] - f[(n-2)a]$$

where $n = 0,1,2,\ldots$; a is the distance between changes from 0 to A. It represents so called quasitelegraphic signal. In these points the existence of the amplitude of the output quantity is assumed. The whole situation is shown in the following figures. In this manner the answer to the dynamic system, called transfer function characteristics, can be calculated $[\zeta(na)]$.

In further study the following questions have to be solved:
a) How to generate the influence of the magnetic field on living structure in the form of quasi telegraphic signals (in time randomely generated).
b) How to define biophysical quantities, the output of which as the reaction of the system we would like to evaluate.

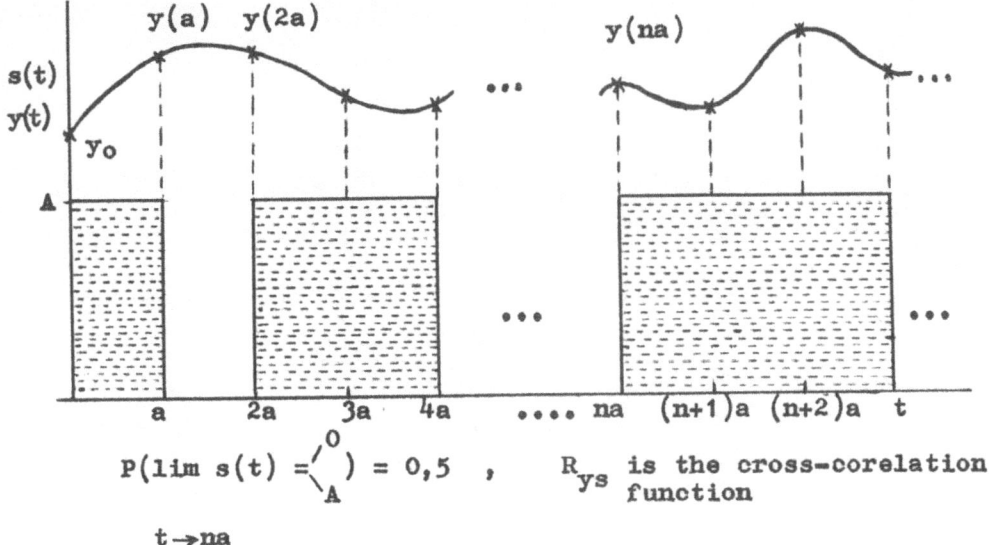

y(a)
y(2a)
y(na)
s(t)
y(t)
yo
A
a
2a
3a
4a
.... na
(n+1)a
(n+2)a
t

$$P(\lim_{t \to na} s(t) = \begin{matrix} 0 \\ A \end{matrix}) = 0,5 \quad , \quad R_{ys} \text{ is the cross-corelation function}$$

Fig.1

s(t)
y(t)
Correlator
$R_{y,s}$
Digital filter
$K(z) = \dfrac{T.z^{-1}}{2Tz^{-1} - Tz^{-2} + 1}$
f(na)

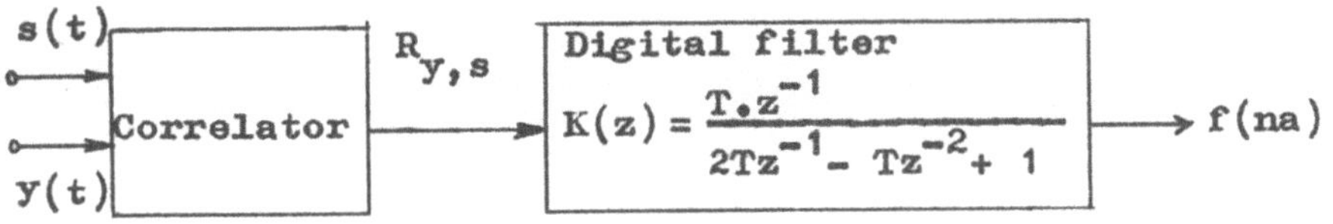

Fig.2

CONCLUSIONS

Applications of the input quantities in the form of quasitelegraphic signal can filtrate and eliminate the side-effects in the study of the answer to effects of magnetic fields on biological structures.

Note: This method has been prooved by the measurements of centre of gravity of human being 1 .

REFERENCES

1 I. Burger, Š. Neuschl, V. Litvinenková, Stochastické procesy v biologických sústavách a vyhodnocovanie dynamických vlastností sústav. Academia, Praha (1969).

2 A. Ottová, J. Baugartner, V. Peter, V. Chrappa, The Effect of Various Inductions of Homogeneous Stationary Magnetic Field on the Hatchability of Fowl and Game Birds, Scientia Agriculturae Bohemoslovaca, Tomus 17 (XXXIV), No.1, p.63-66 (1985).

THE GEOMAGNETIC FIELD AS A LOWENERGETIC FACTOR WITH IMPORTANT BIOLOGICAL SIGNIFICANCE

Dimiter Mikhov

Institute of Neurology, Psychiatry & Neurosurgery
Blvd. "Lenin" - 4 km., 1113 Sofia, Bulgaria

It has been established through many experiments and theoretical analyses that the minimal energy which influencing on the retina and causing light perception is equal to $(2.1 - 5.7).10^{-18}$ joule. It is the custom for this minimal energy to be called "absolute energetic limit" of the ocular perception. It concerns the specific receptors while concerning the other cells (unspecific receptors) the term "absolute energetic limit" of sensivity is bringing into use. It is that minimal stimulus which still could cause any kind of stimulation. According to Plekhanov[1], each cell of the multicellular organism could be an unspecific receptor as the property of excitedness (rising of a local reaction of the bioplasm) is kept in each cell. As a result of many experimental and theoretical studying it is accepted now that "the absolute energetic limit" for the human organism amounts to 10^{-16} J. That means that the minimal power of the energetic flow $\underline{S}^0$ which could induce a transition of the respective structures of the human organism from phase of preventive inhibitions to phase of excitation is equal to 10^{-12} J/s.m². In other words all stimuli whose minimal power of energetic flow $\underline{S}$ exceeds $\underline{S}^0$ irrespective of their nature could cause a state of excitement of some structures.

The influence of the geomagnetic disturbances on biological processes is the object of a great number of investigations. Many monographs and symposiums are devoted to that subject[2,3,4,5,6,7,8,9]. A great part of research works treats the influence of geomagnetic activity on a qualitative base. It is established that there exists a difference between the biological processes during days with geomagnetic disturbances and days without changes of the geomagnetic field. Correlative studies concerning the influence of the geomagnetic activity are made in comparatively few investigations. The basic mistake made in these cases is that the geomagnetic activity is evaluated through the various geomagnetic indices used in geophysics for characterizing and describing the changes of the geomagnetic field. All geomagnetic indices including the most precise three hour interval K -index do not reflect the kind of the geomagnetic fluctuations.

The three hour K -index, defined according to a special scale on the base of the differences between the extreme values of the most stroungly changing parameter of the geomagnetic field, does not give an idea of the number and the form of geomagnetic variations in that three hour interval selected beforhand. According to us the use of the geomagnetic indices for quantitative evaluation of geomagnetic activity is incorrect when its biological effect is being investihated. A similar standpoint completely interprets the different results obtained by some authors[10,11,12,13,14,15] .

During a period of 6 years the monthly death-rate of cerebrovascular disease (98165 dead), ischemic heart disease (91590 dead), children up to the age of 1 year (22618 dead), adults above 80 (17413 dead) has been studied according to the geomagnetic activity. A quantitative evaluation of the geomagnetic activity was obtained by an amplitude and frequency description of all geomagnetic variations whose changes of horizontal component ΔH, in time interval Δt less than 30 minutes, exceed 30 gammas (1 gamma = 10^{-5} Oe). Using the expression for the power of the energetic flow $\underline{S}$

$$S = \frac{1}{4\pi . \Delta t} . (H_o . \Delta H + Z_o . \Delta Z)$$

shown by Musalevskaja[16], it has been found that the power of the energetic flow of the weakest changes of the geomagnetic field, studied in that 6 years period, amounts to 5.10^{-12} J/s.m². That means that S_{min} exceeds S^o and for its part could cause a stimulation. In much the same manner it was found for the strongest fluctuations of the geomagnetic field that $S_{max} = 150.10^{-12}$ J/s.m². That means that all geomagnetic variations which were studied during the investigated period could be perceived by organisms.

With the help of rank correlation method a moderate correlations between the geomagnetic activity and monthly death-rate in persons with cerebrovascular disease and inchemic heart disease, in children up to the age of 1 year and adults above 80 were found. The obtained results could be interpreted with the relatively large percentage of the persons having strongly impaired function of vital importance or diminished compensative capacity which is typical for the above groups.

The causes for the lethal outcome are too many and various. Therefore we should not think that the geomagnetic variations have a decisive importance for the lethal outcome. The geomagnetic disturbances are one of the many hazardous agents, which under certain circumstances (considerably decreased compensative capacity) could contribute to worsen the state of health of the organism.

REFERENCES

1. G. F. Plechanov, "About human perception of weak signals", Dissertation, Medical Institute of Tomsk, (1967)
2. A. L. Chizhevskii, "Earth's echo of solar storms", Mysl', Moskow (1973)

3. A. P. Dubrov, "The geomagnetic field and life", Geo-
   magnitobiology, Plenum Press, New York (1978)
4. "Effect of magnetic fields on biological objects", 3rd
   Allution Symp, Kaliningrad State University, Kaliningrad (1975)
5. "Effect of natural and weak artifical magnetic fields
   on biological objects", M. P. Travkin, ed., Belgo-
   rod (1973)
6. "Effect of some cosmic and geophysical factors on
   earth's biosphere", V. N. Chernygovskii, ed., Nauka, Moscow (1973)
7. "Physicomathematical and biological problems of effect
   of electromagnetic fields and ionization of air",
   Vol. I & vol. II, Nauka, Moscow (1975)
8. A. S. Presman, "Electromagnetic fields and life",
   Plenum Press, New York - London (1970)
9. "Reaction of biological systems to weak mahnetic
   fields", Proc. All-union Symp, 21-23 Sept. 1971.
   Moscow (1971)
10. R. Danneel, "Der Einfluss geophysikalischer Faktoren
    auf die Selbstmordhäufigkeit", Arch. Psychiat. Ner-
    venkr.,219: 153-157 (1974)
11. H. Friedman, R. O. Becker, C. H. Bachman, "Psychiatric
    ward, behaviour and geophysical parameters, Nature
    205: 1050 (1965)
12. I. E. Ganelina, S. K. Shurina, N. V. Savoyarov, "Sta-
    te of environmental physical factors and incidence
    of main complications of acute myocardial infarct,
    Kardiologiya 10: 112-118 (1975)
13. B. J. Lipa, P. A. Sturrock, E. Rogot, "Search for cor-
    relation between geomagnetic disturbances and mor-
    tality", Nature 259: 302-304 (1976)
14. S. R. C. Malin, B. J. Srivastava, "Correlation between
    heart attacks and magnetic activity", Nature 277:
    646 (1979)
15. A. D. Pokorny, R. B. Mefferd, "Geomagnetic fluctua-
    tions and disturbed behaviour", J. Nerv. Mental De-
    sease 143: 140 (1966)
16. N. I. Muzalevskaya, Biological activity of disturbed
    geomagnetic field , In: "Effect of Solar Activity
    on Earth's Atmosphere and Biosphere", M. N. Gnevy-
    shev, I.I.01' ed., Nauka, Moscow (1971)

ELECTROROTATION - INFLUENCE OF PULSE SHAPE AND INTERNAL MEMBRANE SYSTEMS ON ROTATION SPECTRA

J. Gimza, E. Donath, and R. Glaser

Humboldt-Univ., Sektion Biologie, Bereich Biophysik
Invalidenstr. 42, Berlin 1040, G.D.R.

1. Electrorotation in square-topped fields

1.1. The torque in continuously rotating fields

The theory of electrorotation was described by FUHR /1985/ and SAUER&SCHLOEGEL /1985/. The electrorotation spectrum e.g. the spin of a biological particle in dependence on the frequency of a rotating electric field reflects the dielectric properties of the membrane as well as those of the internal and external media of the cell /ARNOLD et al. 1986/ / GLASER&FUHR 1986/ /GLASER&FUHR 1987/.

For the following derivations a plane field is considered. The X- and Y-axis point into the directions of the real and imaginarypart of the complex field vector, respectively.

$$E_x = Re/\overline{E}/ \qquad\qquad E_y = Im/\overline{E}/$$

A continuously rotating field can then be written in the form:

$$\overline{E} = E_o *exp/j\omega t/ \qquad \text{with } w=2\pi f \text{ /f-field frequentcy/ /1/}$$

The complex form of the induced dipole moment m for a spherical model consisting of central symmetrical electrically homogenious shells /see Fig.2/ is given by:

$$\vec{m} = A\overline{E}*/m_x/\omega/ + jm_y/\omega/ \qquad \text{with } A = 4 R_e^3 o e \qquad /2/$$

The torque on the object becomes:

$$\vec{N} = \vec{m} \times \vec{E} = m_x/\omega/E_y - m_y/\omega/E_x \qquad\qquad /3/$$

Equations /1/ and /2/ yield:

$$\vec{N} = AE_o^2 m_y/\omega/$$

The torque is proportional to the imaginary part of the induced dipole moment /see/FUHR 1985/ and /FUHR et al. 1986//.

1.2. The torque in square-topped fields with a key ratio of 1:1

The digital generation of signals having a frequency-independent phase shift of 90° can be accoplished quite easily. Let us analyze a key ratio of 1:1 for the driving signal. The FOURIER-spectrum of the signal within a rectangular 4 electrode arrangement will be:

$$\vec{E} = E_0 * / e^{j\omega t} - e^{-3j\omega t}/3 + e^{5j\omega t}/5 - + .../$$

$$= \quad \vec{E}_1 \quad + \vec{E}_2 \quad + \vec{E}_3 \quad + ... = \underset{m}{\leq} \vec{E}_m \qquad /6/$$

The dipole moment can be expressed in terms of a FOURIER-series, too:

$$\vec{m} = A * /m_x/\omega/+jm_y/\omega// *e^{j\omega t} - /m_x/3\omega/ *e^{-j\omega t}/3 +$$

$$/m_x/5\omega/+jm_y/5\omega/*e^{j5\omega t}/5 - + ... /$$

$$= \vec{m}_1+\vec{m}_2+\vec{m}_3+ ...=\underset{n}{\leq} \vec{m}_n \qquad /7/$$

Using equation /3/ the torque can be written as:

$$\vec{N} = \vec{m} \times \vec{E} = \underset{n}{\leq} \underset{m}{\leq} /\vec{m}_n \times \vec{E}_m/ \qquad /8/$$

The terms $/m_n \times E_m/$ are considered for two cases:

1. $n = m$ This case was considered by FUHR et al. /1986/ /see equation /4//.
2. $n \neq m$ Integration of $N_{n,m}$ over a whole period yields O.
So the torque in a square topped field with a key ratio of 1:1 can be written as:

$$\vec{N} = AE_0*/m_y/w/ - m_y/3w//9 + m_y/5w//25 - + .../ \qquad /9/$$

/see /GIMSA 1987//

In the linear range of friction the torque $\vec{N}$ is proportional to the rotation R. Fig. 1 shows theoretical spectra in dependence on the field frequency in continuously and discontinuously rotating fields, according to equation /10/.

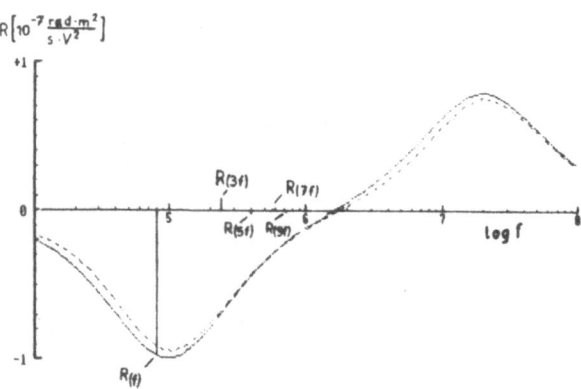

Fig. 1 Electrorotation spectrum of a single shell sphere in a continuously /————/ or discontinuously /- - -/ rotating field
fc1 = 100 kHz ; fc2 = 20 MHz ; R_{min} = $-2*10^{-7}$ radm^2/sv^2 ;
R_{max} = $1.6*10^{-7}$ radm^2/sv^2

2. Electrorotation of protoplasts

For the theoretical description of the rotational behavior FUHR et al. /1985/ and GIMSA et al. /1985/ used a spherical model, consisting of electrically homogenious central symmetrical shells.

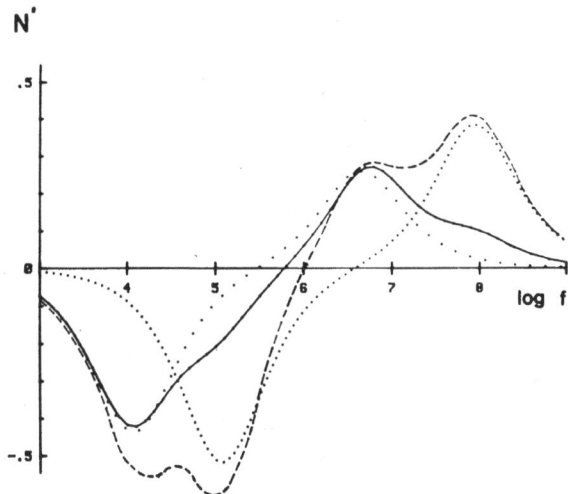

Fig. 2   Single- and three-shell model
         R, d - radii e.g. thicknesses
         G   - conductivities
               - dielectric constants of the shells

The rotational behavior of a songle-shell model reflecting the properties of a simple cell can be decribed by equation /10/:

$$R = R_{min} * \frac{f*fc1}{fc1^2+f^2} + R_{max} * \frac{f*fc2}{fc2^2+f^2}$$ /10/

      R - rotation /rotation speed of the cell per square
          field strengths/
      f - frequency of the external field
fc1, fc2 - 1st e.g. 2nd characteristic frequency
R_{min}, R_{max} - rotations for f = fc1 and f = fc2, respectively

Each addition of a shell results in a new characteristic frequency, which influences the rotational spectrum. FUHR et al./1985/ investigated the theoretical rotational behavior of protoplasts using a three-shell model. Theoretical considerations showed that large errors may be produced using the single-shell approach instead of more complicated models. In our experimental investigations we used 4-electrodechambers with sinusoidal or square-topped pulses applied and field strengths in the measuring chamber between 5000 and 10000 V/m. We found that cell organells and other internal membraneous structures may influence the rotational spectrum, although most of the protoplasts show an electrorotation spectrum similar to that of single-shell models. Deviations from the single-shell behavior were characterized always by a broadening of the rotation spectrum. In our theoretical and practical investigations we never observed two maxima in the range of the first peak. Isolated vacuoles behave like single-shell objects. In most protoplast experiments plasmalemma properties mainly determine the first peak. Calculated specific membrane capacities of vacuoles are as low as those of protoplasts.

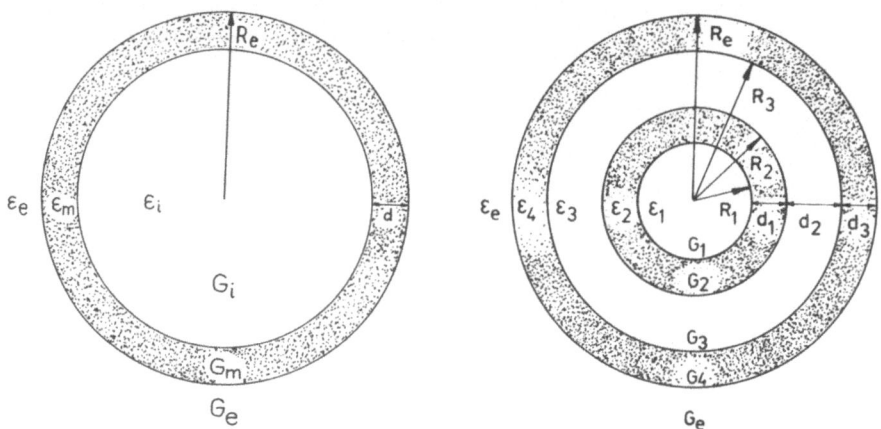

Fig. 3   Electrorotation spectrum of a three-shell sphere
/protoplast/ /————/ and the interpretation of the
resonances using two single-shell models. ... - single-
shell model of the vacuole interior and tonoplast
G_1 = 0.8 S/m, G_2 = 10 μ S/m, G_e = 0.03 S/m, $_1$ = 70,
$_2$ =6, $_3$ =50, vacuole radius: 15 μ m ... -single-shell
model of the cytoplasm and plasmalemma G_1 = 0.03 S/m,
G_2=0.1 μ S/m, G_e=0.01 S/m, $_1$ = 50, $_2$ = 6, $_3$ = 75,
radius: 25 μ m
- - - - superposition of curves / ...../ and / . . ./

REFERENCES

Arnold, W.M., Geier, B.M., Wendt, B. and Zimmerman, U., 1986,
    The change in the electro-rotation of yeast cells effected
    by silver ions, Biochim.Biophys.Acta, 885:35.
Fuhr, G., 1985, Ueber die Rotation dielektrischer Koerper in
    rotierenden Feldern, Dissertation, Humboldt University,
    Berlin
Fuhr, G., Gimsa, J. and Glasser, R., 1985, Interpretation of
    electrorotation of protoplasts. I. Theoretical conside-
    rations, Studia biophysica, 108:149.
Fuhr, G. and Kuzmin, P.I., 1986, Behavior of cells in rotating
    electric fields with account to surface charges and cell
    structures, Biophys.J., 50:789.
Gimsa, J., 1987, Elektrorotation - technische Voraussetzungen
    und biophysikalische Aussagemoeglichkeiten, Dissertation,
    Humbold University, Berlin.
Gimsa, J., Fuhr, G. and Glasser, R., 1985, Interpretation of
    electrorotation of protoplasts. II. Interpretation of
    experiments, Studia biophysica, 109:5.
Glaser, R. and Fuhr, G., 1986, Electrorotation of single cells-
    a new method for assessment of membrane properties, in:
    Electric Double Layers in Biology, M.Blank, ed., Plenum
    Press, N.Y.

Glasser, R. and Fuhr, G., 1987, in: Mechanistic Approaches to Interaction of Electric and Electromagnetic Fields with Living System, M.Blank and E.Findl, ed., Plenum Press, N. Y., in press

Glasser, R., Fuhr, G., Gimsa, J. and HAGEDORN, R., 1985, Electro-rotation - capabilities and limitations, Studia biophysica, 110:43.

Lovelace, R.V.E., Stout, D.G. and Steponkus, P.L., 1984, Proto-plast rotation in a rotating electric field: The influence of cold acclimation, J.Membrane Biol., 82:157.

Mueller, T., Fuhr, G. Hagedorn, R. and Goering, R., 1986, Influ-ence of dielectric breakdown on electrorotation, Studia biophysica, 113:203.

Pastushenko, V.Ph., Kuzjmin, P.I. and Chizmadzhew, Yu. A.,1985, Dielectrophoresis and electrorotation. A unfied theory of spherically symmetrical cells, Studia biophysica, 110:51.

Sauer, F.A. and Schloegl, R.W., in: - Interaction between Electromagnetic Fields and Cells, A.Chiabrera, C.Nicolini and H.Schwan, 1985, Plenum Press, N.Y.

ELECTROROTATION MEASUREMENTS ON MOUSE BLASTOMERS

Torsten Müller, Günter Fuhr, Frank Geißler, and Rolf Hagedorn

Department of Biology,
Humboldt-University of Berlin, GDR

INTRODUCTION

As described earlier[1,2] electrorotation is a new method to determine electrical properties of the constituents of single cells such as membrane or interior. Each cell shows a particular dependence of its rotation speed on the angular frequency of the external applied field frequency . The electrorotation can be used to determine changes in membrane conductivity and capacity[3,4] . It is known, that in many invertebrate and vertebrate species, including mammals, the egg plasma membrane, the egg surface and the egg interior are changed during and after fertilization[5] . This processes[6] should lead to differences in the electrical properties between oocytes and zygotes and consequently should result in different rotation spectra.

MATERIALS AND METHODS

Electrorotation measurements

The experimental chamber consisted of four needle-electrodes fixed by a plastic-ring. The distance between the opposite electrodes was 500 ⁄um and allowing the production of rotating fields with field strengths up to 40kV/m (for measurement of the rotation spectra 5-10 kV/m were used).

The whole chamber was manipulated with a micromanipulator over a fixed microscope slide on which a droplet containing one single cell was placed.

For generating the rotating electric fields four 90° phase shifted square-topped pulses were used .

Cell preparation

The cells were obtained from 6 to 10 week old mice, strain Dummerstorf. The preparation is described in[7].

Before the measurement cells were washed four times in 0.28 M mannitol solution and the conductivity was calibrated with $CaCl_2$ ($5*10^{-3}$S/m). After measuring the electrorotation the eggs were fixed with a 25% acetic alcohol solution.

Staining followed with 2% acetic orcein.

RESULTS

We observed either a rotation of the whole cell or a rotation of the vitellus within the zona pellucida. Only small differences were observed between both the above-mentioned rotation types . Additionally, the rotation spectra of single zona pellucida (the vitellus was destroyed by short field pulses and the zona was washed several times) and vitelli after removing the zona pellucida is given .

At first we were interested in observing differences in the rotational behavior of oocytes, zygotes and embryos in the 2-cell-stage (table 1). The main result was that the mean characteristic frequency (f_{c1}) of three cell populations was different.

Table 1   Maximum rotation (R_{max}) and characteristic
          frequencies (f_{c1}) of several cell types.
          The $\pm$ S.D. is given.
        n-number of cells; w_{cmax}-maximum angular velocity of
        cells; R_{max}-w_{cmax}/E^2;   E-field strength;

Cell-type	$R_{max} * 10^{-7}$ (rad$*$m^2/V^{2*}s)	f_{c1} (kHz)	n
Oocytes	0.45 ± 0.15	11.06 ± 2.45	49
Cells from super-ovulated + mated mice	0.39 ± 0.07	07.82 ± 2.61	79
2-cell-stage	0.47 ± 0.09	13.6 ± 2.22	26

However, the data in table 1 refering to cells from superovulated and mated mice are composed of both fertilized and non-fertilized cells as microscopic observation proved. The distribution of f_{c1} in both populations, oocytes and cells isolated from superovulated + mated mice, respectively (Fig.1).

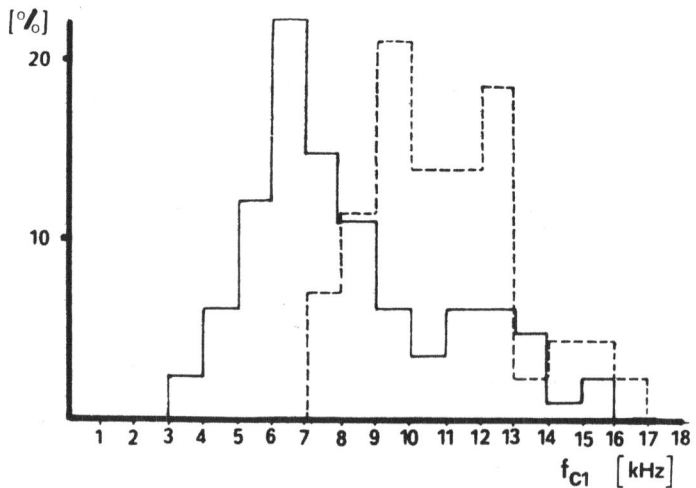

Fig.1: Distribution of the characteristic frequency f_{c1} of oocytes (- - -) and cells isolated from superovulated + mated mice (-----).

There are two groups of cells isolated from superovulated and mated mice differing in their rotational behavior:

(a) A smaller number of cells shows a characteristic frequency almost f_{c1} idential with that of oocytes.

(b) The larger group exhibits a characteristic frequency which is considerable smaller.

Both populations overlap.

For more than 100 cells (>90%) from superovulated and mated mice the observed individual f_{c1} frequency was correlated to the microscopic observation. The mean f_{c1} frequency of cells microscopically classified cells to be fertilized was 6.7± 1.1 kHz. This is approximately half the corresponding oocyte characteristic frequency.

Our experimental chamber allowed us to produce rotating field up to the dielectric membrane breakdown. From this the question arrises at which field strength the resistance behavior of the membrane as measured by electrorotation becomes non-linear. In such a case the relationship between the angular velocity (w_c) and the square of the field strength ((E^2); $w_c \sim E^2$) is changed (see Fig. 2).

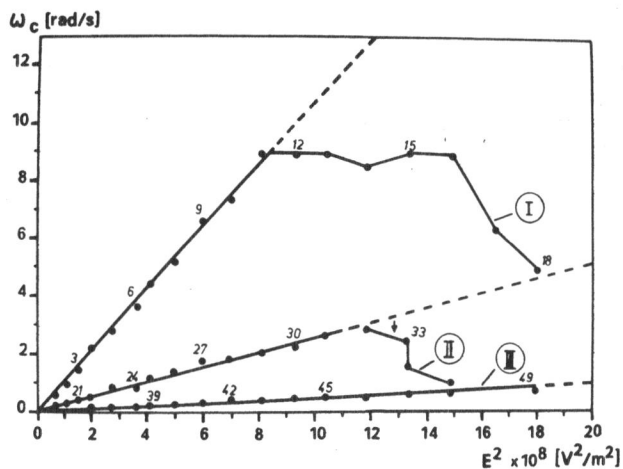

Fig.2: Relationship between the angular velocity (w_c) and the square of the field strength (E^2) measured on single cell in succession of the points (1-49)
arrow - first visible destruction of vitellus

The curves in Fig. 2 represent the typical behavior measured using a single oocyte several times (curve 1,2,3). In curve 1 the linear part ($w_c \sim E^2$; up to 28 kV/m) is followed by a non-linear range. It seems that at field strength > 41 kV/m dielectric membrane breakdown occurs but without visible destruction of the vitellus. The irreversibility of this changes is proved in curve 2. The field strength dependence of w_c was measured again and found to be linear too, but with decreased angular velocity of the egg. At the point marked by the arrow a visible destruction of the vitellus was observed. Therefore curve 3 represents mainly the influence of the zona pellucida.

DISCUSSION

A larger number of cells isolated from superovulated + mated mice differed in their characteristic frequencies in comparison with oocytes. Microscopically these cells were identified to be cells with swollen sperm heads or male pronuclei with sperm tails in the egg cytoplasm or male + female pronuclei. At the present stage a classification between two groups can be made with the electrorotation technique.

Mainly changes of the membrane conductivity influenced the rotation spectrum, whereas changes of the zona conductivity over two decades, which would be necessary to explain the experimentally observed data, are rather improvable. Therefore we concluded that during or after fertilization the membrane conductance was decreased in almost every order of magnitude.

It is necessary to point out that the electrorotation measurements discussed here do not necessarily correlated to short time effects observed during fertilization, since our investigations started several hours after fertilization before and during development of pronuclei. This indicates especially long time changes of the plasma membrane.

Additional it should be possible to investigate cells in the "non-linear range" of electrorotation too, whereas reversible and irreversible changes are unavoidable. The data measured in this field strength range allowed to get informations about the structure, distribution and stability of the egg membrane. This can be important for finding out optimal conditions and strategies of cell preparation, cultivation or analogous processes.

REFERENCES

1. G.Fuhr, R.Glaser, and R. Hagedorn, Rotation of dielectrics in a rotating electric high-frequency field, Biophys.J. 49:395-402, (1986)
2. W.M. Arnold, and U. Zimmermann, Rotating-field induced rotation and measurement of the membrane capacitance of single mesophyll cells of Avena sativa, Z.Naturforsch.Sect.C.Biosci. 37c:908-915, (1982)
3. J.Gimsa, G.Fuhr, and R.Glaser, Interpretation of experiments, II. Interpretation of experiments stud.biophys. 109:5-14, (1985)
4. W.M.Arnold, B.Wendt, U.Zimmermann, and R.Korenstein, Rotation of a single swollen thylakoid vesicle in a rotating electric field, Biochim.Biophys.Acta 813: 117-131, (1985)
5. D.Epel, T. Schmidt, and H.Sasaki,Relationship between cortical granule fusion and transport change at fertilization of sea urchin eggs, in:"Cell fusion, gene transfer and transformation",R.F.Beers Jr. and E.G.Bassett, eds., Raven Press,New York, (1984)
6. L.A. Jaffe, and N.L.Cross, Electrical properties of vertebrate oocyte membranes, Biol.Reprod. 30:50-54, (1984)
7. R.L. Brinster, Studies on the development of mouse embryos in vitro, J.Reprod.Fert. 10: 10-227, (1965)

EFFECT OF THE TEMPERATURE ON THE CELL CONTACT BY DIPOLOPHORESIS AND ELECTROFUSION

R. Pancheva, M. Naplatarova, and P. Antonov

Department of Physics and Biophysics
Scientific Medico-Biological Institute
Medical Academy, 1431 Sofia

INTRODUCTION

Fusion of biological membranes is O.K. stage for a number of physiological processes. In particular, electrofusion consists of two steps. In the first, close membrane contact is achieved by dipolophoresis of the cells in nonuniform alternating electric field. In the second, fusion is triggered between adjacent cells by the application of an intensive field pulse of very short duration. This communication presents results on the temperature dependence of these two events and a model of a single cell-thermo-osmotic monitor.

EXPERIMENTAL

Nicotiana tabacum mesophyll protoplasts were isolated by routine procedures. The protoplasts were peeled at 100 x g for 5 min and washed twice with 0.45 M mannitol. Approximately 10 μl from the suspension at concentration of 10^5 cells/ml was introduced into the electrode gap of a specially designed chamber adapter from microscopic observations. It consists of two parallel thermostated cylindrical electrodes, 600 μm in diameter at a distance of 200 μm. Suspension conductivity was always below $2.5 \times 10^{-5} \Omega^{-1}$ -cm^{-1}. The experiments were performed with fusion /breakdown/ pulses of 15 μs duration at various electric field intensities and temperature from 4 to 37°C.

RESULTS AND DISCUSSION

Previous work has shown temperature dependence of alternating electrical field /AC/ mediated weak and strong cell adhesion.[1] Now the cell contact was studied under strongly non-equilibrium temperature conditions. The AC field was kept at a constant value sufficiently long for alignment of the cells without visible deformation of the interacting surfaces. At 4°C the cells are rotund /Fig.1a/. The effect of the temperature changes /33°C for a few seconds/ was connected with a remarkable deformation of the cell shape and expansion of the

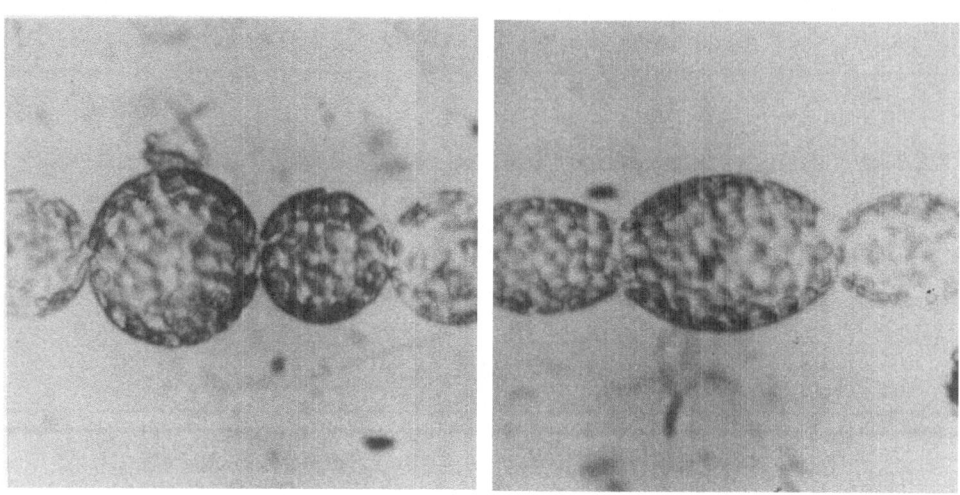

Fig.1 Field-induced cell contact at 4°C /a/ and at 37°C /b/ and
       alternating field of 1 MHz and 400 V cm⁻¹. Magnification
       x 600

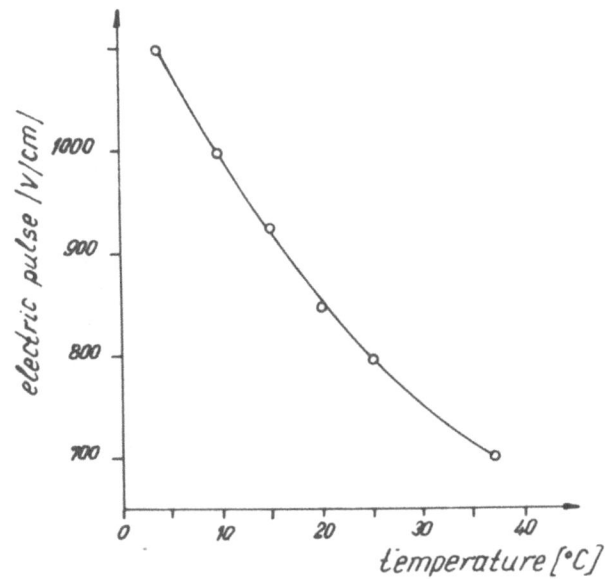

Fig. 2 The temperature dependence of cell fusion for <u>Nicotiana
        tabacum</u> protoplasts.Two pulses of 15 μs duration were
        applied in series

contact area 4-5 times for every 10°C /Fig.1b/, the pattern being completelly reversible.

The temperature dependence of fusion pulses for Nicotiana tabacum protoplasts is given in Fig.2 and agrees in principle with the data reported elswhere.[2]

In the frame of these results it seems resonable to discuss the relationship mainly between three parameters: temperature, contact area and electric field action. The selected experimental conditions permit to neglect AC Joule heating of the solution in the electrode gap and the dielectric perturbation of the membrane permeability due to the AC field.

The role of membrane fluidity receives a priori prominence through similar experiments with native cells in large temperature ranges. As shown by Orgambide et al.,[3] membrane fluidization by ethanol treatment apparently inhibits cell electrofusion. These authors linked this effect to a decrease in the size and lifetime of the induced pores and drew the conclusion that the temperature-induced increase in cell fusion yield /Fig.2/ appears not to be linked to the fluidity of the membrane, but rather to a more general cell reorganization.[4] The cell deformation observed /Fig.1b/ manifested with easier fusion, supports such a speculation.

Figure 1 presents a practical illustration of the temperature jump effect at the level of the individual cell. It has often been demonstrated[5,6] that even a small temperature jump in the suspension changed the osmotic pressure of cells. The magnitude of this thermal osmotic effect[6] δP depends on the energy of transfer /Q/ of water molecules: $\delta P = -$ /Q/VT/δ T, where V is the molar volume of water and δ T is the temperature jump.

According to expectations, the temperature rise leads to an immediate increase in external osmotic pressure and to the appearance of an efflux of water molecules. Indeed, Fig.1 shows how the osmotic pressure gradient is coupled with the expansion of the contact area in adjacent cells. The consequent effect on fusion yield marks at least two contrary tendencies. First, it is a well known fact that a decrease of the osmotic pressure gradient through the cell membrane presupposes higher fusing voltage.[7] Second, the same osmotic effect is coupled with expansion of the contact area and is a good prerequisite for increasing the number of pores, thus facilitating the act of fusion. The available evidence favours the dominating role of the second tendency.

Further, the three-parameter system studied under reversible temperature-jump conditions in fact "worked" as cell thermo-osmotic monitor and this finding is maybe of practical interest.

REFERENCES

1. Antonov, P.P., Pancheva, R.P., Naplatarova, M. and Stoylov, S.P., 1986, Temperature dependence of electrically stimulated red blood cell interactions, Studia Bioph., 113:61.
2. Benz, R. and Zimmermann, U., 1980, Relaxation studies on cell membranes and lipid bilayers in the high electric field range, Bioelectrochem.Bioenerg., 7:723.
3. Orgambide, G., Blangero, C. and Teissie, J., 1985, Electrofusion of Chinese hamster ovary cells after ethanol incubation, Biochim.Biophys.Acta, 820:58.

Blangero, C., 1984, These de 3° cycle, Universite de Toulouse.

5. Tsong, T.Y., 1983, Voltage modulation of membrane permeability and energy utilisation in cells, Bioscience Rep., 3:487.

6. Katchalasky, A. and Curran, P.F., 1965, "Nonequibilibrium thermodinamics in Biophysics", Harvard University Press, Cambridge, Mass.

7. Zimmermann, U., Beckers, F. and Coster, H.G., 1977, The effect of pressure on the electrical breakdown in the membranes of Valonia utricularis, Biochim.Biophys.Acta, 464:399.

THE EFFECT OF SURFACE CHARGED MODIFIERS ON <u>Halobacterium</u>

<u>halobium</u> PURPLE MEMBRANE PHASE TRANSITIONS

Nezabravka Popdimitrova, Jan Gutowicz[+], Andrey Atanassov[*], Andrzej Krawczyk[+], Minko Petrov[++], and Stanislaw Miekisz[+]

Medical Academy, Sofia;[+] Medical Academy, Wroclaw, Poland; University of Sofia;[++*] Institute of Solid State Physics, Sofia, Bulgaria

INTRODUCTION

Being positively charged, spermine and protamine change the electrostatic potential not by exerting a screening effect in the aqueous diffuse double layer, but by adsorbing to the membrane and changing the charge density.[1,2] This interaction with the membrane leads to some spatial rearrangements of the membrane components. The data presented are concerned with microcalorimetric and spectrofluorimetric investigations of the effect of spermine and protamine on purple membrane /PM/ fragments.

MATERIALS AND METHODS

PM from H.halobium strain R_1M_1 were isolated according to Oesterhelt et al.[3] Protamine 5000 /Roche/ was added in a concentration of 50 units/ml and spermine /Roedel-De-Haen/ - 10 mg/ml to the PM, and the samples were incubated at room temperature for 3 hours and then washed with distilled water. Microcalorimetry of the samples was performed in microcalorimeter type 600 /Unipan, Poland/. The screening rate was 0.5 K/min.Lyophilized PM samples were hydrated immediately prior to measurements. All experiments were carried out in the 293-390 K temperature range. Spectrofluorimetric investigations of water suspensions of PM treated with 1-anilino-8-naphthalene sulphonate /Fluka/ were carried out on spectrofluorimeter Yvon Jobin JY-3D. The samples were irradiated at 360 nm and the emission spectra were taken in the 400-550 nm interval with maximum at 480 nm.

RESULTS AND DISCUSSION

Microcalorimetric studies of untreated PM samples reveal two peaks in the temperature ranges of 310-318 K and 350-365 K /Fig.1/. These results correlate with our data obtained by studying PM by Rayleigh light scattering.[4,5] According to them, three anomalies are registered, with the first peak in the 306-310 K range being reversible. It is assumed to be connected with some changes in the phase state of lipids. The second peak, occuring in the narrow range of 339-344 K, is partially reversible and is probably due to certain structural rearrangements in the bacteriorhodopsin. These results of ours correlate with the data obtained by other authors[6] that regard the second peak

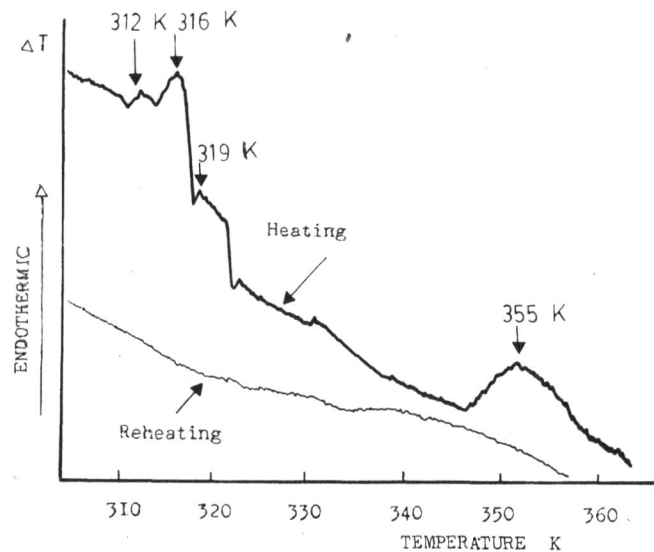

Fig. 1   Theromogram of unmodified purple membranes
          After first scanning the samples were cooled and
          heated once again
          On the abscissa - temperature /in Kelvins/; on ordinate-
          heat uptake

Table 1

Effect of the surface charge modifiers on purple membrane
thermotropic transition temperatures

Material	Midpoints of peaks in thermograms /Kelvins/		
Purple membranes unmodified	312-316	319	355
Purple membranes unmodified 2nd scan		NO PEAKS	
Purple membranes+spermine		NO PEAKS	
Purple membranes+protamine	362	368	
Purple membranes+protamine 2nd scan		NO PEAKS	

Weight of samples /dry matter/ - 1-3 mg

188

in the thermograms as a result of membrane fragmentation with
the formation of smaller vesicles. The transition at 365-369 K
corresponds to irreversible protein denaturation due to the
decrease of its helicity. The treatment with spermine leads to
the elimination of the peaks in the thermogram, while protamine
treatment is characterized by the appearance of two irreversible
peaks: one observed at 360-364 K and the other - at 366-370 K
/ Table 1 /.

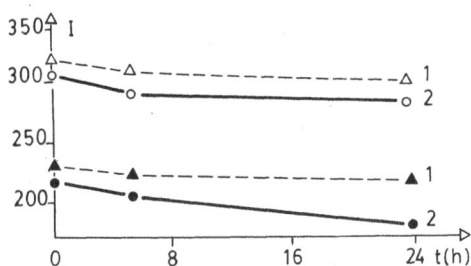

Fig. 2   Dependance of the fluorescence intensity of purple
         membranes in time
         On the abscissa - time / in hours /; on the ordinate -
         fluorescence intensity / in arbitrary units /
         Concentrations : purple membranes - 30 μg/ml ;
         protamine - 10 μg/ml ; ANS - 20 μg/ml. Measurements were
         carried out at 20°C and pH 6.5
         1 - untreated membranes ; 2 - protamine - treated
         membranes

Spectrofluorimetric investigations of ANS-membrane complex in
the range of 277-310 K show also temperature dependance of its
fluorescence intensity /Fig. 2 /. Partial solidification of the
lipid phase is observed at lower temperatures, while the rise
of the temperature to 310 K leads to a decrease in the density
of the packing of the lipid bilayer and increase in the mobility
of the fatty acids. The treatment of PM with the polyamines
investigated reduces significantly the fluorescence intensity.
It is assumed that there is a competition between ANS molecules
and the polycations for some charged sites on the membrane surface specific for ANS. These data are in agreement with the
results from microelectrophoresis of native and polyamines-treat-
ed membrane fragments.[2,4] It is assumed that the addition of
spermine and protamine favours confirmational transition of the
bacteriorhodopsin which is accompanied by unfolding of the membrane with the polycation added. The latter promotes the formation of multiglobular salt aggregates, stabilized by weak
electrostatic forces with the polycation laying flat on the
surface. In this way the full electrostatic potential at the
surface would minimize the flexibility of the membrane and,
therefore, its deformability.

ACKNOWLEDGEMENT

The authors thank Prof.V.P.Skulachev for the kindly supplied samples of purple membranes.

REFERENCES

1. McLaughlin, S., Mulrine, N., Gresalfi, T., Vaio, G., McLaughlin, A.C., 1981, Adsorption of Divalent Cations to Bilayer Membranes Containing Phosphatidylserine. J.Gen. Physiol. 77:445
2. Chung, L., Kaloyanidis, G., McDaniel, R., McLaughlin, A., McLaughlin, S., 1985, Interaction of Gentamicin and Spermine with Bilayer Membranes Containing Negatively Charged Phospholipid. Biochemistry, 24:442
3. Oesterhelt, D., Stoeckenius, W., 1974, Isolation of the Cell Membrane of Halobacterium halobium and Its Fractionation into Red and Purple Membrane. Methods in Enzymology /Part A.Biochemistry/ 31:667
4. Atanassov, A., Popdimitrova, N., Petrov, M., 1986, The Effect of Protamine and Spermine on the Electrophoretic Mobility of Purple Membranes from H.halobium. XVII FEBS Meeting, Abstracts, W.Berlin, 07.02.69
5. Petrov, M., Popdimitrova, N., Atanassov, A. 1987, Depolarized light Scattering Intensity Temperature Anomalies in Thin Oriented Purple Membrane Film. In Press: Mol.Cryst.a.Liqu. Cryst.
6. Shnyrov, V., Zakis, V., Borovyagin, V., 1984, A Study of Effect on the Purple Membrane Structural Characteristics. Biologicheskii Membrany, /in Russian/ 1:349.

SURFACE-INDUCED ORIENTATION AND ELASTICITY IN PURPLE MEMBRANE THIN FILM MEDIA

Minko Petrov, Nezabravka Popdimitrova*, and
Andrey Atanassov**

*Institute of Solid State Physics
**Medical Academy, Sofia
University of Sofia, Bulgaria

INTRODUCTION

The membrane fragment media elasticity and the fragments interaction with the orienting solid surfaces, as well as the interfragment interaction, are connected with the orientation of the fragment . Electrical and magnetic fields, applied parallel and perpendicular to the normal of the surface plate, have a definite contribution to the system's free energy, surface-induced orientation and its penetration into the volume. The competition between the mechanical orientation and the external electrical or magnetic fields is the object of investigation of this work. The contemporary understanding of the liquid crystal polymers can be a basis for interpretation of the results presented here.

MATERIAL AND METHODS

An effective method for obtaining oriented sample preparation is the method of drying purple membrane /PM/ suspension[1], inserted by the capillary forces between two substrate plates, accompanied by an applied electric or magnetic field. The PM suspension was used at high temperature /O.D.10 to 20/. Such a medium is compact enough to be studied as being continuous / Landau and De Genes considerations/. As a fundamental liquid crystal unit we used a fragment having pronounced anisotropic form-disc,[2] with a diameter D/Height H$\sim$0.1D/.

RESULTS AND DISCUSSION

Figures 1 and 2 present some cases of orientation of PM fragments sandwiched between two oriented substrates. Fugure 1a presents the PM suspension without any orientation with[3] permanent dipole moment perpendicular to the disc's plane, unit vector $\vec{n}$ coinciding with $\vec{P}$.[4] As a result of termal fluctuations of the local orientation, presented by $\vec{n}$, the latter always forms an angle Θ with the substrate's normal $\vec{N}$. The mechanical orientation of PM fragments is shown on Fig.1b. In the experiments microchannels parallel to the Y-direction were formed onto the substrate surface by the known method of SiO oblique evaporation.[1] As it is assumed, this mechanical way of orientation results in a definite position of the membrane fragments, i.e. close contact of the largest side of the membrane disc with the topology of the orienting boundary. Among the factors determining in general the position of the PM fragment in the medium, the following can be mentioned: The elastic adaptation of the fragments to the surface topology, the Van-der--Vaals long range interaction and the membrane/membrane interaction in the medium. The full energy of the system has two components : volume and surface energy. When the surface energy dominates, the orientation penetrates deeper in to the medium. The penetration length depends on the elasticity of the PM fragment's system:

$$\lambda = (K/B)^{1/2} ,$$

where K is an effective elastic constant of the medium ($K \sim 10^{-6}$ dyn/cm) and B is a constant characterizing the elasticity of the compression of the interlayer distance in smectic A approximation ($B \sim 10^8$ erg/cm^3).[2] Usually λ is of the order of an interlayer distance. The balance between the two components of the energy is strongly dependent on the temperature, which can provoke some shifts in the known PM phase transitions points.[5,6] The ordering induced by the boundary is characterized by the parameter S_B, which differs from S_V, therefore when a strong anchoring of the PM fragment to the orienting surface exists, the phase transition close to the boundary and into the volume will occur at different temperatures. Using electric or magnetic field, we can increase the degree of orientation. For the purpose constant electric field parallel to the substrate's normal was used /Fig.2/ In order to avoid instability of the system, an electric field of low intensity /2.10^4 V/cm/ was applied Since the applied electric field coincides with $\vec{P}$ /and $\vec{n}$/, it stabilizes the initial orientation and depresses the thermal fluctuations. Under polarizing microscope, when $\Theta \rightarrow 0$, such a complex seems to be uniaxial, with optical axis Z coinciding with the substrate's normal direction. The results mentioned above are confirmed by means of polarization investigations under laser irradiation. The electric field, applied perpendicular to the substrate's normal, has a destabilizing effect on the fragment's orientation. A rotation of fragments around the X-axis is possible. In the course of the competition between the surface-induced orientation and the electric field, the latter is predominant and close to the surface the fragments ought to contact with a narrow part of the disc. This is energetically unfavourable for system, which is destabilized, so again under the microscope a disordered system with large clusters, formed by separate PM, is obseved. On the contrary, the magnetic field applied parallel to the substrate's normal, destabilizes the orientation.Clusters similar to the case of perpendicularly applied electric field are seen. Meanwhile the magnetic field applied perpendicular to the substrate's normal produces stable orientation, meaning that the PM fragments can be considered as diamagnetics, oriented so as to minimize the total energy. Different external actions, including the temperature variations, introduce additional effects on the system, increasing its total energy. As the system has the tendency to elastic adaptation to the surface and maintains a constant distance between the

192

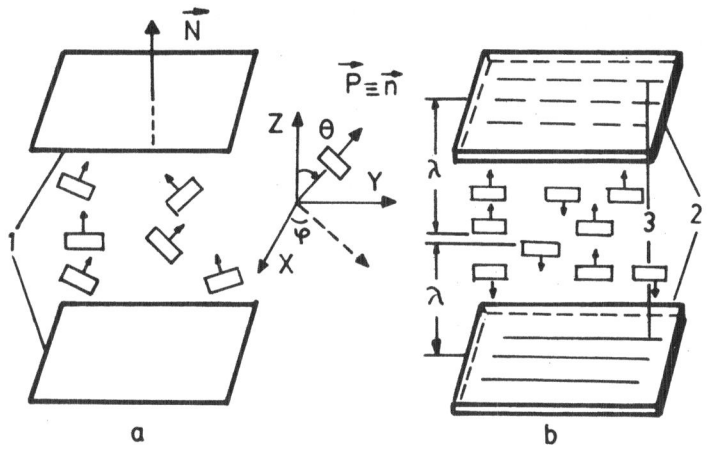

Fig.1
    a-Purple membrane fragment cell without any orientation:
      1 - glass plates; XYZ - laboratory coordinate system
      connected with the cell. PM fragments are presented like
      discs with permanent dipole moment $\vec{P}.\vec{N}$ - substrate's
      normal, Θ - and φ - polar angles of PM fragments direc-
      tor n.
    b-Ordered PM fragment state induced by surfaces. λ - pene-
      trating length of orientation induced by surfaces. 2 -glass
      plates; 3 - microchannels formed on the substrate surface

193

layers /according to the smectic A approximately/, the energy
is minimized by forming different dislocations, which are
observed under the polarizing microscope.

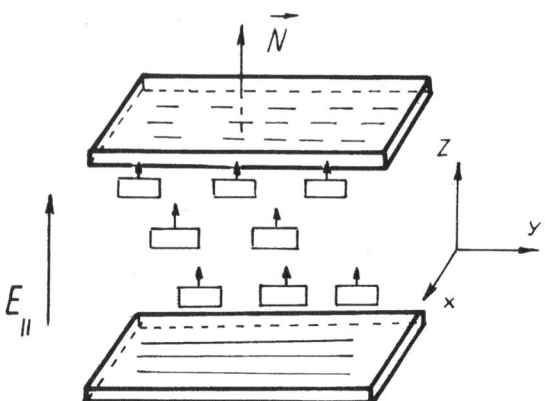

Fig.2 Ordered PM fragment's state induced by surfaces and
electric field parallel to the substrate's normal direction /E_{II}/

ACKNOWLEDGEMENT

     The authors thank Prof.V.P.Skulachev for the kindly supplied
samples of purple membranes.

REFERENCES

1. Petrov, M., Popdimitrova, N., 1983, Concerning Ion Con-
    ductivity of Purple Membranes Thin Oriented Films, Physio-
    logie, 20:89.
2. Clark, N., Rotschield, K., Luippold, D., Simon, B., 1980,
    Surface-induced Lamellar Orientation of Multilayer Membrane
    Arrays. Theoretical Analysis and New Method with Application
    to Purple Membrane Fragments, Biophys.J, 31:65.
3. Keszthely, L., 1982, Orientation of Purple Membranes by
    Electric Field, Methods in Enzymology, 88:287.
4. DeGenes, P.G., 1974, Physics of Liquid Crystals, Clarendon,
    Oxford.
5. Jackson, M., Sturtevant, J.M., 1978, Phase Transitions of
    the Purple Membranes of Halobacterium halobium, Biochemis-
    try, 17:911.
6. Hiraki, K., Hemanaka, T., Mitsui, T., Kito, I., 1981, Phase
    Transitions of the Purple Membrane and the Brown Holo-Mem-
    brane. X-Ray Diffraction, Circular Dichroism and Absorption
    Spectrum Studies, Biochem.Biophys.Acta, 647:18.

A STUDY OF THE FLUORESCENCE SPECTRA OF NATIVE AND MONOMERIC

BACTERIORHODOPSIN IN PURPLE MEMBRANES

Tzvetana Lazarova-Skumrieva and Alexander Ivanov

Central Laboratory of Biophysics
Bulgarian Academy of Science
Sofia 1113, Bulgaria

INTRODUCTION

A major advance in the investigation of the functional activity of the proton pump bacteriorhodopsin (BR) has been the determination of a completely amino acid sequence[1]. On the basis of electron diffraction density maps, lengths of inter-helix linking segments the most probable sequences of arrangements of polipeptide α-helicals has been proposed[1]. In conformity with the best of these structural models a transversal location of tryptophan (Trp) residues in the membrane tend to distribute fairly uniformly. There is compelling evidence that Trp residues are of a great importance both for the structure and the function of BR[2,3]. The fluorescence of purple membranes (PM) and apomembranes has been investigated with a purpose of qualifying the location of Trp residues in the membrane[4,5,6,7]. In the present paper are reported results about fluorescence spectra of native (nBR) and monomeric (mBR) bacteriorhodopsin and are analysed their parameters. A fluorescence probe pyrene was used in an attempt to characterize the position of Trp residues responsible for the protein fluorescence.

MATERIALS AND METHODS

Purple membrane fragments were prepared as described in[1]. The monomer form of BR was obtained by a solubilization with β-octyl-D-glucoside according to[8]. BR concentration was determined using an extinction coeficient $\varepsilon_{570}=6,3.10^4 M^{-1} cm^{-1}$. The membranes were suspended in 25 mM phosphate buffer (pH 6,8), so that the protein concentration to be suitable for fluorescence measurements, i.e. the BR absorption at 280 nm to remain less than 0.1 cm^{-1}. All steady-state fluorescence and polarization measurements were performed using a Jobin-Yvon 3YD and Perkin Elmer MPF 44A spectrofluorimeters equipped with a thermostated sample holder at $20^{o}C$ (slits width 4 nm). The quantum yield was determined as previously described in[9]. Excitation energy transfer from Trp residues of BR (donor) to the fluorescent probe pyrene (acceptor) was measured by quenching the protein fluorescence at 315 nm. A small aliquots of twice recrystalized pyrene in ethanol (stock solution $10^{-3}M$) were added to the membrane suspensions. The final concentration of the fluorescent probe was determined spectrofotometrically using an extinction coeficient $\varepsilon_{337}=5,4.10^4 M^{-1} cm^{-1}$. The efficiency of energy transfer (E) was calculated using equation: $E=1-J/J_o$, where J and J_o are the fluorescence intensities of the donor in the presence or absence of the acceptor respectively.

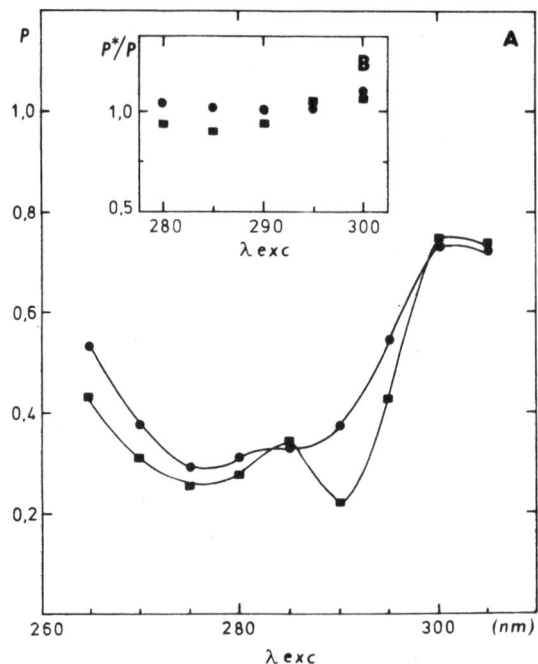

Figure 1. A. Excitation polarization spectra of nBR
($\bullet$) and mBR ($\blacksquare$). λ_{fluo}=340 nm. B. Polarization ratio P*/P as a function of excita-
tion wavelength. P*-polarization at 315 nm,
P-polarization at 340 nm. Fluorescence
measurements were performed as in [10].

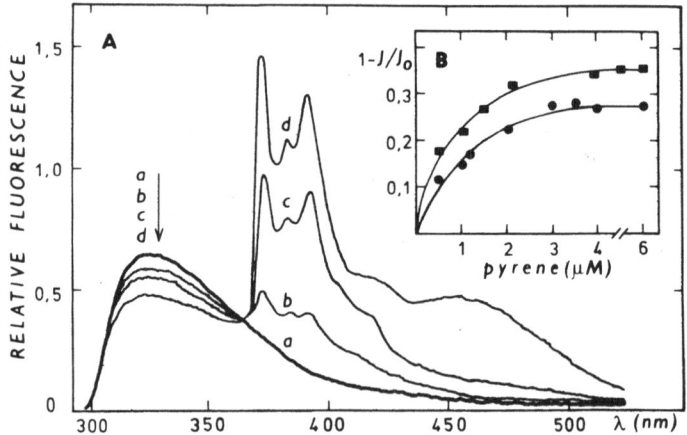

Figure 2. A. Fluorescence spectra of nBR in the absence
(a) and in the presence of increased concentrations of pyrene (b-0.5 μM,c-1.2 μM,d-2.1 μM).
B. Efficiency of energy transfer from Trp to
pyrene in nBR ($\bullet$) and mBR ($\blacksquare$) as a function
pyrene concentration.

RESULTS AND DISCUSSION

The fluorescence spectrum of nBR shows a maximum ($\lambda=315$ nm) which is considerably shortwavelengths shifted, as compared to the one for L-Trp in water ($\lambda=350$ nm), as well as the fluorescence spectra of most Trp containing proteins /Table 1/. The measured spectrum of nBR could be related to Trp residues located predominantly in a hydrophobic environment. Bleaching of PM results in a red shift of the fluorescence /see Table 1/, which suppose a participation of Trp in a more polar environment as compared to those in native PM. In case of covalent linking of the retinal to apomembranes long wavelength component in the fluorescence spectrum vanishes due to quenching of some Trp by retinal[3]. Changes in the fluorescence of PM has been observed also after treatment with $NaBrH_4$, heat-denaturation[4], β-octylglucoside.

Table 1. Parameters of the UV fluorescence spectra of L-Trp and BR in different states

sample	L-Trp	nBR	mBR	apomembranes
λ_{max} (nm)	350	315	324	326[6]
$\Delta\lambda$ (nm)	62	50	56	-
Q	0.20	0.013±0.001	0.025±0.003	0.170±0.030[6]

λ_{max}-position of the fluorescence maximum, $\Delta\lambda$-spectral halfwidth

Heterogeneity of the fluorescence of PM has been observed by means of derivative spectroscopic techniques[5]. Obviously, the main contribution to the fluorescence maximum of nBR have Trp located in hydrophobic environment, while Trp residues located in a more polar environment are quenched. It should be noted that such short wavelength fluorescence maximum of the protein in PM implies the participation of tyrosine (Tyr) residues. There is a marked inconsistency with regard to this view[5],[4]. In order to clarify a likely contribution of Tyr to the fluorescence of PM polarization data of BR were analyzed. The excitation polarization spectra of nBR and mBR correspond to those for multi-Trp proteins[10], though different values of polarization of the two samples are measured /Fig. 1A/. Since the ratio of polarizations P^*/P is almost independent of the excitation wavelength, it might be assumed that either there is no Tyr contribution to the fluorescence of nBR and mBR or that it is negligible /Fig. 1B/. On the other hand, the high values obtained for P_{280}/P_{max} for both nBR (0.423) and mBR (0.375) can be interpreted in term of a significant energy transfer between Tyr and Trp in protein molecule. The stronger quenching of Tyr in the spectrum of nBR /Fig. 1B/ most probably is a result of more effective energy transfer from Tyr to Trp compared to mBR. Hence the main contribution in the fluorescence spectra of nBR and mBR have Trp residues with an indirect contribution of Tyr (by energy transfer).

The fluorescence parameters: fluorescence maximum, quantum yield and spectral halfwidth of BR in different states are listed in Table 1. The position of the fluorescence maximum of nBR in PM presented in Table 1 is in good agreement with the data in [4],[6], and quite different from those in[5]. Native BR has the lowest values of Q (0.013) and spectral halfwidth (50 nm). Bleaching of PM produces not only a red shift in the maximum of fluorescence but an increase in quantum yield also /see Table 1/. The decrease of a quantum yield in nBR more than 10 fold confirms the view that between Trp and the retinal exist strong energy transfer[3]. The solubilization of BR with β-octylglucoside induces changes in the structure leading to changes in the fluorescence. The fluorescence maximum is slightly red shifted (324 nm) and the quantum yield is markedly increased (0.025) with respect to those in nBR.

At the same time spectral halfwidth goes up to 56 nm. It seems likely that during a conformational transitions Trp residues became more external and their quantum yield and spectral halfwidth grow much higher.

An effect of quenching of the protein fluorescence was measured upon an incorporation of the hydrophobic fluorescent probe pyrene in PM. An increase of the amount of pyrene leads to a decrease of the protein fluorescence and an appearance of a pyrene fluorescence spectrum, respectively /Fig. 2A/. As can be seen from Fig. 2B the efficiency of the energy transfer from Trp residues to pyrene is more pronounced in mBR. The effect of pyrene in PM with mBR is very similar to the one for 9-(9-antroyloxy) fatty acids reported in[1]. According to the neutron diffraction data[2], the position of the retinal is close to the centre of the membrane and it quenches Trp which are at distance $R_o = 25$ Å^3. Therefore, a decrease of protein fluorescence is a result of quenching of Trp residues presumably near to the membrane surface. It is reasonable to assume that more effective energy transfer in mBR is mainly due to the structural changes on the level of the aggregation state of BR molecule. Obviously, accessibility to Trp residues is larger for more open configuration. Energy transfer experiments imply that Trp which are not involved in quenching by retinal are probably burried in the membrane but not too far from the external surface.

REFERENCES

1. W. Stoeckenius, R. H. Lozier, and R. A. Bogomolni, Bacteriorhodopsin and the purple membrane of Halobacteria, Biochim. Biophys. Acta 505:215 (1979).
2. J. P. Hardy, A. E. W. Knight, K. P. Ghiggino, T. A. Smith, and P. J. Rogers, Effects of tryptophan oxidation on bacteriorhodopsin structure and function, Photochem. Photobiol. 39:81 (1984).
3. O. Kalisky, J. Feitelson, and M. Ottolenghi, Photochemistry and fluorescence of bacteriorhodopsin excited in its 280-nm absorption band, Biochemistry 20:205 (1983).
4. A. E. Permyakov and V. L. Shnyrov, A spectrofluorometric study of the environment of tryptophans in bacteriorhodopsin, Biophys. Chem. 18: 145 (1983).
5. B. J. Plotkin and W. V. Sherman, Spectral heterogeneity in protein fluorescence of bacteriorhodopsin: Evidence for intraprotein aqueous regions, Biochemistry 23:5353 (1984).
6. A. U. Acuna, J. Gonzalez, M. P. Lillo, and J. M. Oton, The UV protein fluorescence of purple membrane and its apomembrane, Photochem. Photobiol. 40:351 (1984).
7. R. C. Chatelier, P. J. Rogers, K. P. Ghiggino, and W. H. Sawyer, The transverse location of tryptophan residues in the purple membranes of Halobacterium halobium studied by fluorescence quenching and energy transfer, Biochim. Biophys. Acta 776:75 (1984).
8. N. A. Dencher and M. P. Heyn, Preparation and properties of monomeric bacteriorhodopsin, Methods Enzym. 88:5 (1982).
9. F. W. J. Tealer and G. Weber, Ultraviolet fluorescence of the aromatic amino acid, Biochem. J. 65:476 (1957).
10. G. Weber, Fluorescence polarization spectrum and electronic energy transfer in proteins, Biochem. J. 75:345 (1960).

HCO_3-STIMULATED ATPase ACTIVITY OF BRAIN MITOCHONDRIA:

THE ROLE OF Ca^{2+} AND Ca^{2+} AND Mg^{2+} IONS COMBINATION

Radoy I. Evanov and Nina Lambadjieva

Department of Human and Animal Physiology
Faculty of Biology, University of Sofia
Sofia, Bulgaria

INTRODUCTION

Anion-dependent ATPase[1] are localized in almost all animal tissues, plants and microorganisms.[2] Such ATPases are found and characterized in the brain tissues as well.[3] One assumption is that anion ATPase are of mitochondrial nature, but that can hardly be accepted as a general principle.[4,5,6]

Mg^{2+}-ions and above all the Mg^{2+} ATP complex are natural stimulators of the activity of almost all types of ATPases. In the literature there are scarce data on the effect of Ca^{2+} and the combination between Mg^{2+} and Ca^{2+} on HCO_3-ATPase activity in brain tissue. It is shown[6] that Ca^{2+} /1-10 mM/ inhibits significantly the ATPase activity of liver mitochondria, nuclei and plasma membranes.

In the present investigation we study the effect of Ca^{2+} and Ca^{2+} and Mg^{2+} ions on the HCO_3-ATPase activity of brain mitochondria and on the Na^+, K^+-ATPase activity of nerve tissie.

MATERIALS AND METHODS

Brain mitochondria were isolated according to Stahl et al.[7]

HCO_3-ATPase activity was astimated in a medium containing: 50 mM tris-buffer with pH=8.6, 1.5 mM $MgCl_2$, 1.5 mM vanadium--free ATP, 30 mM KCl, 50 mM NaCl or 50 mM $NaHCO_3$, 30 mM NaSCN / in control samples/, 0.1-1.0 mM ouabain, 50-100 μ g/sample enzyme protein. The reaction was initiated by the addition of ATP and completed by the addition of 0.5 ml TCA. HCO_3-ATPase activity / in μ mol or μ g P_i/mg protein/hour/ was determined as the difference between the activities of the total /Mg,K, HCO_3-/ and SCH^- - insensitive ATPase.

The enzyme was solubilized by Triton X-100 / detergent: protein ratio=2:1/.[8]

Inorganic phosphorus /P_i/ was determined by the method of Fiske and Subbarow,[9] the protein content was determined by the method of Lowry et al.[10]

Cat or rat brain membrane samples with a high Na, K-ATPase activity were isolated by the method of Guerra and Skou.[11]

RESULTS AND DISCUSSION

Brain mitochondria are among the main systems controlling the endogeneous Ca^{2+} content in nerve cells. On the other hand, divalent cations are important and obligatory regulators of the main mitochondrial processes. The combined effect of Ca^{2+} and Mg^{2+} ions on the enzyme kinetics of anion ATPases of brain tissue is one of the unsufficiently investigated links of the mechanisms of regulation of the anion ATPase activity.

The data presented on Fig.1 show that Ca^{2+} in the absence of Mg^{2+} can not be accepted as physiological regulators of the activity of mitochondrial anion ATPase. Stimulation of enzyme activity with the increase of Mg^{2+} concentration is registered with a maximum at 0.1-1.0 mM. The HCO_3-ATPase activity in the presence of Ca^{2+} only is four times lower than that of the

Table 1. Influence of Ca^{2+} and EGTA on the activity of different ATPase forms of brain mitochondria / n=6; $p^x < 0.001$/

Ion-modulated ATPase forms /ions in mM /	ATPase activity, % of Control			
	Ca^{2+}- 0.1 mM	Ca^{2+}- 0.1 mM and EGTA - 0.1 mM	Ca^{2+}- 1 mM	Ca^{2+}- 1 mM EGTA - 1 m M
Mg^{2+} / 1.5/	105	112	68^x	112
Mg, K /30 /				
HCO_3 / 50 /	75^x	82	43^x	100
SCN^- - sensitive- ATPase	65^x	50^x	28^x	78

Mg-dependent HCO_3-ATPase activity. The regulatory effect of Ca^{2+} ions has physiological importance only at optimum concentration of Mg^{2+} and optimum Mg^{2+}/ATP ratio. Under these conditions the HCO_3-ATPase activity increases parallel with the reduction of the Ca^{2+} -concentration from 1.0 mM to 1.0 μM. The HCO_3-ATPase activity is optimal at ratio $Mg^{2+}/Ca^{2+}=$ 1.5 mM/1.0 μM.

In the case of ion modelling of mitochondrial ATPase activity it is shown /Table 1/ that Ca^{2+} -ions /1.0 mM/ inhibit significantly all forms of ATPase /40-70 per cent/, while EGTA, respectively EDTA, in equimolar concentration eliminates totally /Mg^{2+}-dependent ATPase/ or partially /HCO_3-stimulated ATPase/ the inhibitory effect of Ca^{2+} ions.

The effect of Ca^{2+}-ions on the Na, K-ATPase activity
of purified membrane preparation from brain tissue is expressed
by dose-dependent inhibition of the enzyme activity in the
range of Ca^{2+}-concentration from 10 μM to 1.0 mM /Fig.2/.
The kinetic curve of the inhibitory effect of Ca^{2+} is similar
to that of the ouabain effect.[13] The points of application,
the mechanism of action and the inhibitory effectivity of
these substances are, however, different.

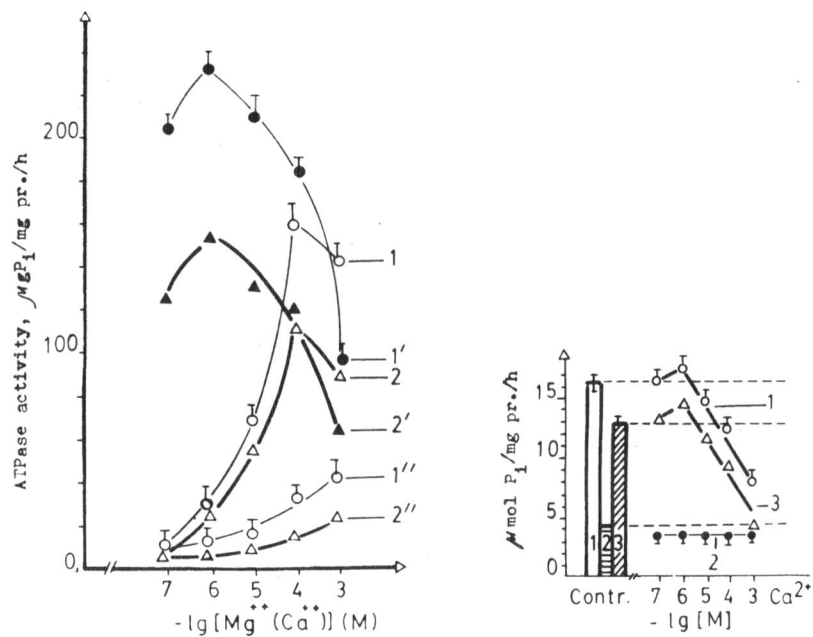

Fig.1 Influence of Mg^{2+} /1, 2/, Ca^{2+} / 1″, 2″/ and Mg^{2+}
/ 1.5 m M / plus Ca^{2+} / 0.1 μM - 1.0 mM/ /1′, 2′/ on the
Mg, K, HCO_3-ATPase /1/ and HCO_3-ATPase /2/ activities
of brain mitochondria /ATPase activity - in μ mol
P_i/mg protein/hour, n=8/

Fig. 2 Influence of Ca^{2+} - ions on the Na, K-ATPase activity
of membrane enzyme preparation of brain tissue
/ 1 - Mg, Na, K-ATPase activity, Na^+ : K^+ = 120:30 mM;
2 - ouabain-sensitive /Na,K-/-ATPase activity; n=6/

The effect of Ca^{2+}-ions on Na, K-ATPase is expressed
at the first and third step of the five-step enzyme reaction.
The Ca^{2+}.ATP complex is negative allosteric modulator,
while the Mg^{2+}.ATP complex is a positive allosteric modulator.[11] The inhibitory effect of Ca^{2+} on Na, K-ATPase is shown

on Fig.2, explaining the participation of cation-transport ATPases in the mediatory process, the effect of biologically active substances and the cell response. The dose-dependent effect of Ca^{2+}-ions on the mitochondrial ATPase activity reveals some mechanisms of their participation in the supply of energy for nerve functions.

REFERENCES

1. Kasbekar, D.K., Durbin, R.P., 1965, An adenosine triphosphates from frog gastric mucosa, Biochim.Biophys.Acta, 105, 3:472-482.
2. Ivaschenko, A.T., 1977, Adenosintriphosphataznaja sistema jivotnih kletok, stimuliruemaja ionami HCO_3^-, Biol.nauki, 1:25-33 /in Russian/
3. Kimelberg, H.K., Bourke, R.S., 1973, Properties and localization of bicarbonate-stimulated ATPase activity in rat brain, J.Neurochem, 20:347-359.
4. Amelsvoort, V.J.M.M., De Pont, J.J.H.H.M, Bonting, S.L., 1977, Is there a plasma membrane-located anion-sensitive ATPase?, Biochim.Biophys.Acta, 466, 2:283-301.
5. De Pont, J.J.H.H.M. and Bonting, S.L., 1981, Anion-sensitive ATPase and /K^+ + H^+/-ATPase. In: "Membrane Transport", Elsevier, North-Holland, Biomed.Press, 209-234.
6. Ivaschenko, A.T., 1980, Svojstva aniontchuvstvitelnoj ATP-azi plasmaticheskoj membranaj kletok petcheni krajsaj, Biochimia, 45, 3:424 /in Russian/.
7. Sttahl, W.L., Smith, J.C., Napolitano, L.M., Basford, R.E., 1963, Brain mitochondria. I.Isolation of bovine brain mitochondria, J.Cell Biol., 19, 2:293-307.
8. Blum, A.L., 1971, Properties of soluble ATP-ase of gastric mucosa. II. Effect of bicarbonate ion, Biochim.Biophys.Acta, 249:101-113.
9. Fiske, C.H., Subbarow, Y., 1925, The colorimetric determination of phosphorus, J.Biol.Chem., 66:375-400.
10. Lowry, O., Rosenbrough, M., Farr, A. and Randall, R.,1951, Protein measurement with the folin reagent, J.Biol.Chem., 193:265-275.
11. Guerra, L., Skou, J.C., 1974, A inibicao pelo Ca^{2+} da ATPase activada pelo Na^+ e pelo K^+ extraida do cortex cerebral de boi na presenca e na ausencia da oligomocina, Rev.Cien.Med., /Maputo/, 7:1-23.
12. Ivanov, R., Velichkova-Markova, S., Simova, B., 1984, Cloxazepine and chlorpromazine:Effect on the activity of Na, K- and HCO_3-ATPases of rat brain, Med.-Biol.Inform., /Sofia/, 1:16-23.
13. Ivanov, R.I., 1987, Acetylcholinat kato regulator i modulator na funktionalno-metabolitnoto sastojanie na cholinergichnite structuri, Aftoreferat na dissertatzia, s.47, Sofia /in Bulgarian/.

THE INFLUENCE OF ACETYLCHOLINE AND SEROTONIN ON THE HCO_3-ATPase ACTIVITY OF BRAIN MITOCHONDRIA

Radoy I. Ivanov and Dobrinka K. Dobreva

Department of Human and Animal Physiology
Faculty of Biology, University of Sofia
Sofia, Bulgaria

INTRODUCTION

In some of our works[1,2,3] it was shown that adrenaline and noradrenaline in concentration of 1×10^{-3} M stimulated the HCO_3-ATPase activity of cat brain mitochondria by 300 and 216 per cent, respectively, and in concentration of 1×10^{-8} M - by 57 and 12 per cent. Chlorpromazine and cloxazepine /1×10^{-3} M/ dose-dependently inhibit the activity of this enzyme.[4]

The present study, and those mentioned above, are among the first studies devoted to the neurotransmitter regulation of the activity of anion-dependent ATPase in nerve tissue.

Investigations on the role of transmitter as regulatores of the activity and function of membrane ATPase are a trend in modern neurobiology.[3,5,6,7]

Obvious is the necessity of a complex study of the effects of neurotransmitters, neuropeptides, neuroleptics, Tranquillizers and antidepressants upon different ATPase and on bioenergy exchange in the nerve tissue as possible mechanisms of the physiological and theraupeutic action of these substances and druds.[4,8,9,10]

The present study is a part of this programme. The data presented on the influence of acetylcholine, carbamylcholine, atropine and aerotonin reveal a new aspect in the effect of biologically active substances on the activity of a comparatively poorly studied enzyme system.

MATERIAL AND METHODS

Brain mitochondria were isolated according to Stahl et al.[11]

The procedure for determining the HCO_3-ATPase activity in brain mitochondria is described elsewhere /Ivanov, Lambadjieva - in this volume/.

Inorganic phosphorus was determined by the method of Fiske and Subbarow,[12] the protein content - by the method of Lowry et al.[13] ATPase activity was calculated in μ mol Pi/mg protein/hour.

RESULTS AND DISCUSSION

Acetylcholine /Ach/ and carbamylcholine /Cch/ in concentrations 10 μM - 1.0 mM dose-dependently inhibit HCO_3-stimulated ATPase activity of brain mitochondria and in concentration 0.1-1.0 μM moderately stimulate the activity of this enzyme /Fig. 1/.

Essential inhibition of the HCO_3-ATPase activity at higher concentrations of Ach and Cch is mainly due to the reduction of the stimulating action of HCO_3^- and the inhibitory effect of SCN^- on the activity of anion ATPase.

The retained native structure of the mitochondria and of the lipid environment of the enzyme molecule are necessary conditions for expressing the dose-dependent influence of Ach. For solubilized forms of the enzyme this effect is not presented /Table 1/.

In the case of delipidated membranes the effect of specific anions, neurotransmitters and some drugs decreases considerably.[4]

Atropine does not affect significantly the effects of Ach on the HCO_3-ATPase activity /Table 1/. These data call in question the presence of an M-choline receptor unit in the mechanisms of action of Ach upon anion-dependent ATPases.

The presence of Ca^{2+}-ions in samples reduces and modifies the effect of Ach on the HCO_3-ATPase activity of brain mitochondria, while EDTA and EGTA /1 mM/ restore the dose-dependent action of this substance /Table 1/.

Table 1. Influence of acetylcholine /Ach/, Ca^{2+} and atropine on the HCO_3-ATPase activity of brain mitochodria - membrane-bound from /m.b.f./ and solubilized /Triton X-100/ from /s.f./ n=10

SERIES	ATPase activity/ mol Pi/mg protein hour/					
	Mg^{2+},K^+,HCO_3^--ATPase			HCO_3^--ATPase		
	Control	Ach		Control	Ach	
		10 μM	10 mM		10 μM	10 mM
Control /m.b.f./	11.60	10.45	9.90	7.30	8.70	2.20
Atropine /1x10^{-4} g/ml/	10.60	9.40	7.00	6.40	4.10	0.95
Ca^{2+} - 1 μM	12.50	8.90	10.60	6.10	3.20	4.60
Ca^{2+} - 1 mM	2.40	2.70	2.40	1.50	1.40	1.20
Control /s.f./	10.15	9.50	9.10	5.10	4.70	4.60

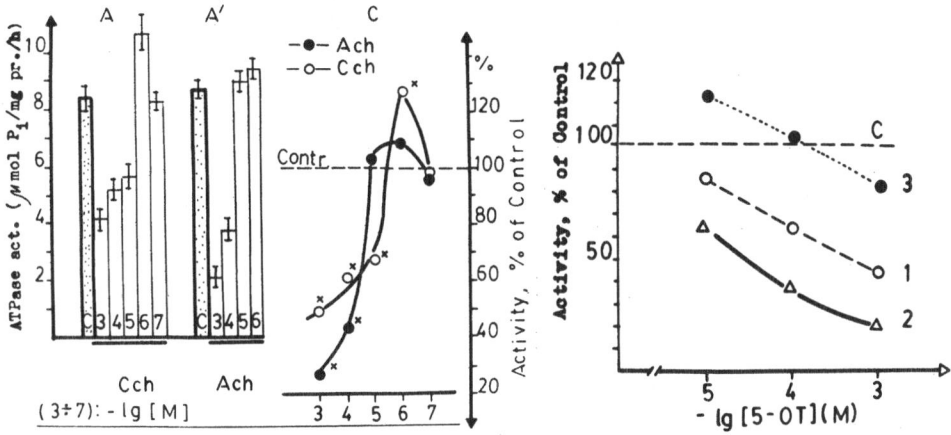

Fig.1   Influence of acetylcholine /Ach/ and carbamylcholine
         /Cch/ on the Mg, K, HCO$_3$-ATPase /A/ and SCN$^-$ - sensitive
         ATPase /C/
         /n=10; x- p< 0.001; Control /C/=100 per cent/

Fig.2   Influence of serotonin /5-OT/ on the Mg, K, HCO$_3$-/I/,
         SCN$^-$ -sensitive- /2/ and SCN$^-$ -insensitive- /3/-ATPase
         activities of brain mitochondria /n=6/

     Among the main groups of mechanisms of the membrane-active
effect of cholinergic agents on the activity of anion ATPases
are the influence of transport processes, of conformational
state of membrane and membrane-bound enzyme, and the existence
of competition between Ach /Cch/, HCO$_3^-$, SCN$^-$ and Ca^{2+} -ions.
     The effect of serotonin /5-OT/ on the HCO3-ATPase activity
of brain mitochondria is expressed by dose-dependent inhibition
of enzyme activity /Fig.2/. The presence of Ca^{2+} in the samples
does modify the action of 5-OT /data are not shown/.
     In another report it has already been shown that 5-OT
stimulates the respiratory activity in the range of 30-35 per
cent / in concentration of 0.01-0.1 mM /, as well as the succi-
nate-dehydrogenase activity of brain mitochondria /in conc.
10 μM/ and Na, K-ATPase activity of membrane brain preparation
/in conc. 0.1 mM/.[3]
     The data presented reveal another aspect of the biochemical
mechnisms of the physiological effect of neurotransmitters,
expressed mainly by the influence of bioenergetic processes in
the nerve tissue.

REFERENCES

1.   Atanassova, Y.G., Ivanov, R.I., 1981, Effect of adrenaline
     and noradrenaline on HCO$_3$-ATPase activity of brain mitochon-
     dria, Comp.rend.Acad.bulg.Sci., 34, 9:1305-1306

2. Atanassova, Y.G., Ivanov, R.I., 1982, Effect of histamine and serotonin on HCO_3-ATPase activity of brain mitochondria., Compt.rend.Acad.bulg.Sci., 35, 2:215-216.

3. Ivanov, R.I., 1987,"Acetylcholinat kato regulator i modulator na functionalno-metabolitnoto sastojanie na cholinergichnite structuri". Aftoreferat na dissertatzia, s.47, Sofia /in Bulgarian/.

4. Ivanov, R., Velichkova-Markova, S., Simova, B., 1984, Chloxazepine and chlorpromazine: Effect on the activity of Na, K^- and HCO_3-ATPases of rat brain., Medico-Biol.Inform., Sofia, 1; 16-23.

5. Meyer, E., Cooper, J., 1981, Correlations between Na^+ -K^+--ATPase activity and acetylcholine release in rat cortical synaptosome., J.Neurochem., 36/2/:467-475.

6. Rodriguez, L.A., Antoneli, M., 1981, The effect of several neurotransmitter substances on nerve ending membrane ATPases. Acta Physiol.Latinoam., 31:39-44

7. Vizi, E.S., Török, T., Seregi, A., Sereözö, P. and Adam-Vizi, V., 1982, Na, K - activated ATPase and the release of acetylcholine and noradrenaline., J.Physiol. /Paris/, 78:399-406.

8. Tomov, T., Velichkova-Markova, S., Russanov, E., 1981, Effect of cloxazepine on mitochondria., Acta Physiol. Pharmacol.bulgarica., 7 /4/:65-71.

9. Sawas,A., Gilbert, J., 1981, Effects of adrenergic agonists and antagonists and of the catechol nucleus on the Na^+ -K^+--ATPase and Mg^{2+}-ATPase activities of synaptosomes. Biochem. Pharmacol.,30 /13/:1799-1803.

10. Sicora, J., Krulic, R., 1974, Psychotropic drugs and differently stimulated ATPase in CNS. Activ.nerv.sup./Praha/, 16, 3 ; 222.

11. Stahl, W.L., Smitt, J.C., Napolitano, L.M. Basford, R.E., 1963, Brain mitochondria. 1. Isolation of bovine brain mitochondria., J.Cell.Bioł., 19, 2:293-307.

12. Fiske, C.H., Subbarow, Y., 1925, The colorimetric determination of phosphorus., J.Biol.Chem., 66:375-400.

13. Lowry, O., Rosenbrough, M., Farr, A., Randall, R., 1951, Protein measurement with the folin reagent., J.Biol.Chem., 193:265-275.

EFFECT OF CHLORAMINE-T ON SINGLE SODIUM CHANNELS IN

NEUROBLASTOMA CELLS

Ilko Iliev* and Karoly Nagy

I. Physiological Institute, University of the Saarland
D-6650 Homburg/Saar, German Federal Republic and *Central
Laboratory of Biophysics, Bulgarian Academy of Sciences
Acad, G. Bontchev str., Bl. 21, Sofia 1113, Bulgaria

ABSTRACT

Single sodium channel currents were measured in cell-atached patches
of native and chloramine-T treated neuroblastoma cells, N1E 115. Channels
modified by chloramine-T can open and close several times during depolarization. The open time of these channels was never longer than that of the
native channels. Therefore the slower decay of the averaged current must be
explained by the flickering behaviour of the modified channels.

INTRODUCTION

Externally applied chloramine-T markedly inhibits sodium channel inactivation in myelinated nerve fibres[12] and squid axons[13]. A brief exposure
to a milimolar concentration of chloramine-T is sufficient to inhibit inactivation irreversibly without reducing the size of the sodium current. The
observations on myelinated nerve fibres can be interpreted by a three state
inactivation model with one open and two closed states[11]. Whereas Wang[12]
reported no effect of chloramine-T on the activated process, later investi-
gators described a small positive shift of the activation curve by chloramine-T[6,10]. The present paper describes the effect of chloramine-T on single
channel sodium currents of mouse neuroblastoma cells. The effects of chloramine-T are compared with those of pronase and N-bromacetamide (NBA) which
also inhibit inactivation of the microscopic current and have been studied
with the patch clamp technique[9,1].

MATERIALS AND METHODS

Single sodium channel currents were measured in cultured neuroblastoma
cells in cell-attached configuration[4]. Cells were grown under standard conditions[7]. The experimental methods have been described earlier[8]. Cells were
incubated in bath solution containing 0.5 to 1 mM chloramine-T. After 5 to 8
min the chloramine-T solution was washed out and experiments were carried
out in normal bath solution. Pipette and bath solutions contained (in mM):
NaCl 140, KCl 5.0, CaCl$_2$ 1.8, MgCl$_2$ 0.8, HEPES 20, glucose 20. pH was adjusted to 7.3 and temperature (between 8 and 15°C) was kept constant during
each experiment. Both with native cells (in control experiments) and with
chloramine-T treated cells patch pipettes formed seals with a resistance of
30-50 Gohm. The cell membrane was hyperpolarized by 40-60 mV to remove resting inactivation. From this constant holding potential depolarizing pulses

were applied with one per second repetition. Pulse lenth was 20-40 ms. A DEC/LSI 11/23 microcomputer generated the voltage pulses, collected the data with 10 kHz sampling rate and was used for off-line analyses[5]. Analog signals were filtered at 2 kHz (-3 dB) by a four-pole low-pass Bessel filter. The major part of the leakage and capacitive currents was compensated by an analog circuit. The half amplitude threshold detection[3] was used for construction of the open, close and wurst open histograms. During averaging single channel currents, records without openings were averaged separately and subtracted from the mean of records with openings to eliminate the rest of the capacitive artifact. Due to the large variation in the cell's resting potential[7,8] results obtained with different cells cannot be compared directly. The fast time constant of the first delay histograms was used to recalculate the pulse potential and to compare the single channel parameters of different patches. Chloramine-T does not alter the activation time course of sodium channels[11,12]therefore this method is suitable for the comparison of results obtained with both native and chloramine-T treated cells

RESULTS AND DISCUSSION

Single sodium channel current records measured with native and chloramine-T treated cells are shown in Fig. 1a and b respectively. Clear diffe-

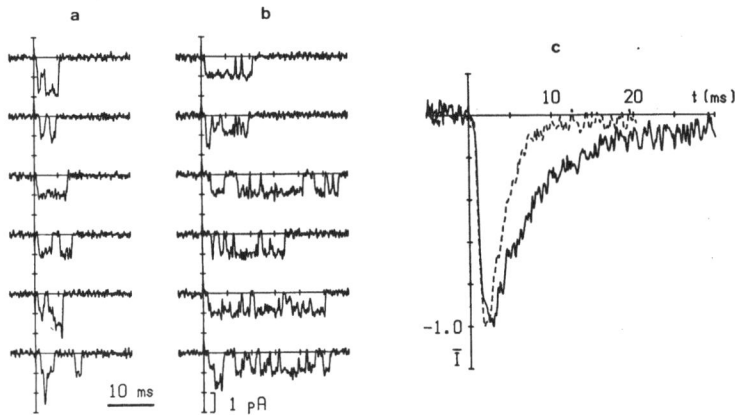

Figure 1.   Single sodium channel currents recorded in a cell-attached patch of a native (a) and chloramine-T treated (b) neuroblastoma cell. c) The average of the single channel currents for native (dashed line) and modified channels (continuous line) are normalized and compared. Curves are from the same patch as a) and b). V_m = RP+20 mV, V_H = RP-60 mV and T = 10 °C.

rences between the two plots can be observed. Channels in the native cells (Fig. 1a) usually open once during a depolarization and no openings appear later than about 10 ms. In chloramine-T treated cells (Fig. 1b) channels open several times before they shut. Open channel currents are interrupted by brief closings, which are sometimes so short that the base line cannot be reached. Openings frequently occur after 20 ms. Chloramine-T did not influence the single-channel conductance (not shown). In chloramine-T treated cells inactivation of the macroscopic sodium current is markedly delayed (see Introduction). The same effect occurs in neuroblastoma cells as demonstrated by the averages of the single channel current records in Fig. 1c. For better comparison the current records have been normalized to the peak value. The mean current measured on a chloramine-T treated cell (continuous line) decays much more slowly than the current of the native cell (dashed line). Open time histograms were constructed for both native (Fig. 2a) and chloramine-T treated patches (Fig. 2b). The mean open time of chloramine-T modified channels measured in 14 patches was never longer than that of the native channels. Therefore the slower decay of the current in treated cells can only be explained by the repetitive opening or bursting. Closing time histograms could be best fitted by the sum of two exponentials (Fig. 2c), corresponding to the brief nonconducting states within bursts and the longer nonconducting states between bursts (see third record in Fig. 1b). Nonconducting states longer than a critical duration t_c were treated as gaps between bursts. We chose $t_c = 0.5$ ms for Fig. 2c becouse at 0.5 ms the proportion of long closed intervals erroneously classified as gaps between bursts was equal to the proportion of misclassified short closed intervals[2]. Ignoring all nonconducting states which are shorter than 0.5 ms the burst time histogram in Fig. 2d was constructed. The mean burst time (BT = 2.35 ms) is difinitely longer than the mean open time of the native channels

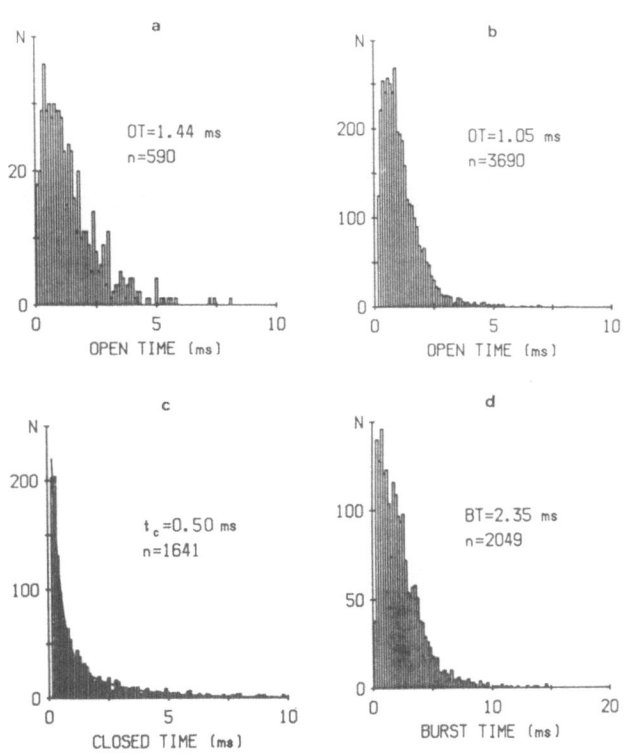

Figure 2.   Open time, closed time and burst time histograms. a) Open time histograms of native and modified (b) channels, OT is mean open time, n is the sum of events, c) closed time histogram for CT treated cell and was fitted by the equation $N(t) = 243 \exp(-t/0.32) + 44 \exp(-t/2.06)$, t_c- specific closed time used for the construction of the BT histogram shown in plot d (see text), n is the sum of events, d) birst time histograms for the modified channels, BT - mean burst time, n is the sum of the events, plot a is from the same patch as Fig. 1a, b, c, and d are from the same patch as Fig. 1b

(OT = 1.44 ms; compare Fig. 2d and a). From the ratio of the open and burst events the number of mean openings during a burst can be obtained. It was 3690/2049 = 1.8 for Fig. 2d and varied between 1.4 and 3.6 in different patches (14 patches measured).

The effect of chloramine-T described here differs from that of the other substances which modify the inactivation of channels. NBA prolongs the open time of channels by about a factor of ten[9,14]. Pronase induces similar effect in rat myotubes[9] but in chick dorsal root neurones it causes both a three-fold increase in the open time and the bursting of channels[1]. The effect of both substances can be explained by the block of inactivation. The nonincreased open time and the bursting of the chloramine-T treated channels suggest another mechanism. These results could be explained by supposing that the open-inactivated pathway is not blocked but the absorbing property of the inactivated state is abolished by chloramine-T. Therefore the channel may open from this modified inactivated state several times.

REFERENCES

1. E. Carbone, H. D. Lux, Sodium channels in cultured chick dorsal root ganglion neurones, Eur. Biophys. J. 13:259 (1986).
2. D. Colquhoun, B. Sakmann, Fast events in single-channel currents activated by acetylcholine and its analogues at the frog muscle endplate J. Physiol. 369:501 (1985).
3. D. Colquhoun, F. Sigworth, Fitting and statistical analysis of single channel records, In "Single Channel Recording", B. Sakmann, E. Neher, eds, Plenum Press, New York, 191 (1983).
4. O. Hamill, A. Marty, E. Neher, B. Sakmann, F. Sigworth, I±proved patch clamp techniques for high-resolution current recording from cells and cell-free membrane patches, Pflug. Arch. Eur. J. Physiol. 391: 85 (1981).
5. D. Hof, A pulse generating and data recording system based on the microcomputer PDP 11/23, Comput. Method. Program. Biomed. in press (1987).
6. H. Meves, N. Rubly, Kinetics of sodium current and gating current in the frog node of Ranvier, Pflug. Arch. 407:18 (1986).
7. W. Moolenaar, I. Spector, Ionic currents in cultured mouse neuroblastoma cells under voltage clamp conditions, J. Physiol. 278:265 (1978).
8. K. Nagy, J. Kiss, D. Hof, Single Na channels in mouse neuroblastoma membranes. Indication for two open states, Pflug. Arch. 399:302 (1983).
9. J. Patlak , R. Horn, Effect of N-bromacetamide on single sodium channel currents in excised membrane patches, J. Gen. Physiol. 79:333 (1982).
10. M. Rack, N. Rubly, C. Waschow, Effects of some chemical reagents on sodium inactivation in myelinated nerve fibres of the frog, Biophys. J. in press (1986).
11. J. Schmidtmayer, Behaviour of chemically modified sodium channels in frog nerve supports a three-state model of inactivation, Pflug. Arch 404:21 (1985).
12. G. Wang, Irreversible modification of sodium channel inactivation in toad myelinated nerve fibres by the oxidant chloramine-T, J. Physiol (Lond) 346:127 (1984).
13. G. Wang, M. Brodwick, D. Eaton, Removal of sodium channel inactivation in squid axon by oxidant chloramine-T, J. Gen. Physiol. 86:289 (1985).
14. D. Yamamoto, J. Z. Yeh, Kinetics of 9-aminoacridine block of single sodium channels, J. Gen. Physiol. 84:361 (1984).

A RECTIFYING POTASSIUM CHANNEL IN HUMAN NEUROBLASTOMA CELLS

Tuula Jalonen

Department of Biology
University of Turku
20500 Turku, Finland

INTRODUCTION

Mouse neuroblastoma cells have extensively been used in studies on the function of neural cells. Differentiation of cells in culture may be achieved with various substances including butyrate, c-AMP or dimethylsulphoxide (Kimhi et al., 1976; Palfrey et al., 1976, 1977). More recently similar studies have been made using human neuroblastoma cell lines (Åkerman et al., 1984; Littauer et al., 1979; Påhlman et al., 1981, 1983; Scott et al., 1986a,b; Spinelli et al., 1982). In this study the human ganglion cell derived neuroblastoma cell line SH-SY5Y has been used for characterizing the single ion channels in non-differentiated and differentiated cells. Induction of differentiation in this cell line with the phorbol ester, 12-o-tetradecanoyl-phorbol-13-acetate (TPA) leads to an increase in the average resting membrane potential as well as to a depolarisation-induced Ca^{2+} ion uptake and transmitter release (Åkerman et al., 1984; Scott et al., 1986a,b) and an increase in catecholamine synthesis (Påhlman et al., 1981, 1983). There is little or no electrophysiological data available on human neuroblastoma cells, though sodium, calcium and two types of potassium currents as well as a nonselective Ca^{2+}-activated current have previously been demonstrated in the mouse neuroblastoma cell line N1E-115 (Moolenaar and Spector, 1977; Spector, 1981; Yellen, 1982).

MATERIALS AND METHODS

Cell Culture. The SH-SY5Y cells were grown in Dulbecco's MEM medium supplemented with 7 % foetal calf serum (Gibco, U.K.) at 37°C in CO_2 air ventilated, humified incubator. Media were changed twice a week. Cells used for experiments were subcultured on cover slips in petri dishes by seeding approximately 10^5 cells/ml. The cells were used after 3-4 days in culture. To induce cell differentiation 20 nM TPA (Sigma, USA) was added into the culture medium. Cell differentiation was judged from the appearance of long neuritelike processes with frequent varicosities described by Påhlman et al. (1981).

Solutions. For the electrophysiological experiments the cover slips were moved from the growth medium into the experimental media. The extracellular solution contained (mM): NaCl 137; KCl 5.4; $MgCl_2$ 1.2; KH_2PO_4 0.44; TES 20; pH 7.4, the intracellular solution contained: NaCl 3; KCl 150;

MgCl$_2$ 1; EGTA 1; HEPES 10; pH 7.2 and the Ba-solution contained: BaCl$_2$ 90; KCl 5; MgCl$_2$ 1.2; KH$_2$PO$_4$ 0.44; TES 20; pH 7.4. All solutions were filtered before use through 0.22 μm Millipore-microfilters. All measurements were done at room temperature, 22$\pm$2 °C.

Recordings. The single channel currents were recorded using the improved patch-clamp method introduced by Hamill et al. (1981) with the L/M EPC-5 amplifier (List-Medical, West-Germany) and stored on tape (TEAC MR-10/30 Recorder, Japan). The glass microelectrodes were drawn from haematocrit capillaries (Hirschmann EM Techcolor, West-Germany) and coated with Sylgard^R. The recordings were made from inside-out membrane patches of the soma part of the cells.

RESULTS AND DISCUSSION

The results show that there are several different channel types in these human neuroblastoma cells (Table 1.). The rectifying K$^+$ channel is the most common K$^+$ channel in both the control and TPA-treated cells and is found in about half of the membrane patches of the TPA-treated cells. This channel invariably activates after the formation of an excised inside-out patch and stays active for long periods (even over 1 hour) with no sign of inactivation. The channel tends to stay in the open configuration with very fast flickering closures. Extracellular Ba^{2+} (90 mM) reduces the probability of the channel opening but does not affect the single channel current amplitude. The addition of EGTA (0.1 mM) or verapamil (70 μM) into the bath (inside the membrane) increases the tendency of the channel to flicker between the closed and open states (Figure 1. A to D). At higher concentrations verapamil also seems to cause total closure of the channels. The current-voltage relationship for this channel is not linear but shows clear outward rectification. The single channel conductance changes from about 80 pS at negative membrane potentials to 150 pS at positive potentials (Figure 1.). Because of the large single channel conductance and the high frequency of the channel openings this channel is able to let through large amounts of K$^+$ ions and so strongly influence the cell resting potential. However, in this study the channel activity has been studied in excised cell-free membrane patches and does not give information about the channel behaviour in a whole intact cell. These cultured human neuroblastoma cells might prove to be a useful object for studies on the role of the ion channels in cell differentiation.

Table 1. The Frequency of Different Channel Types in the Membrane Patches of the Control and TPA-treated Human Neuroblastoma Cells

Channel Type	Control(%)	TPA-treated(%)
Delayed Rectifier K$^+$	29	50
Transient K$^+$	12	27
Ca^{2+}-activated K$^+$	–	15
Chloride	35	32
Na$^+$, Nonselective and Other channels	6-10	9

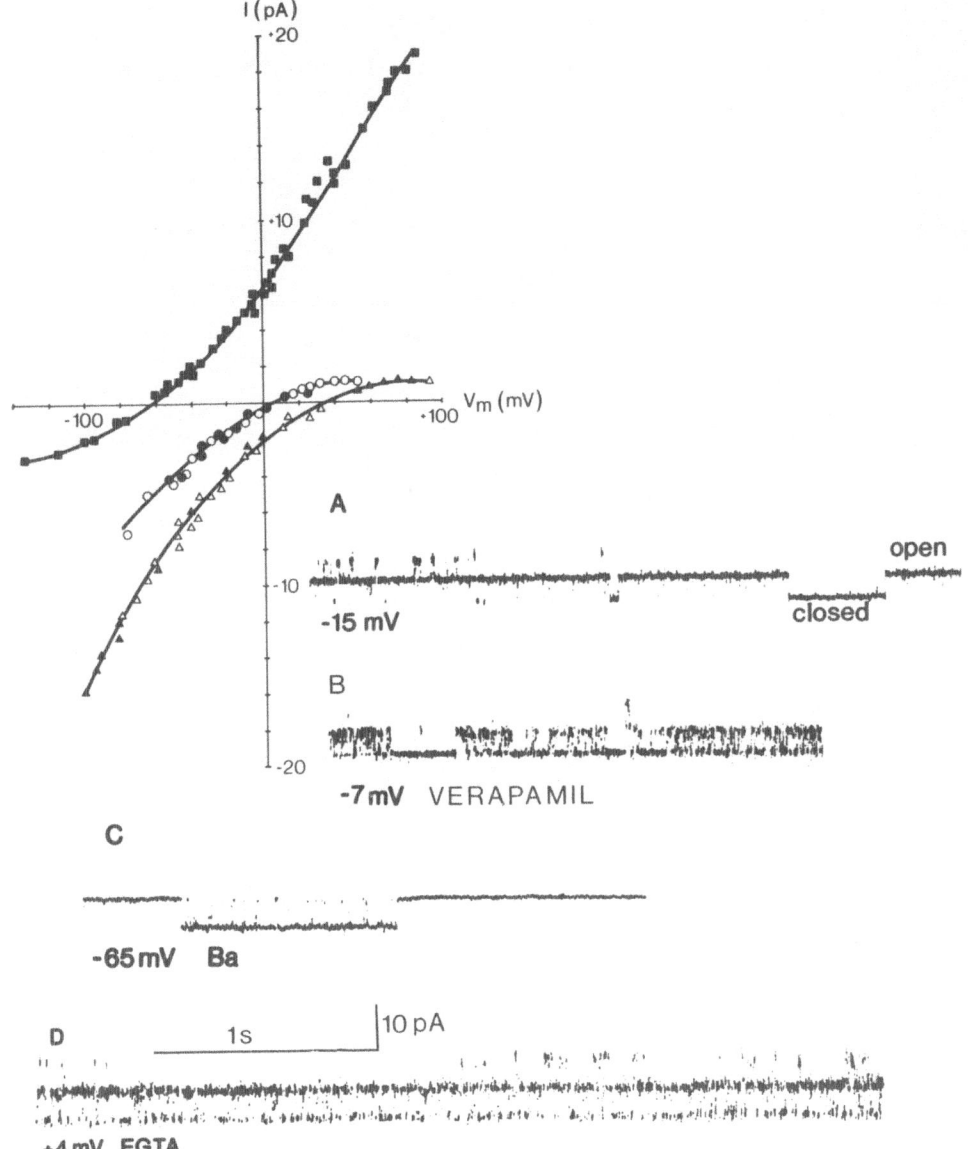

Fig. 1. The current-voltage relationships of the rectifying K^+ channel and
samples of channel activity. The I-V curve: In ($\blacksquare$, TPA-treated
cells) there is 150mM K^+ in the bath and 5.8mM K^+ in the pipette;
in ($\blacktriangle$,control) and ($\triangle$,TPA) the K^+ concentrations are reversed;
in ($\bullet$,control) and ($\bigcirc$,TPA) there is 5.8mM K^+ in the bath and 5.4mM
K^+ together with 90mM Ba^{2+} in the pipette. The reversal potentials
for the currents are about -60, +5 and +40 mV, respectively (the
calculated reversal potentials for pure K^+ currents would be -82,
+2 and +82 mV). A. Channel activity with 150mM K^+ in bath and 5.8
mM K^+ in the pipette. B. Same recording after addition of 70 µM
final concentration of verapamil into the bath. C. The decreased
channel activity with 90mM Ba^{2+} in the pipette, channel openings
downwards. D. Channel activity after the addition of 0.1mM EGTA
into the bath. In this case the channels open upwards.

213

ACKNOWLEDGEMENTS

T.J. would like to thank Prof. K.E.O. Åkerman for providing the cells and the Academy of Finland for financial support.

REFERENCES

Åkerman, K. E. O., Scott, I. G., and Andersson, L. C., 1984, Functional differentiation of a human ganglion cell derived neuroblastoma cell line SH-SY5Y induced by a phorbol ester (TPA), Neurochem. Int., 6:77.

Hamill, O. P., Marty, A., Neher, E., Sakmann, B., and Sigworth, F. J., 1981, Improved patch-clamp techniques for high-resolution current recording from cells and cell-free membrane patches, Pfluegers Arch., 391:85.

Kimhi, Y., Palfrey, C., Spector, I., Barak, Y., and Littauer, U. Z., 1976, Maturation of neuroblastoma cells in the presence of dimethylsulphoxide, Proc. Natl. Acad. Sci. USA, 73:462.

Littauer, U. Z., Giovanni, M. Y., and Glick, M. C., 1979, Differentiation of human neuroblastoma cells in culture, Biochem. Biophys. Res. Commun., 88:933.

Moolenaar, W. H., and Spector, I., 1977, Membrane currents examined under voltage clamp in cultured neuroblastoma cells, Science, 196:331.

Påhlman, S., Odelstad, L., Larsson, E., Grotte, G., and Nilsson, K., 1981, Phenotypic changes of human neuroblastoma cells in culture induced by 12-o-tetradecanoyl-phorbol-13-acetate, Int. J. Cancer, 28:583.

Påhlman, S., Ruusala, A. I., Abrahamsson, L., Odelstad, L., and Nilsson, K., 1983, Kinetics and concentration effects of TPA-induced differentiation of cultured human neuroblastoma cells, Cell Differentiation, 12:165.

Palfrey, C., and Littauer, U. Z., 1976, Sodium-dependent efflux of K^+ and Rb^+ through the activated sodium channel of neuroblastoma cells, Biochem. Biophys. Res. Commun., 72:209.

Palfrey, C., Kimhi, Y., and Littauer, U. Z., 1977, Induction of differentiation in mouse neuroblastoma cells by hexamethylene-bis-acetamide, Biochem. Biophys. Res. Commun., 76:937.

Scott, I. G., Åkerman, K. E. O., Heikkilä, J. E., Kaila. K., and Andersson, L. C., 1986a, Development of a neural phenotype in differentiating ganglion cell-derived human neuroblastoma cells, J. Cell. Physiol., 126:285.

Scott, I. G., Heikkilä, J. E., and Åkerman, K. E. O., 1986b, Muscarinic receptor-linked Ca^{2+} mobilisation in differentiating neuroblastoma cells, Biochem. Soc. Trans., 14:587.

Spector, I., 1981, Electrophysiology of clonal nerve cell lines, In: "Excitable Cells in Tissue Culture," P. G. Nelson and M. Lieberman, eds., Plenum Press, New York.

Spinelli, W., Sonnenfeld, K. H., and Ishii, D. N., 1982, Effect of phorbol ester tumour promoters and nerve growth factor on neurite outgrowth in cultured human neuroblastoma cells, Cancer Res., 42:5067.

Yellen, G., 1982, Single Ca^{2+}-activated nonselective cation channels in neuroblastoma, Nature, 296:357.

EFFECT OF IONS AND ATP ON THE /^{3}H/QNB BINDING TO MUSCARINIC CHOLINERGIC RECEPTORS IN RAT BRAIN SYNAPTIC MEMBRANES

Nevena Mouleshkova and Petja Naumova

Department of Human and Animal Physiology
University of Sofia
Sofia, Bulgaria

INTRODUCTION

Muscarinic cholinergic receptors /mChRs/ play a key role in mediating synaptic activity in cholinergic synapses. They are wide spread in the central nervous system.[1,2] The initiation and the development of synaptic events are characterized by essential changes in the ionic composition and concentration in the synaptic region.[3,4] On the other hand, ATP is stored and realised together with the neurotransmitter acetylcholine when exocytosis takes place.[5]

In view of these observations it seemed of interest to investigate the effect of various concentrations of Na^+, K^+, Ca^{2+} and Mg^{2+}, alone or in combination with ATP, on the binding of the specific antagonist /^{3}H/-quinuclidinyl benzilate, /^{3}H/QNB, to mChRs in rat brain synaptic membranes.

MATERIALS AND METHODS

Male albino rats weighing 150-180 g were decapitated and the brains were rapidly removed. The forebrains were isolated on Petri dishes cooled with crushed ice and homogenized in 0.32 M sucrose solution to obtain 10 % /w/v/ homogenate.

Preparation of crude synaptosomal fraction. The tissue homogenate was centrifuged at 1.000 g for 10 min. The pellet consisting of nuclei and cell debris was discarded and the supernatant was centrifuged at 17.000 g for 20 min. The pellet was resuspended in 0.32 M sucrose and recentrifuged at 17.000 g for 20 min. The pellet defined as crude synaptosomal fraction was resuspended in the incubation buffer.

Cholinergic muscarinic receptor binding assay. Muscarinic receptor binding was carried out according to the rapid filtration method of Kobayashi et al.[6] using the specific ligand /^{3}H/QNB. The samples, prepared in triplicate, were incubated for 45 min at 37°C. A duplicate set of samples containing additionally I μM atropine sulphate were likewise incubated to determine nonspecific binding. The protein content, estimated according to Lowry et al.[7] was in the range of 50-60 μg per sample. The amount of specifically bound ligand was calculated as the difference in bound /^{3}H/QNB in samples with and without atropine. All the estimations were carried out both in absence /control sample/ or presence of NaCl, KCl, CaCl$_2$ and MgCl$_2$ / all were from Merck, Darmstadt/. In some samples ATP /Boehringer, Mannheim/ was added.

RESULTS AND DISCUSSION

The in vitro effect of variations concentrations of Na$^+$, K$^+$, Ca^{2+} and Mg^{2+} on /^{3}H/QNB binding in the crude synaptosomal fraction are presented on Fig. 1. The addition of increasing concentrations of Na$^+$ or K$^+$ to the incubation medium did not affect significantly the binding of the ligand to the mChRs. However, the presence of Ca^{2+} or Mg^{2+} in the medium reduced

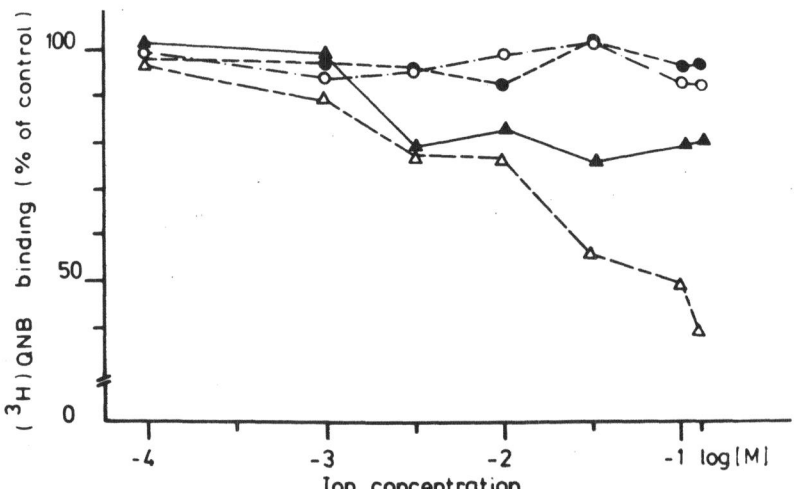

Fig. 1   Effect of Na$^+$ /O/, K$^+$ /●/, Ca^{2+} /△/ and Mg^{2+} /▲/ on /^{3}H/QNB binding to the mChRs in rat brain.

the binding. This influence was especially well expressed when studying the effect of Ca^{2+}. The inhibitory effect increased dose-dependently with the rise of the Ca^{2+} concentration /more than 50 % inhibition at 200 mM/. Mg^{2+} showed a similar effect, though less pronounced. Inhibition was first manifested at a higher concentration /5 mM/, increasing till 10 mM and remaining unchanged up to 200 mM.

Table 1.   Effect of ATP on the /^{3}H/QNB Binding to mChRs

Addition	/^{3}H/QNB binding /fmol.mg protein^{-1}/	% of control
Control	1468 ± 54	100
ATP - 0.1 mM	1393 ± 23	95
1 mM	1430 ± 48	100
3 mM	1403 ± 26	96
5 mM	1425 ± 30	97

All values are the mean ± S.E.M. of three separate experiments.

It is of interest to note that ATP itself did not affect the /^{3}H/QNB binding /Table 1/, but when added along with Ca^{2+} or Mg^{2+} it potentiated their inhibitory action /Table 2/.

Table 2.   Effect of ATP in the Presence of Ca^{2+} or Mg^{2+} on the /^{3}H/QNB Binding to mChRs

Addition	/^{3}H/QNB binding /% of control/
0.1 mM Ca^{2+}	93 ± 9
0.1 mM Ca^{2+} + 3 mM ATP	88 ± 6
1 mM Ca^{2+}	90 ± 5
1 mM Ca^{2+} + 3 mM ATP	72 ± 6
10 mM Ca^{2+}	77 ± 6
10 mM Ca^{2+} + 3 mM ATP	40 ± 3
0.1 mM Mg^{2+}	96 ± 6
0.1 mM Mg^{2+} + 3 mM ATP	88 ± 3
1 mM Mg^{2+}	95 ± 7
1 mM Mg^{2+} + 3 mM ATP	86 ± 5
10 mM Mg^{2+}	84 ± 5
10 mM Mg^{2+} + 3 mM ATP	50 ± 4

All values are the mean ± S.E.M. of three separate experiments.

This effect was most expressed at the highest Ca^{2+}/Mg^{2+}/ concentration studied /10 mM/.

Among the mechanisms which may explain the effect of ATP in the presence of divalent cations /Ca^{2+}, Mg^{2+}/ on the /^{3}H/QNB binding, the following could be suggested: /a/ a direct effect on the muscarinic receptor molecule /e.g. phosphorylation and/or binding of Ca^{2+} or Mg^{2+} to cation binding sites/, and /b/ an effect on a molecule /protein or lipid/ adjacent to the receptor resulting in a change of the receptor binding properties. In this aspect it is worth mentioning the results of Corpus and Sun[8] on the stimulation of the Ca^{2+} binding to synaptic membranes by ATP.

REFERENCES

1. Yamamura, H.J., Kuhar, M.J., Greenberg, G. and Snyder, S.H.,
    1974, Muscarinic cholinergic receptor binding: Regional
    distribution in monkey brain, Brain Res., 66:541.
2. Salvaterra, P.M. and Foders, R.M., 1979, /^{125}I/α-Bungaro-
    toxin and /^{3}H/-quinuclidinylbenzilate binding in Central
    nervous system of different species, J.Neurochem., 32:1509.
3. Katz, B. and Miledi, R., 1969, Spontaneous and evoked acti-
    vity of motor nerve ending in calcium Ringer, J.Physiol,
    203:689.
4. Hutter, O.F. and Kostial, K., 1954, Effect of magnesium
    and calcium ions on the release of acetylcholine,
    J.Physiol., 124:234.
5. Dowdall, M.J., Boyne, A.F. and Whittaker, V.P., 1974, Adeno-
    sine triphosphate, a constituent of cholinergic synaptic
    vesicles, Biochem.J., 140:1.
6. Kobayashi, R.M., Palkovits, M., Rothshild, R. and Yamamura,
    H.J., 1978, Regional distribution of muscarinic cholinergic
    receptors in rat brain, Brain Res., 154:13.
7. Lowry, O.H., Rosenbrough, N.J., Farr, A.L. and Randall,R.
    J., 1951, Protein measurement with the Folin phenol rea-
    gent, J.Biol.Chem., 193:265.
8. Corpus, V. and Sun, A., 1983, Effect of ATP on calcium
    binding to synaptic plasma membrane, Neurochem.Res., 8:501.

THE ROLE OF SOME CRYOPROTECTANTS ON THE SURVIVAL OF RAM

SPERMATOZOA. A STUDY ON THE MEMBRANE FLUIDITY

Antoaneta Ilieva, Alexander Ivanov*, and Pavel Marinov

Institute of Biology and Immunology of Reproduction
and Development of Organisms, *Central Laboratory of
Biophysics, Bulgarian Academy of Sciences, Sofia 1113
Bulgaria

INTRODUCTION

The degree of the cell damages depends from the molecular organization
of the cell membranes in the different animals, from the composition of the
protective mediums and from the regime of freezing/thawing[1,2]. The composition of the extendors influences on the sensitivity of the sperm cell again:
cold shock and have reflected on their survival, motility and morphology
after freezing. It is supposed that the cold shock induces primary damages
which ocurrs in the plasma membranes, are expressed sensibilities to the
processes of freezing/thawing and are determined by their complicated membrane organization. The mechanism by which protectants prevent membrane damages during freezing or dehydratation is not well understood. In the previous papers have been reported that the action of some cryoprotectants
took place into biological membranes[3,4]. It is belived that the protective
action of many of the natural compounds are associated with a capacity to
bound hydrogen strongly with water, even at low water contents[5]. The goal
of the present paper is to study the changes of the hydrophobic interaction:
at different temperatures in the presence of some cryoprotectants. It was
examined their protective role under deep freezing of the sperm cells from
ram and the concequence changes of the parameters of membrane fluidity of
the spermatozoa, refrigerated in the presence of milk dilutors.

MATERIALS AND METHODS

The influences of the cryoprotectants on the sperm motility and survival after deep freezing were investigated using the following diluents:
1- 0.3 M solution of raffinose, 2- 0.3 M solution of inositol, 3- 0.3 solution of sorbitol combined with 5% v/v glycerol and 20% v/v egg yolk, presented in equal concentrations in another extendors, 4- lactose-egg yolk-
glycerol, 5- raffinose-milk-egg yolk-glycerol diluent. The semen was divided to the portions, diluted 1:3, equilibrated for 4h at 0-4 °C and frozen
according to the method of Nagase-Niwa and stored at -196 °C. The pellets
were thawed at 39 °C in 2.8% Na citrate. Sperm motility was estimated microscopically in percentage at 0-90. The spermatozoa diluted with milk-glycerol diluent and in raffinose-milk-glycerol diluent were frozen by pellet
method on solid CO_2 (at -79°C) for 3 min and at -196°C in liquid nitrogen.
Fluorescent spectra of frozen and unfrozen pyrene labeled samples were compared. The morphological experiments were carried out by means of a differential staining procedure with reactive procion according to[6]. A total of

200 cells from each sample were analyzed at 100 x 8 magnification. Student's test was used for statistical analysis. The cells were incubated at room temperature with 20 µM pyrene (10 mM stock solution in ethanol). After washing of the fluorescent probe, a 0.3 M solutions containing raffinose, inositol, sorbitol, lactose were added to the cell suspension. (It must be noted that ram spermatozoa were obtained from fresh ejaculates with normal parameters, suspended and washed with Ringer-fructose buffer, according to[7], centrifugated at 800 x g for 15 min, rewashed again in PBS buffer[8], pH-7.4 and were diluted up to 1.8-2.2 x 10^7 cell/ml). All measurments were carried out at 20° C or $0-4^\circ$ C. Fluorescence emission spectra of pyrene was excited at $\lambda = 327$ nm and was recorded from 350 to 550 nm using Jobin Yvon 3YD spectrofluorimeter (slits width-4 nm). All spectra were normalized to the fluorescence intensity of control samples at 375 nm. Excimer/monomer ratio (I_e/I_m) were calculated from the fluorescence as in[9]. The data were presented as a percentage of I_e/I_m ratio in control samples.

RESULTS AND DISCUSSION

The results presented in Table 1 show that after deep freezing and thawing the higest post-thaw motility percentage remain unchanged using diluents containing raffinose and lactose. It must be noted that no any protective effect has been observed by using sorbitol even in the presence of glycerol and egg yolk. The percentage of motility spermatozoa is similar to that in cold-shocked cells. The cryoprotective effect of the raffinose was increased adding milk lipoproteins.

Table 1. Effect of storage medium of ram spermatozoa on sperm motility, survival and acrozome morphology.

medium	motility 0-4°C/4h in (%)	post-thaw motility	survival	abnormal acrosome			
				without acrosome	swallen	brocken	remote
1	68±1.8	39±0.9 (p<0.01)	280±13.3 (p<0.001)				
2	68±2.7	30±1.0 (p<0.001)	200±8.9 (p<0.001)				
3	22±1.4	10±1.1 (p<0.001)	61±0.9 (p<0.001)				
4	60±3.4	38±0.7 (p<0.001)	315±7.3 (p<0.1)	10.4±0.5	13.1±1	8±0.4	1.6±0.2
5	70±1.5	42±0.7	345±9.8	6.2±0.3 (p<0.001)	10.5±1 (p<0.1)	6±0.6 (p<0.1)	2±0.5

Mean values ± standard deviation are calculated from 10 undependent observations and statistically significant differences are found to the value in fifth mediums.

Table 1 shows the morphological damages in the acrosomal membrane in that extendors which ensure good protection. Usually the following morphological changes were registered: spermatozoa without acrosome, high number of cells with swallen and brocken acrosomal membrane, etc. (Fig. 1). Ussing raffinose-milk diluent the total acrosome changes were 7.4% less that in the control. The loss of the motility and survival under the same conditions probably reflect some membrane effects (damages, destabilization) induced by decreasing of the temperature and closely connected with changes in the medium. Fig. 2A and B and Table 2 show the effects of some components of the tested cryoprotective mediums on the fluidity properties of ram spermatozoa membranes under different temperature conditions. At 20° C the influence of all components tested is related to the considerable in-

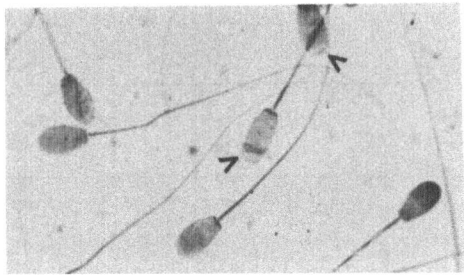

Fig. 1.
Tipical morphological anomalies in deep freezing ram spermatozoa

crease of pyrene excimerization compared to the control (Fig. 2 A). Lowering the temperature up to 0-4° C (without freezing) leads to decrease of the eximer formation and this effect does not depend from the nature of the protectants used. Nevertheles it is interesting to note that I_e/I_m ratios in all samples with cryoprotectants remain higher than in control. Some

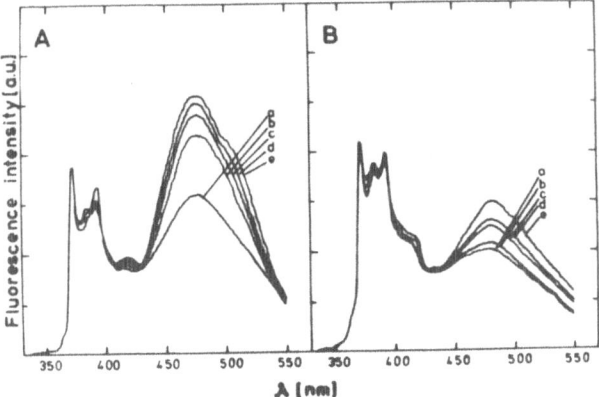

Fig. 2.
Pyrene fluorescence emmision spectra in ram spermatozoa membrane, A- at 20° C and B- at 0-4° C. a)control, b)sperms+sorbitol, c)sperms+inositol d)sperms+raffinose, d) sperms+lactose

sugars and alcohols, which favourable action is ilustrated in Table in Table 1 led to increase of the membrane fluidity at 20° C which is in agreement with their possible role as a protectants. By decreasing the temperature to 0-4° C protectants maintained a wide variety of conformational mobilities of phospholipids. The action of lactose and raffinose is better expressed with the respect to the increasing and keeping of the membrane fluidity comparing to inositol and sorbitol which is in agreement to their cryoprotective action (Table 1). We supposed that the tested nonelectrolites could be considered as a compounds which may interact with the polar head groups of the phospholipids thus stabilizing aqueous phase during dehydratation processes concominated freezing. Decreasing of the membrane fluidity at 0-4° C when no phase transition was occured could be destabilized the membrane structure and may have serious consequence on the hydrophobic interactions in adequate cryoprotection and freezing. Previous results[10] show that cryodamages of the cells were determined from the changes of the lipid phase state from the membrane hydration states.

The degree of the pyrene excimerization tends to decrease after the exposure to ultralow temperatures and warmed to the physiological temperatures undependent from usable medium (Table 2). Forming the more rogaraus

Table 2. Effect of some cryoprotectants on the fluidity of the spem membrane

Cryopro- tector	20° C	0-4° C	F I_e/I_m in (%) -79/39° C	-196/39° C
1	154±2.8	75.8±5.3		
2	165±4.7	87.5±4.4		
3	143±6.6	76.6±5.4		
4	127±2.9	61.0±6.0		
5	100	59.6±3.4		
6			78.9±3.2	77.6±2.1
7			95.5±5.6	87.6±0.5

1-sperms+raffinose, 2-sperms+lactose, 3-sperms+inositol, 4-sperms+sorbitol 5-control, 6-milk-glycerol diluent, 7-raffinose-milk-glycerol diluent

membrane structure after ultralow temperatures influences might be due to a phase transitions and lateral phase separation of the lipids in the bi-layer[10]. Combined raffinose-milk-glycerol diluent retains the membrane in more fluidity states than the milk-glycerol extendors. We assume that the milk lipoproteins may interact with the phospholipid molecules with the following changes in the processes of the lipid phase transitions. The integration into the lipid matrix probably due to the restriction of the pyrene excimerizations. The changes of membrane fluidity after cooling to 0-4° C and deep freezing in the condition of particular protection find out reflection in the displaying of irreversible structural and functional changes, appearence of the injury areas into the membrane or mechanical destructions, swelling, vacuolization etc.

REFERENCES

1. P. Fiser, L. Ainsworth and G. Langford, Effect of osmolarity of skim-milk diluent and thawing rate of ram spermatozoa, Cryowiology, 18: 399 (1981).
2. F. Tasseron, D. Amir and H. Schidler, Acrosome damage of ram spermatozoa during dilution, cooling and freezing, J. Reprod. Fert. 51:461 (1977).
3. P. Watson, The interaction of egg yolk and ram spermatozoa studied with fluorescent probe, J. Reprod. Fert. 42:105 (1975).
4. H. Hammerstedt, A. Keith, W. Snipes, R. Amann, D. Arruda, L. Griel, Use of spin labels to evaluate effects of cold shock and osmolarity on sperm, Biol. Reprod. 18:686 (1978).
5. F. Franks, Solute-water interactions:dopolyhydroxy components after the properties of water, Cryowiology. 20:355 (1983).
6. E. Chakarov and M. Mollova, A one-act differential stain of the acrosome with active dyes, J. Reprod. Fert. 48:245 (1976).
7. T. Mann and C. L. Mann, "Male Reproductive Function and Semen", Springer-Verlag, Berlin-Heidelberg/New York (1981).
8. G. Edelman and C. F. Millette, °olecular probes of spermatozoan structure, Proc. Natl. Acad. Sci. USA 68:2436 (1971).
9. H. Galla and E. Sackmann, Lateral diffusion with hydriphobic region of membranes use pyrene excimers as optical probes, Biochim. Biophys. Acta 339:103 (1974).
10. P. Quinn, A lipid-phase separation model of low temperature damage to biological membranes, Cryobiology 22:128 (1985).

REGULATION OF THE QUANTUM YIELD OF PHOTOSYNTHESIS IN CHLORELLA DURING NUTRIENT LIMITED GROWTH

Pavel S. Venediktov

Biophysics Department, Faculty of Biology
M.V.Lomonosov State University
Moscow, 119 899

INTRODUCTION

Remarkable progress has been done in recent years in under-standing of general features of structural and functional orga-nization of the primary reactions of photosynthesis in the chloroplast /1/. Many of the features of such a typical chloroplast can however dramatically be affected by environmental factors during plant growth. The most known are changes in photosynthetic apparatus during adaptation to light conditions /2,3/. Much less is known about changes in the organization of primary photosynthetic reactions induced by other factors.

The data presented in this work show that in nutrient-limited Chlorella cells the quantum yield of photosynthesis is lowered and the fluorescence yield increases due to modification of the reaction centre of PS2 probably leading to blocking of electron transport from pheophytin to primary quinone acceptor.

RESULTS AND DISCUSSION

The suppresion of Chlorella cell division in growth medium containing no nitrogen or phosphorus induces the same set of changes in the photosynthetic apparatus. The content of active reaction centers of PS1 and 2 decreases and the Chl a/b ration decreases from 3 to 1.5-2.0 indicating the erichment of resting cells in light-harvesting Chl a/b-protein complex. Inspite of the greater apparent size of antenna, photosynthesis at low light intensities is decreased thus impling a lower quantum efficiency of photosynthesis in resting cells.

Variable fluorescence $/F_v/$ from such cells is suppressed because in resting cells with open PS2 reaction centers the fluorescence yield $/F_o/$ increases to a level of closed reaction centers $/F_{max} = F_v + F_o/$ /Fig.1/. This enhanced level of F_{max} is quenched to its initial level when the resting cells are illu-minated with intense light /photoinhibition/ /Fig.2/ which is known to induce the radiationless deactivation of excitons in the reaction center Chl /4/. Therefore the high F_o level in resting cells is not associated with the disconnection of the

223

light-harvesting antenna from the PS2 reaction centers but can be related to inefficient trapping of excitons in the reaction centers. The contribution of the slow component with half-time of 2.5 ns to the decay kinetics of picosecond fluorescence

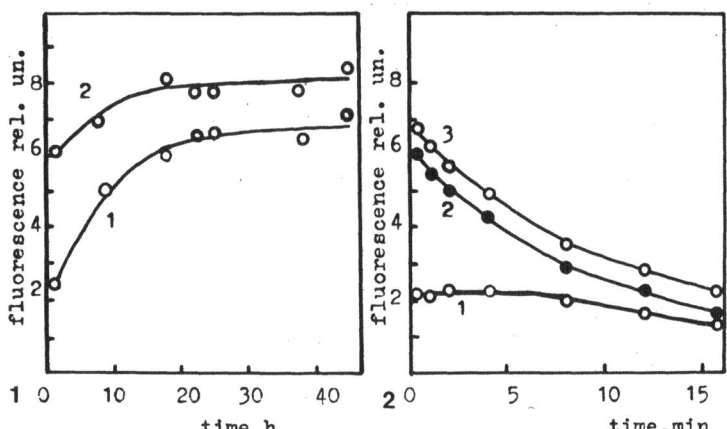

Fig. 1   Changes of Chl fluorescence yield during nitrogen star-
tion of Chlorella. /1/ F_O - fluorescence yield probed
with weak exciting flashes of 1 s duration; /2/F_γ -fluo-
rescence yield probed with the same flashes in cells
illuminated by conditions light in the presence of
10 μM diuron.

Fig.2   Kinetics of fluorescence yield changes during intense
illumination of Chlorella cells. /1/ F_O - normal cells;
/2/ F_O, after 48 h of nitrogen starvation; /3/ F_{max},
normal cells.

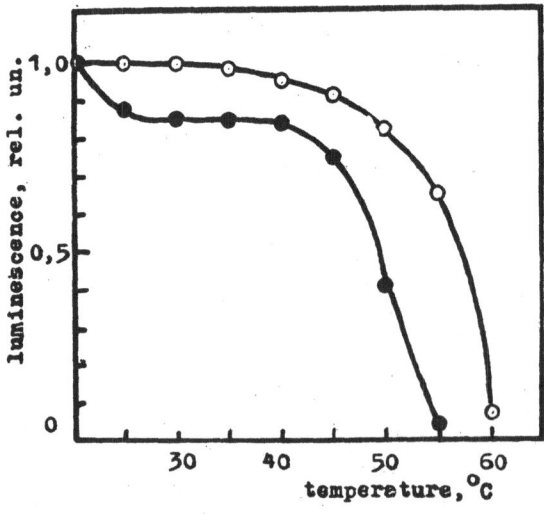

Fig.3   Temperature dependence of the long-lived delayed fluorescence of Chlorella cells. /o/ normal cells; /•/ cells
after 48 h of nitrogen starvation.

increased at the level F_o from 0.17 to 0.67. This component originates from the recombination of reduced pheophythin and the oxidized reaction center Chl of PS2 /5/. Apparently the electron transfer from pheophytin to the primary quinone acceptor is blocked in non-diving cells.

The changes in fluorescence characteristics are reminescent of free fatty acid action on chloroplast membranes /6/. The progressive decrease of thermostability of the oxygen-evolving sytem characteristic of fatty acid action /7/ and reflected in decrease of the heat inactivation tempereture of the long-lived delayed fluorescence of Chl /Fig.3/ make the similarity even more pronounced.

These changes in the photosynthetic apparatus are not a result of cell death for they are readily reverted when the starved cells are placed in the full medium and, moreover, can be seen in continuously dividing cells whose growth is limited by low concentration of nutrients. Drop of the light intensity to the growth-limiting level reestablishes high efficiency of PS2 photochemistry in such nutrient-limited cells.

Increase of recombination fluorescence yield can be prevented if photosynthesis is suppressed due to lack of CO_2 or by addition of diuron. On other hand, the changes induced in the resting state become stronger in CO_2-enriched air. Therefore, the changes are triggered by the accumulation of CO_2 reduction products under conditions when they can not be used for growth.

It is evident that in algal cells the quantum yield of PS2 photochemistry is regulated to adjust photosynthetic electron transport to growth conditions. The accumulation of intermediates of carbohydrate synthesis during nutrient-limited growth does not only represses synthesis of new chloroplast proteins /7,8/ but triggers also events leading to the inactivation of the existing PS2 reaction centers in a process analagous to catabolite inactivation of enzymes in yeasts /9/. The inactivation is of adaptive significance, preventing light saturation of electron transport due to limitation by reactions using photoreductans in dark carbon matabolism. Such a condition may lead to accumulation of reduced acceptors of PS1 and to reduction of O_2, which initiates a cascade of hazardous oxidative reactions /10/.

REFERENCES

1. Murphy, D.J., 1986, The molecular organization of the photo-synthetic membranes of higher plants, Biochim.Biophys.Acta 864:33.
2. Bjorkman, O, 1981, Responses to different quantum flux densities, in: "Encycl.Plant Physiol." , O.L.Lange et al., ed., Springer, Berlin etc.
3. Fork, D.C. and Satch, K., 1986, The control by state transitions of the distribution of excitation energy in photosynthesis, Ann.Rev.Plant Physiol. 37:355.
4. Powels, S.B., 1984, Photoinhibition of photosynthesis induced by visible light, Ann.Rev.Plant Physiol.35:15.
5. Klimov, V.V. and Krasnavsky, A.A., 1981, Pheophytin as a primary electron acceptor in photosystem 2 reaction centers, Photosynthetica 15:592.
6. Warden, J.T. and Csatorday, K., 1987, On the mechanism of linolenic acid inhibition in photosystem 2, Biochim.Biophys. Acta 890:215.
7. Shihira-Ishikawa, I. and Hase, E., 1965, Effects of glucose

on the process of chloroplasts development in Chlorella pro-
tothecoides, Plant CellPhysiol.6:101.

8. Semenenko, V.E., 1978, Molecular biological aspects of photo-
    synthesis regulation, Sov.Plant Physiol.25:903.
9. Holzer, E.F., 1976, Catabolic inactivation in yeast, Trends
    Biochem.Sci.1:178.
10. Elstner, E.F., 1982, Oxygen activation and oxygen toxicity,
    Ann. Rev. Plant Physiol.33:73.

SURFACE CHARGE DENSITY AND ROTATIONAL DIFFUSION OF LIPOSOME RECONSTITUTED PHOTOSYSTEM I REACTION CENTRES STUDIED BY SPIN-LABEL TECHNIQUE

Tsvetan Gantchev

Faculty of Biology
Sofia University
8. D.Tsankov blvd., 1421 Sofia, Bulgaria

Due to certain methodical and technical complexity, very little is known about the dynamic behaviour of integral proteins within native membranes. One promising approach to studying membrane macromolecular mobility is the isolation, purification, selective labelling with molecular probes such as spin-labels and final reconstitution in artificial membranes. This techniques was applied to study the rotational diffusion of photosynthetic reaction centres in liposome membranes with respect to membrane surface and light-induced transmembrane potentials. The results obtained at this stage of experiments imply the existence of certain relations between rotational correlation times /τ_c/ of protein moiety and membrane potentials.

MATERIAL AND METHODS

Photosystem I Reaction Centres /PSI RC/ were prepared from spinach chloroplasts according to Bengin and Nelson.[1] Proteoliposomes were obtained by sonication of soybean phospholipids /azolectin/ and PSI RC for 15-30 min under Ar flow or by detergent-removal technique using gelfiltration on AH-Sepharose 4B column.[5] Some of these preparationa were able to generate transmembrane electrochemical potential /$\Delta\mu H^+$/ in the presence of the cyclic electron carrier phenazine methosulphate /PMS/.[2] The generation of ΔpH was measured by quenching of 9-aminoacredine fluorescence. Membrane surface potentials were determined using spin-labelled amphiphile CAT$^+$9.[3] Isolated PSI RC were covalently modified by maleimide spin-label in media containing: 20 mM MOPS /pH 7.8/, 50 mM NaCl, 1 mM EDTA, 0.02 % Triton X-100. Under these conditions 2-3 cystein SH-groups of the protein complex werealkylated with the spin-label. Rotational correlation times were measured using saturation energy transfer technique according to Thomas et al.[4]

RESULTS AND DISCUSSION

Reconstruction of PSI RC into liposomes by sonication procedure leads to formation of small unilamellar proteoliposomes containing 4-5 mol.RC per vesicle, while gel-filtration technique results in 10-15 mol.RC per vesicle and as it was shown by negative staining electron microscopy, the average diameter is three times greater than that of the former ones. The orientation of reconstituted RC in respect to the vesicle exterior was investigated by ferricyanide oxidation of the primary donor P_{700}. As it is seen from Table 1, sonicated proteoliposomes have a tendency for "right" orientation /PSI RC oriented with their acceptor side to the bulk aqueous phase/. An opposite orientation predominates for liposomes prepared by detergent-removal technique and no detectable transmembrane electrochemocal potential /positive inside/ was observed for this preparations /Table 1/.

TABLE 1

LIPOSOMES	% OXIDAZED P700 by FeCy	$\Delta \tau$/uC/cm/^{2} per mol.RC	Δ pH INDUCED
SONICATED 2R=45nm	12	-0.43	2.5
GEL - FILTR. 2R=150nm	87	+0.12	-

The dynamic partition of the spin-labelled amphiphile CAT 9 between the membrane and aqueous phases was used to determine the membrane surface potentials /Ψ/ and surface charge density /σ/ of poteoliposome preparations /Fig.1/. The outer surface charge density of sonicated poteoliposomes was found to be $-2.3 \mu C/cm^2$. The reconstruction of PSI RC increases this value and this increase depends on the reconstruction stoichiometry and orientation of the RC complex within the membrane. It can be easily calculated that for sonicated preparations the increase of the surface charge density is $-0.43 \mu C/cm^2$ per reconstituted RC complex. As is seen from Fig.1, when reconstitution is made by gel-filtration technique a decrease of the negative surface potential occurs. This fact can be explained taking into account that on the acceptor side of PSI RC complex positively charged aminoacid side chains are exposed to the bulk phase. ESR spectra /first and second derivatives/ of liposomes containing spin-labelled PSI RC are shown on Fig.2. Maleimide spin-labes are tightly immobilized and correlation time /τ_c/ of about 350-400 ns was measured in the absence of light-induced tramsmembrane potential. A certain increase of rotational freedom was observed when Δ pH was created by light and PMS. Parallel decrease of τ_c of 12-doxyl steric acid, used as a spin-probe to monitor the fluidity of the phospholipid bilayer, was observed /not shown/. Thus the decrease of the SL-RC τ_c can be related to the increased fluidity of the phospholipid double layer upon generation of Δ pH. When 3 mM MgCl was present in the reaction media and upon longer light illumination /30 min/, a very strong immobilization of the spin-labelled RC complex is obvious /Fig.2/. The calculated τ_c in this case is about 300 μs /one order of magnitude

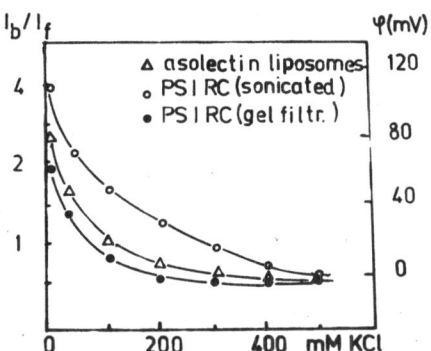

Fig.1   Surface membrane potentials /Δψ/ of:Δ - asolectin liposomes; Ο - PSI RC proteoliposomes prepared by sonication;
● - PSI RC proteoliposomes prepared by gel-filtration.
Measurements were done using the phase distribution of
the charged quanternary spin-labelled amine CAT^+9.

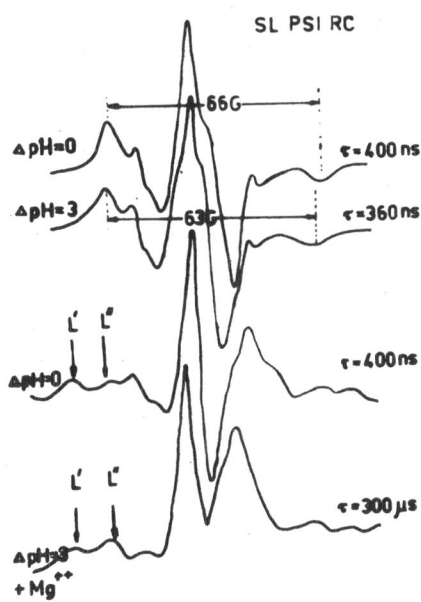

Fig.2   ESR spectra - first and second /Saturation energy tran-
sfer/ derivatives of SL PSI RC in liposomes. Generated
pH and additives as indicated.

larger than when divalent cations are lacking/. The only expanation is that RC complexes undergo irreversible aggregation or cross-linking under these conditions. It is notable that this strong immobilization of membrane-imbedded PSI RC is prevented by addition of 5 % cholesterol to the preparation before the sonication and such increase of the rotational correlation time was not observed with proteoliposomes prepared by gel-filtration.

From the experiments described above it is clear that changes of surface and transmembrane potentials together with bilayer fluidity can moderate PSI RC macromolecular dynamic behaviour in membrane media. At this stage of the experimental work it is not clearly understood whether the surface charge screening by divalent cations or the formation of coordination bonds with the exposed protein negative charges is the main reason for protein aggregation. It is also doubted whether these strong macromolecular mobility changes are proper to the native thylakoid membrane-embedded PSI RC or whether they are a reconstitution artefact.

REFERENCES

1. Bengis, C., Nelson, N., 1977, Subunit structure of chloroplasts photosystem I reaction center, J.Mol.Biol., 252, pp. 4564-4569.
2. Orrich, G., Hauska, G., 1980, Reconstitution of photosynthetic energy conservation; Proton movements in liposomes conatining reaction center of photosystem I from spinach chloroplasts, Eur.J.Biochem., 111, pp.525-533.
3. Castle, D., Hubbell, W., 1976, Extimation of membrane surface potentials and charge density from phase equilibrium of a paramagnetic amphyphile, Biochemistry, 15, pp. 4818-4831.
4. Thomas, D., Dalton, L., Hyde, J., 1976, Rotational diffusion studied by passage saturation transfer EPR, J.Chem. Physics, 65, pp.3006-3024.
5. Taguchi, T., Kasai, M.m., 1983, A new method of vesicle formation by salting-out and its application to the reconstitution of sarcoplasmatic reticulum, Biochem.,Biophys.Acta, 729, pp.229-236.

STRUCTURAL AND FUNCTIONAL CHARACTERIZATION OF THE EFFECTS OF

DETERGENT TREATMENT ON CHLOROPLAST MEMBRANES

Emilia Apostolova and Alexander Ivanov

Central Laboratory of Biophysics
Bulgarian Academy of Sciences
Sofia 1113, Bulgaria

INTRODUCTION

Detergents are widely used for the isolation and investigation of biological membranes and their protein components. Treatment of the membrane with detergents leads to structural changes which are associated with a specific detergent concentrations. The detergent-induced changes in chloroplast organization involve the rearrangement of membrane components and are accompanied by the lose of certain polypeptides.[1] The photochemically active particles from photosynthetic membranes obtained by detergent treatment have a different photochemical activity depending on the detergent used.[2] It can be supposed that the detergent in concentration below CMC intercalates in the membrane and alters both the membrane organization and the photochemical activity of the chloroplasts. In this study the influence of ionic /SDS/ and nonionic /digitonin and Triton X-100/ detergents on the chloroplast membrane properties and function is compared.

MATERIAL AND METHODS

Intact pea chloroplasts were isolated as described in.[3] The chloroplasts were incubated for 30 min in a reaction medium containing: 66 mM phosphate buffer /pH 7.3/, 0.4 M sucrose, 100 mM NaCl, 10 mM $MgCl_2$, 1 mg chl/ml and detergent at different concentrations. The photochemical activity of PS II / H_2O to DCPIP and DPC to DCPIP / was measured spectrophotometrically at 580 nm in a medium: 66mM phosphate buffer /pH 7.3/, 0.4 M sucrose, 100 mM NaCl, 10 mM $MgCl_2$, 0.01 mM DCPIP and 20 μg chl/ml. In some samples DPC was added to a final concentration of 0.5 mM. Chlorophyll fluorescence emission was excited with 436 nm monochromatic light and registrated at 685 nm using interference filter /IF 685/ and red sensitive photomultiplier. The membrane microviscosity was measured by ESR spectroscopy using spin-labelled stearic acid /SSL/ as described in.[3]

RESULTS AND DISSCUSION

Our previous results show that detergent treatment of the chloroplast membranes leads to selective solubilization of the membrane components /lipids and proteins/, depending on the type of the detergent used.[4] Rearrangement of the membrane components as a result of the solubilization and/or incorporation of the detergent molecules is accompanied by changes of the membrane properties.

Table 1. Effect of detergents on Mg^{2+} - induced chlorophyll fluorescence increase in pea chloroplasts. The relative size of the fluorescence increase is presented as a percentage of the $\Delta F/F_O$ level in control chloroplasts/F_O - fluorescence intensity before Mg^{2+} addition;ΔF - change observed after salt addition/. The $t_{1/2}$ corresponds to the time for increase to $\Delta F/2$.

Sample	Concentration /μmoles/mg chl/	Fluorescence	
		$\Delta F/F_O$	$t_{1/2}$/sec/
Control	0	100	14.0
Digitonin	0.081	95	15.0
	0.570	59	18.0
Triton X-100	0.081	81	20.0
	0.570	64	20.0
SDS	0.081	119	16.0
	0.570	93	21.0

From Fig. 1 it can be seen that the mobility of SSL decrease in the chloroplasts treated with both nonionic and ionic detergents, which indicates an increased membrane microviscosity. These changes are observed at concentrations below critical

Table 2. Influence of the detergent treatment on DCPIP photoreduction in pea chloroplasts

Concentration / moles/mg chl/	DCPIP reduced / % of control /					
	Digitonin		Triton X-100		SDS	
	-DPC	+DPC	-DPC	+DPC	-DPC	+DPC
0.	100.0	117.6	100.0	117.6	100.0	117.6
0.24	101.3	117.8	130.0	135.2	81.7	106.1
0.81	104.9	122.1	65.0	34.6	22.7	28.2
1.60	63.8	80.5	44.7	72.9	9.9	14.2
4.00	49.8	68.2	30.9	61.2	0	0

100 % correspond to 44.3 moles reduced DCPIP/mg chl per hour.

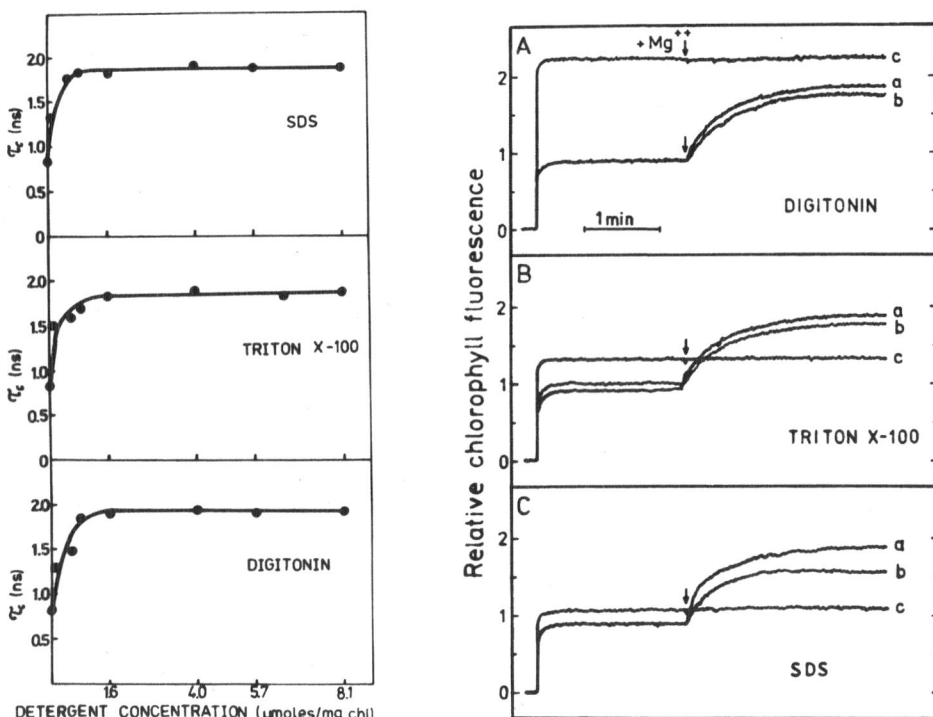

Fig.1 Effect of digitonin, Triton X-100 and SDS on the behaviour
of SSL incorporated in the chloroplasts. The spectra were
recorded at 9.05 GHz with 27 min scan time over 200 G,
magnetic field - 3440 G, time constant 0.5 sec, at room
temperature

Fig.2 Effect of digitonin /A/, Triton X-100 /B/ and SDS /C/ on
the salt-induced chlorophyll fluorescence. For all.deter-
gents: /a/ control, /b/ below CMC, /c/ above CMC.Chloro-
plasts were suspended in low salt medium of 0.01 M tricine
/pH 8.0/, 0.33 M sucrose and 0.0005 M EDTA. Chloroplast
concentration in the sample was equivalent to 5 g chl/ml.
Mg^{2+} was added to a final concentration of 5 mM as
indicated.

micellar concentration /CMC/, above this concentration furthure changes are not observed. Earlier observations show that the incorporation of relatively rigid molecules leads to a decrease in the membrane fluidity.[5,6] It is supposed that the loss of membrane components and incorporation of detergent molecules lead to changes in the membrane fluidity. The observed reduced size and kinetic values of Mg^{2+} -induced increase of the chloro- phyll fluorescence in all cases of chloroplasts treated with detergent concentration below CMC / Fig.2 and Table 1 / most probably correlate with the decreased fluidity under the same condition / see Fig. 1/. These results are in full agreement with the data in[5,6], which indicate that the salt-induced in- crease of fluorescence seems a result of a stacking process strongly depending on the fluidity of thylakoid membranes. On the other hand, the observed much higher initial levels and completely suppressed ability to increase the fluorescence on addition of Mg^{2+} in digitonin- and Triton X-100 treated chlo- roplasts /at concentrations above CMC - 1.6 μ moles per mg chl/ allow us to assume that under these conditions the membrane fractions isolated at 3000xg for 12 min contain mainly the appressed areas of the thylakoids. This assumption is further supported by the decrease of the chl a/chl b ratio[4] and by the reported remarkable decrease in the fluidity of the appressed grana stack.[7] The latter observation is in full agreement with our fluidity data for all detergents used at concentrations above CMC.

The results in Table 2 show that the detergents used /digitonin, Triton X-100 and SDS/ have a different effects on the PS II activity /H_2O to DCPIP/ at concentration below CMC. This activity is not changed by digitonin, it is stimulated by Triton X-100 and inhibited by SDS. The data of Vernon and Shaw[8] are in agreement with our experimental results concerning the stimulation of DCPIP reduction in Triton X-100 treated chloroplasts, while Benesova et al.[9] have not observed a stimu- lation of this process. This contradiction with the data of Benesova et al. could be explained by the higher initial detergent/chlorophyll ratio in their experiments than in our study. All detergents used at concentrations above CMC lead to a considerable inhibition of electron transport from water to DCPIP /Table 2/. The addition of DPC leads to an increase in the PS II reducing capacity by 20-30 per cent. It could be supposed that the reduction of the PS II activity is due to the destabilization of electron donor site/s/ /water split- ting complex/ of PS II. In SDS-treated chloroplasts the addition of DPC does not cause an increase in the rate of DCPIP reduction. In this case the changes in the PS II acti vity are most probably due to the destruction of the PS II reaction centre. The different effect of the detergent used on the PS II act-vity can be explained by the selective solu- bilization of the membrane components.[4] On the other hand some direct influence of the fluidity /below CMC/ on the PS II activity cannot be excluded. Bearing in mind that a diffusion- -limited step between an O_2-evolving complex and PS II reac- tion centre has been proposed,[10] it is possible to assume that the effectively restricted lateral mobility of protein complexes could be responsible to some extent for the observed inhibition of the DCPIP-photoreduction rate.

On the basis of the experimental results in this study it could be supposed that: /1/ the detergents in concentrations below CMC induce a change in PS II activity, which is a result of destabilization of the membrane components and decreased membrane fluidity, /2/ the changes above CMC are caused by detachment of membrane fragments.

REFERENCES

1. Lyon, M.K. and Miller, K.R., 1984, Ultrastructural characterization of the effects of detergent treatment on stacked thylakoids, J.Ultrastruct.Res. 88:229.
2. Dunahay, T.G., Staehelin, L.A., Seibert, M., Ogilvie, P.D. and Berg, S.P., 1984, Structural, biochemical and biophysical characterization of four oxygen-evolving photosystem II preparations from spinach, Biochim.Biophys.Acta 764:179.
3. Apostolova, E.L. and Kafalieva, D.N., 1985, Changes in the microviscosity of chloroplasts after glutaraldehyde treatment, Compt.rend.Acad.bulg.Sci. 38:1061.
4. Apostolova, E. and Kafalieva, D., 1985, Photochemical activities of chloroplasts after detergent treatment,in:"Proceedings of Fifth International Seminar on Energy Transfer in Condensed Matter", P.Pancoska and J.Pantoflicek, eds., Prague 183.
5. Jamamoto, J. Ford, R.C. and Barber, J., 1981, Relationship between thylakoid membrane fluidity and the functioning of pea chloroplasts, Plant Physiol 67:1069.
6. Pedersen, I.J. and Cox, R.P., 1984, Relationship between thylakoid membrane fluidity and the kinetics of salt induced fluorescence changes: A spin label study, Adv.Photosynth.Res. 3:51.
7. Ford, R.C., Chapman, D.J., Barber, J., Pedersen, J.Z. and Cox, B.P., 1982, Fluorescence polarization and spin-label studies of the fluidity of stromal and granal chloroplast membranes, Biochim.Biophys.Acta 681:145.
8. Vernon, L.P. and Shaw, E., 1965, Photochemical activities of spinach chloroplasts following treatment with the detergent Triton X-100, Plant Physiol.40:1269.
9. Benesova, H., Hladik, J., Kaplanova, M. and Sofrova, D., 1984, Photochemical activity of photosynthetic pigments in thylakoid membranes and Triton X-100 micelles, Photobiochem. Photobiophys.7:15.
10.Packham, N.R. and Barber, J., 1984, Stimulation by manganese and other divalent cations of the electron donation reactions of photosystem II, Biochim.Biophys.Acta 764:17.

THE HEAT STABILITY OF THE PHOTOSYNTHETIC APPARATUS DURING THE PROCESS OF HIGH TEMPERATURE ACCLIMATION

Atanaska Andreeva, Mariela Marinova and I. Yordanov

Faculty of Physics, Sofia University, 1126 Sofia, Bulgaria
Institute of Plant Physiology, Bulg. Academy of Sciences
1113 Sofia, Bulgaria

INTRODUCTION

The high temperature acclimation of higher plants results in an improved photosynthetic performance at high temperatures[1]. To evaluate the extent of this acclimation under controlled temperature conditions[2], we investigate the heat stability of the photosynthetic apparatus of plants by measuring the dependence of the chlorophyll a fluorescence kinetics at room temperature on high temperature treatments. As it is known, the non-monotonic behaviour of the chlorophyll a fluorescence kinetics at room temperature in dark adapted native photosynthetic system reflects mainly the state of the photosynthetic apparatus and the changes in electron-transport reactions[3-6].

MATERIALS AND METHODS

The high temperature acclimation of the investigated plants (french bean) was performed under controlled conditions described in Table 1.

Chlorophyll a fluorescence was measured using a spectrofluorimetric set-up eqquiped with a halogen lamp. The exciting and analysing monochromators were grating Specol 11 (Carl Zeiss, Jena). The fluorescence was excited with light with wavelength of 480 nm and measured in a right-angle arrangement with a photomultiplier M12F VC51 (RFT). The chlorophyll a fluorescence kinetics were recorded using compensative recorder K101 (Carl Zeiss, Jena) at wavelength of 685 nm (the maximum in the emission spectrum at room temperature) after dark adaptation for 10 minutes. The high temperature treatments before the measuring of the chlorophyll a fluorescence kinetics last 3 minutes at the following temperatures: $40^{\circ}C$, $45^{\circ}C$, $50^{\circ}C$, and $55^{\circ}C$.

237

Table 1. Scheme of controlled temperature conditions of plant growth

N	Plant	Conditions of plant growth at days			
		I day	II day	IIIday	IVday
1	control (K)	6h at 25°C			
	non-acclimated (NA)	6h at 45°C			
2	control (K)	at 25°C	at 25°C		
	non-acclimated (NA)	at 25°C	6h at 47,5°C		
	acclimated (A)	6h at 45°C	6h at 47,5°C		
3	control (K)	at 25°C	at 25°C	at 25°C	
	non-acclimated (NA)	at 25°C	at 25°C	6h at 50°C	
	acclimated (A)	6h at 45°C	6h at 47,5°C	6h at 50°C	
4	control (K)	at 25°C	at 25°C	at 25°C	at 25°C
	non-acclimated (NA)	at 25°C	at 25°C	at 25°C	6h at 52,5°C
	acclimated (A)	6h at 45°C	6h at 47,5°C	6h at 50°C	6h at 52,5°C

RESULTS

As a measure for the heat stability of the photosynthetic apparatus of the investigated plants we use the ratio between the value of the highest maximum in the non-monotonic chlorophyll a fluorescence kinetic curve (F_p) and the value of the steady state in the same curve (F_t) - F_p/F_t. The dependence of the F_p/F_t ratio for control (K), non-acclimated (NA), and acclimated (A) plants on the high temperature treatment are shown in the Figs 1-4. Figs 1-4 are for plants grown under temperature regimes №1-4 described in Table 1, respectively.

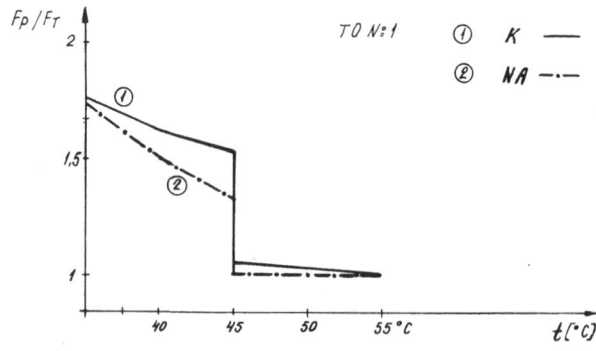

Fig. 1. The temperature dependence of the ratio F_p/F_t for plants grown under temperature regime №1 described in Table 1.

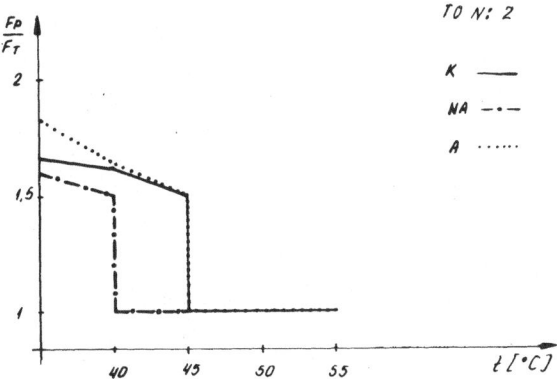

Fig. 2. The temperature dependence of the ratio F_p/F_t for plants grown under temperature regime N 2 described in Table 1.

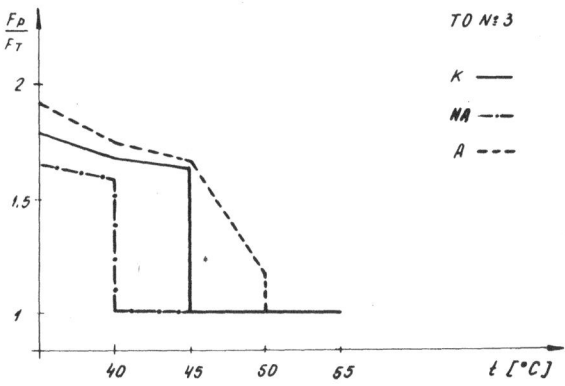

Fig. 3. The temperature dependence of the ratio F_p/F_t for plants grown under temperature regime N 3 described in Table 1.

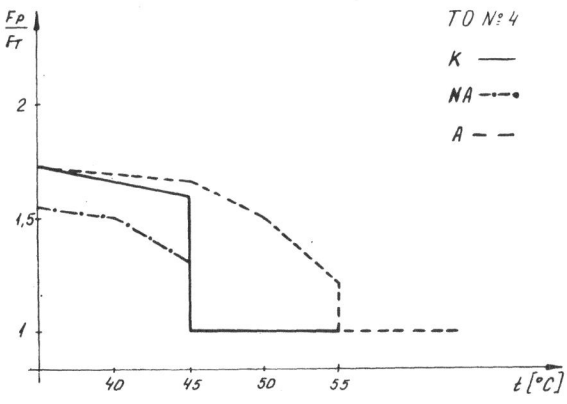

Fig. 4. The temperature dependence of the ratio F_p/F_t for plants grown under temperature regime N 4 described in Table 1.

CONCLUSIONS

Our experimental results show that the chlorophyll a fluorescence
kinetics maintains its non-monotonic behaviour in high temperature
acclimated plants in the temperature range up to 50 - 55°C, while for non-
acclimated plants this limiting temperature value is 40°C. Thus, the
proposed temperature acclimation regime leads to an increased heat stability
of the photosynthetic apparatus of higher plants.

REFERENCES

1. J. Berry and O. Björkman, Photosynthetic response and adaptation
   to temperature in higher plants, Ann. Rev. Plant Physiol.31:491
   (1980).

2. I. Yordanov, Resistance of the photosynthetic apparatus of intact
   bean plants to raised temperatures and their effect on the
   metabolism of photoassimilated ^{14}C, Physiol. rast. V:3 (1979).

3. H. Kautsky, W. Appel and N. Amann, Chlorophyllfluorescenz und
   Kohlensäureassimilation. XIII. Mitteilung., Biochem. Z. 332:277
   (1960).

4. J. Lavorel and A. Etienne, In vivo chlorophyll fluorescence, in:
   "Primary Processes of Photosynthesis", J. Barber, ed., Elsevier,
   Amsterdam-New Jork- Oxford (1977).

5. J. Barber, Ionic regulation in intact chloroplasts and its
   effect on primary photosynthetic processes , in: "The intact
   Chloroplast", J. Barber, ed., Elsevier, Amsterdam -New Jork-
   Oxford (1976).

6. A. Andreeva, Chlorophyll fluorescence and EPR signal I kinetics
   during dark-light transition in whole leaves of higher plants,
   Photobiochem. Photobiophys. 4:17 (1982).

TEMPERATURE DAMAGE OF THE OXYGEN-EVOLVING

COMPLEX IN THYLAKOID MEMBRANE PARTICLES

Mira Busheva and Antoaneta Popova

Central Laboratory of Biophysics
Sofia, 1113, Bulgaria

INTRODUCTION

The temperature sensitivity of photosynthetic processes has been known for a long time. Its complexity is also well known. In the physiological range 10 to 40° C the temperature response reflects Boltzmann's factor of kinetic process and is reversible (1). Above these temperatures irreversible effects result due to thermal denaturation (2).

Electron transport mediated by PS II is particularly susceptible to heat stress. It was initially attributed to damage of oxygen evolution system (3). However, Gounaris (4) observed a partial restoration of electron transport on adding a substitute electron donor as DPC to heat chloroplasts, indicating the involvement of more fundamental changes of PS II. Analyses of changes in chlorophyll a fluorescence (5) showed that increased temperature led to a separation of antennae-protein complex from the central core of the LHC II. A detachment of LHC II from PS II has been proved recently by freeze fracture studies (6).

As a rule, the chloroplast thylakoids are more sensitive to heat than are either the chloroplast envelope, or the leaf (7). During the last few years, preparations of PS II particles have proved to be very useful for the studies of PS II, especially the mechanism of water oxidation.

Here we focus on heat stress upon the oxygen-evolving function and corresponding changes in the Chl fluorescence of PS II oxygen-evolving sub-chloroplast particles (PS II OEP).

MATERIALS AND METHODS

Chloroplast and PS II OEP were isolated from 14 days old pea seedlings.
Broken chloroplasts were isolated in a grinding medium containing 0,4 M sucrose, 4 mM MgCl$_2$, 10 mM MES (pH 6,5), 2 mM ascorbate. After grinding 3 times for 30 sec the muss was filtered through 8 layers of cheese cloth and was centrifugated for 1 min at 1 000 x g. The supernatant was centrifugated for 5 min at 2 000 x g. The pellet was washed with the same medium but without ascorbate.

ABBREVIATIONS: PS II OEP = oxygen-evolving subchloroplast particles; BQ = 1,4 benzoquinone; DPC = diphenylcarbazide; PS I, II = photosystem I, II; LHC II = light harvesting chl a/b complex of PS II; chl = chlorophyll.

PS II OEP were isolated by the method of Yamamoto et al. (8).

The activity of PS II was measured by the rate of oxygen evolution in the reaction (water-BQ) recorded by a Clark-type electrode at room temperature.

The light was white and saturating. The concentration of BQ was 0,45 mM, DPC - 0,5 mM. The concentration of chloroplasts and PS II OEP was 50 µg chl/ml.

Fluorescence emission and excitation spectra were measured at room temperature on Jobin - Ivon spectrofluorimeter. Fluorescence emission was excited with 480 nm light. The excitation spectra were recorded at 685 and 722 nm. Chlorophyll concentration was 15 µg chl/ml. Determination of chlorophyll was done by (9). Assay medium was 330 sucrose, 4 mM $MgCl_2$, 10 mM MES (pH 6,5).

RESULTS AND DISCUSSION

The Yamamoto et al. (8) PS II particles showed high steady-state O_2 evolution (150 - 400 mol O_2 mg/chl.h) but contained a small contamination of PS I according to studies in (10). We measured PS I activity (dichlorphenol indophenol to methylviologen) in PS II OEP. It was only 0,3 % of the one in chloroplasts, which suggests that particles have some inactive (in terms of oxygen uptake) tightly bound PS I.

Incubation of pea chloroplasts and PS II OEP for 5 min at different temperatures leads to a marked loss of PS II mediated electron transport. The rate of oxygen evolution in the reaction water - BQ for both preparations is shown on Fig. 1. Chloroplasts have a typical plot of inactivation with heat range 25 - 50° C (4,11). The similarity of the kinetic and threshold temperature in chloroplasts and PS II OEP indicates that destruction of thylakoid membrane does not affect the primary electron flow through PS II. When DPC is used as an electron donor for PS II, it restores the O_2 evolution at 45 ° C in a close range 640 % for chloroplasts, 30 % for PS II OEP; Table I), which shows higher temperature sensitivity of PS II OEP.

The fluorescence emission spectra at room temperature of PS II OEP is similar to the one of chloroplasts well known from literature data (12). They have their maximum at 685 nm and a broad band at 720 nm (not shown). While at 20° C PS II OEP show an enhancement of F 685, due to inactive PS I, in heat stress particles chl fluorescence is 20 % weaker than in chloroplasts (Table 1).

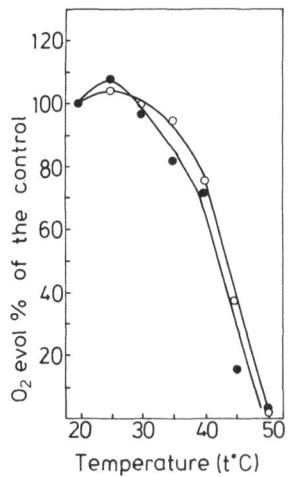

Fig.1. Effect of heat treatment on the rate of O_2 evolution in chloroplasts (o) and PS II OEP (•).

Table I. PS II response on thermal stress 5 min at 45 - 50° C, cooled and measured at 20° C. Values are given in % of control (20° C).

	PS II rates [a] with DPC	F_{685} [b]	F_{720}	Absorbance at 680 nm
chloroplasts	40 (%)	65	63	92
PS II OEP	30	45	52	85

[a] Values are calculated for heat treatment at 45° C.

[b] F 685, F 720 = rel. intensity of fluorescence emission at 685 and 720 nm, respectively.

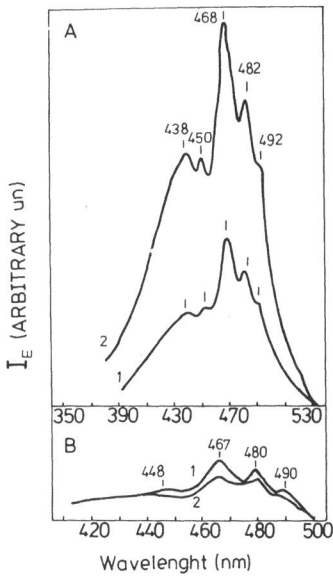

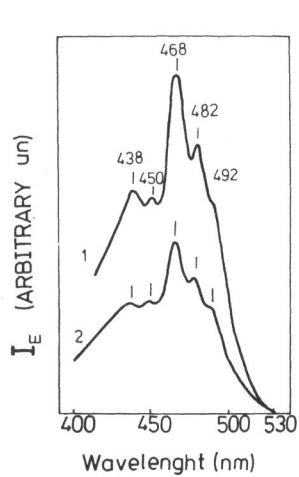

Fig. 2. Fluorescence excitation spectra at room temperature of control chloroplasts (1) and PS II OEP (2). The fluorescence was monitored at 685 nm (A) and at 722 nm (B).

Fig. 3. Fluorescence excitation spectra at room temperature of PS II OEP at 20° C (1) and 50° C (2).

The rate of decreasing of F 720 emission is slower than F 685 and chlorophyll molecules responsible for F 720 band are more stable at heat stress. The general agreement is that at room temperature chl a fluorescence is largely emitted by PS II core and by attached LHC (12). Probably more affected by heat stress would be LHC II (1,4,6) and for PS II OEP we could suggest that decreasing of chlorophyll fluorescence with heat treatment was a result of destacking of chlorophyll-protein complexes (4).

Heat stress induced no changes in absorption spectra in both preparations. They resemble those reported for chloroplasts (not shown). There are slight losses of chl absorbance after 50° treatment (Table 1). Similar results have been introduced for cyanobacteria (13).

Excitation spectra of chloroplasts and PS II OEP when light is monitored at 685 nm are shown on Fig. 2 A. The absorption band responsible for strong fluorescence is centered around 468 nm. Changes in excitation spectra of heat stress (50° C) PS II OEP, Fig. 3 revealed lower chl a_{438} intensity, which is not characteristic of heat stressed chloropasts (i.e. chl$_{438}$ is connected mainly with the antennae size of LHC II7.

Scanning of the excitation for the broad band at 722 nm for both preparations possessed four bands (Fig. 2B). The heat treatment does not change the relation between the maxima, which again speaks of thermal stability of long wavelength chlorophylls.

REFERENCES

1. C. Sundby, A. Melis, P. Mäenpää and B. Anderson, Temperature-dependent changes in the antennae size of Photosystem II. Reversible conversion of Photosystem II to Photosystem II, Biochim.Biophys.Acta, 851: 475-483 (1986).
2. N. I. Bishop, R. Lumry and J. D. Spikes, The mechanism of the photochemical activity of isolated chloroplasts. I. Effect of temperature, Arch.Biochim.Biophys., 58:1-18 (1955).
3. S. Katoh and A. San Pietro, Ascorbate-supported NADP photoreduction by heated Euglena chloroplasts, Arch.Biochim.Biophys., 122:144-152 (1967).
4. K. Gounaris, A. Brain, A. J. Quinn and W. P. Williams, Structural and functional changes associated with heat-induced phase separations of non-bilayer lipids in chloroplast thylakoid membranes, FEBS lett, 153:47-52 (1983).
5. U. Schreiber and P. A. Dumond, Heat-induced changes of chlorophyll fluorescence in isolated chloroplasts and related heat damage at the pigment level, Biochim.Biophys.Acta, 502:138-151 (1978).
6. K. Gounaris, A. Brain, P. J. Quinn and W. P. Williams, Structural reorganization of chloroplast thylakoid membranes in response to heat stress, Biochim.Biophys.Acta, 766:198-208 (1984).
7. C. H. Krauce and K. A. Santarius, Relative thermostability of chloroplast envelope, Planta, 127:285-299 (1975).
8. Y. Yamamoto, T. Ueda, H. Shinkai, M. Nishimura, Preparation of O_2-evolving photosystem II from spinach, Biochim.Biophys.Acta, 679:347-350 (1985).
9. D. I. Arnon, Copper enzymes in isolated chloroplasts. Polyphenol oxidase in Beta vulgaris, Plant Physiol., 24:1-15 (1949).
10. T. G. Dunahay, L. A. Staehelin, M. Seibert, P. D. Ogilvie and S. P. Berg, Structural, biochemical and biophysical characterization of four oxygen-evolving photosystem II preparations from spinach, Biochim.Biophys. Acta, 764:179-193 (1984).
11. E. Weis, The influence of metal kations and pH on the heat sensitivity of photosynthetic oxygen evolution and chlorophyll fluorescence in spinach chloroplasts, Planta, 154:41-47 (1982).
12. C. H. Krause and E. Weis, Chlorophyll fluorescence as a tool in plant physiology. II.Interpretation of fluorescence signals, Photosynth.Res. 5:139-157 (1984).
13. P. Mohanity, S. Hoshina and D. Fork, Energy transfer from phycobilins to chlorophyll a in heat stressed cells of Anacytis nidulans, Photochem. Photobiol., 41:589-596 (1986).

LIGHT INTENSITY DEPENDENCE OF P700 PHOTOOXIDATION IN

HEAT-STRESSED PEA CHLOROPLASTS

Maya Velitchkova and Alexander Ivanov

Central Laboratory of Biophysics
Bulgarian Academy of Sciences
Sofia 1113, Bulgaria

INTRODUCTION

The short heat stress-induced alterations of photosynthetic membrane structure are well established[1,2,3]. A considerable decrease of PS II activity[1,4,5] and an increase of PS I-associated reactions (PS I-mediated electron transport[1,5], P700 photooxidation[3,6]) have been observed within the temperature range where heat-induced damages in the chloroplasts organization occur. Whereas the inhibition of PS II activity could be attributed to the heat damage of the PS II pigment-protein complex, the nature of PS I activity increase remains still controversial. It is generally accepted that PS I-related activity increase after heat-treatment could be due to the redistribution of excitation energy in favor to PS I[7], and/or to the exposure of new electron acceptor sites in the electron transport chain[5]. It has been recently supposed that the increase of P700 photooxidation efficiency in heat - treated chloroplasts could be partially due to the increase of absorption cross section of PS I via the heat - induced rearrangements of the thylakoid protein complexes[3,6]. In the present study the ESR spectroscopy of the light - induced ESR signal I (P700$^+$) and the kinetics investigations of P700$^+$ formation under limiting light conditions are applied to asses the effects of heat - treatment on the efficiency of P700 photooxidation and PS I antenna size.

MATERIAL AND METHODS

Pea chloroplasts were isolated as described in[6], in final medium containing 0.33 M sucrose, 5 mM $MgCl_2$ and 50 mM Tricine (pH 8.0). ESR measurements of the light - induced ESR signal I and its formation rates were performed on an ESR 220 spectrometer as described in[6], except that neutral density filters (Carl Zeiss, Jena) were used to vary the light intensity. 77°K chlorophyll fluorescence emission was registered by an Jobin Yvon JY3 spectrofluorimeter equipped with a red sensitive photomultiplier (Hamamatsu R 928) and a low temperature device (slits width = 4 nm). Excitation wavelenght was 436 nm.

RESULTS AND DISCUSSION

The curves of the light - intensity dependence of the ESR I amplitude (respectively P700$^+$ amount) are presented in Fig.1. Both curves follow the similar pattern as the temperature dependences of the steady state PS I -

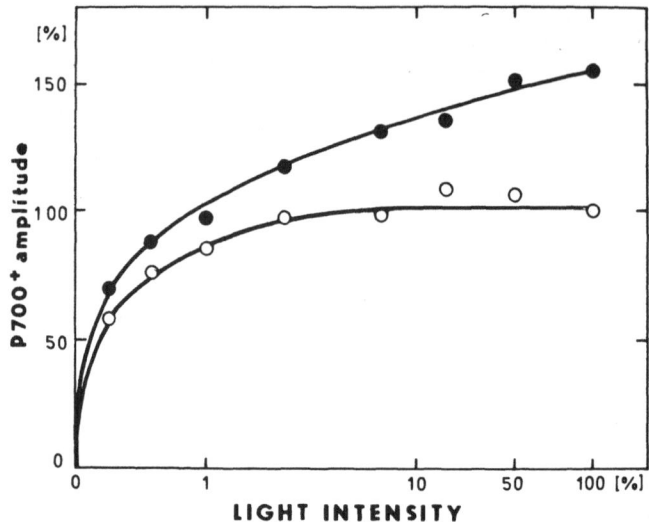

Fig. 1. Light intensity dependence of the ESR signal I (P700$^+$)
amplitude in control (O) and heated for 5 min at 50°C
(●) pea thylakoids. The ESR spectra were recorded at
20°C in g=2.0025 region with the following instrument
setting: frequency 9.5 GHz, microwave power 12.5 mW,
modulation amplitude 3,0 G. Suspending media contained:
0.01 M Tricine (pH 8,0), 0.33 M sucrose, 0.005 M MgCl$_2$,
0.1 mM methylviologen and 3.0 mg Chl/ml. 100% on the
light intensity scale corresponds to 45 W.m^{-2}.

mediated electron transport rates reported in[5]. This allows to suppose that
different response of light – induced P700$^+$ to the temperature variations
reflects the same events considered to take place in the heat – induced
changes of PS I activity under the similar experimental conditions. The
amplitudes of ESR I (P700$^+$) are higher in heat – treated chloroplasts than
those of the control chloroplasts at all light intensities tasted. It is
important to note, that P700$^+$ values are higher in heat – treated samples

Table 1. Half times ($t_{\frac{1}{2}}$) of P700$^+$ formation in control and heat-
treated (50°C, 5 min) chloroplasts as a function of the
light intensity. Kinetics were monitored at the low-field
peak of ESR I. All other conditions as in Fig.1.

Sample	$t_{\frac{1}{2}}$ formation (s)					
	Relative light intensities (%)					
	100	50	20	1.0	0.5	0.2
Control chloroplasts	0.50	0.67	1.20	6.40	7.60	8.60
Heated chloroplasts	0.50	0.57	0.90	4.90	6.70	7.90

heat – stressed chloroplasts. The most intriguing feature is that the $t_{\frac{1}{2}}$

even at very low light intensities, thus indicating that the efficiency of P700 photooxidation increases as a result of heat treatment.

The half times of P700$^+$ formation are presented in Table 1. The decrease in light intensity leads to the increase of $t_{\frac{1}{2}}$ values both in control and values for heat - treated chloroplasts are lower as compared to those for non - treated chloroplasts (Tabl.1.). Taking into consideration that: i/ P700 photooxidation kinetics under limiting light conditions is sensitive to the structural and organization state of PS I[8,9], and that: ii/ the rate of P700 photoconversion is directly related to the functional antenna size of PS I (under similar conditions) we suppose that the increase of P700 photooxidation efficiency in heat - stressed chloroplasts might be due to an increase of PS I absorption cross section. The heat - induced disruption of chloroplast granal structure[2,3] and subsequent rearrangement of protein complexes of both photosystems are similar to those observed after destacking of the thylakoids under low ionic strenght. Therefore, it could be expected that the heat - induced increase of P700$^+$ is partially due to increase of the spillover. Moreover, the spillover type mechanism has been proposed to explaine the heat - induced redistribution of excitation energy in favor to PS I[7].

In order to check at what extent the spillover changes can be responsible for P700$^+$ increase in heat - treated chloroplasts, measurements of 77°K chlorophyll fluorescence were performed[10]. As can be seen in Fig.2.

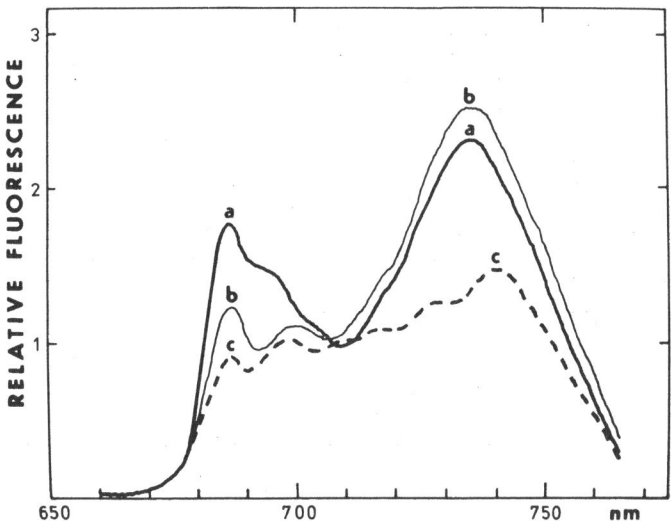

Fig. 2. 77°K fluorescence emission spectra of non-heated stack
       [+ 5 mM MgCl$_2$] - (a) and unstack [+ 10 mM KCl] - (b)
       chloroplasts. (c) - heat -treated [50°C, 5 min] stack
       chloroplasts. Chlorophyll concentration in the probe was
       7 µg/ml.

resuspending the chloroplasts in low - salt medium (unstack chloroplasts) results in strong increase of F735/F685 ratio (1.91) in comparison with that for stack chloroplasts (F735/F685=1.21). Furthermore, the 20% increase of P700$^+$ amount has been accounted for the increase of spillover caused by the unstacking of chloroplasts[6]. Heating of stack chloroplasts does not cause remarcable changes in the F735/F685 ratio (1.32), although the overall fluorescence intensity emission is reduced, which is in agreement with earlier observations[11]. Obviously, the strong increase of P700 photooxidation efficiency observed can not be explained only by the spillover changes.

The results presented here allow us to assume, that the nature of heat-induced increase of PS I activity is rather more complicated than the mechanism proposed by Thomas et al.[5]. The possible heat - induced exposure of new electron donor sites located within the cyt f/b_6 complex of the electron transport chain can offer a good explanation of the PS I activity increase, particularly under saturating light conditions and in the presence of an artifitial electron donor. On the other hand our data provide an evidence for changes in the PS I antenna size, which could be partialy responsible for the heat - induced increase of $P700^+$. The enlargement of PS I absorption cross section could be explained by the heat-induced physical dissociation[2-4] of LHC II from PS II reaction center complex[2,4], and the subsequent migration of the dissociated (free) LHC II from the grana to the PS I - enriched stroma membranes, although Sundby et al.[4] postulated that the dissociated LHC II remains free and separated from both photosystems. Our assumption is supported by recent data[3,6], indicating that glutaraldehyde fixation of chloroplast membranes (which prevents their structural changes) abolishes the heat - induced increase of P700 photooxidation efficiency.

REFERENCES

1. K. Gounaris, A.P.R. Brain, P.J.Quinn and W.P. Williams, Structural and functional changes associated with heat-induced phase-separations of non-bilayer lipids in chloroplast thylakoid membranes, FEBS Lett. 153:47 (1983).
2. K. Gounaris, A.P.R. Brain, P.J. Quinn and W.P. Williams, Structural reorganization of chloroplast thylakoid membranes in response to heat-stress, Biochim. Biophys. Acta 766:198 (1984).
3. A. G. Ivanov, M. Velitchkova and D. Kafalieva, Multiple effects of trypsin- and heat-treatments on the ultrastructure and surface charge density of pea chloroplast membranes. Influence on $P700^+$ parameters,

    in: "Progress in Photosynthesis Research", vol. II, J. Biggins ed., Martinus Nijhoff Publishers, Dordrecht, p. 741 (1987).
4. C. Sundby, A. Melis, P. Maenpaa and B. Andersson, Temperature-dependent changes in the antenna size of Photosystem II. Reversible conversion of Photosystem II to Photosystem II , Biochim. Biophys. Acta 851: 475 (1986).
5. P. G. Thomas, P.J. Quinn and W.P. Williams, The origin of photosystem I-mediated electron transport stimulation in heat-stressed chloroplasts, Planta 167:133 (1986).
6. A. G. Ivanov, M.Y. Velitchkova and D.N. Kafalieva, Heat-induced changes of photosystem I reaction centre in pea chloroplast membranes, Compt. rend. Acad. bulg. Sci. 39:123 (1986).
7. E. Weis, Light- and temperature-induced changes in the distribution of excitation energy between Photosystem I and Photosystem II in spinach leaves, Biochim. Biophys. Acta 807:118 (1985).
8. A. Melis and R.A. Ow, Photoconversion of chloroplast photosystem I and II. Effect of Mg^{2+}, Biochim. Biophys. Acta 681:1 (1982).
9. W. Ortiz, E. Lam, M. Chirardi and R. Malkin, Antenna function of a chlorophyll a/b protein complex of photosystem I, Biochim. Biophys. Acta 766:505 (1984).
10. G. H. Krause and E. Weis, Chlorophyll fluorescence as tool in plant physiology. II Interpretation of fluorescence signals, Photosynth. Res. 5:139 (1984).
11. U. Schreiber and P.A. Armond, Heat-induced changes of chlorophyll fluorescence in isolated chloroplasts and related heat-damage at the pigment level, Biochim. Biophys. Acta 502:138 (1978).

EFFECTS OF ALPHA-TOCOPHEROL DERIVATES ON THE PHOTOSYNTHETIC

ACTIVITY OF THYLAKOID MEMBRANES

Vassilii Goltsev, Ognjan Popov, Virginia Doltchinkova, Ivan Yordanov[+], and Valerian Kagan[++]

Department of Biophysics and Radiobiology, Sofia University, [+]Institute of Plant Physiology, Bulgarian Academy of Sciences, [++]Institute of Physiology, Bulgarian Academy of Sciences, Sofia, Bulgaria

INTRODUCTION

The light-harvesting and electron transport processes of photosynthesis are localized in lamellae which consist of about 50 % protein, 40 % acyl lipid and 10 % pigment and other neutral lipids /by weight/ /Gounaris et al., 1986/. Thylakoid membranes contain about 45-50 % of their acyl lipids as mono-galactosyldiacylglycerol, 25-30 % as digalactosyldiacylglycerol and up to 10 % as sulpholopids /Williams et al.,1984/. It is generally assumed that the major role of lipids in membranes is the formation of a bilayer which will act as a structural matrix and will limit the transmembrane movement of most hydro-philic solutes. In addition, the lipids in the bilayer prevent the appearance of non-specific protein-protein aggregation and conformational changes. A number of assumptions have been made, ascribing specific roles to the thylakoid neutral lipids in the functioning of the membrane. In particular, alpha-toco-pherol /AT/ is considered as one of the essential neutral li-pids of the thylakoid membrane. The main function of alpha-toco-pherol is the protection of thylakoid membrane lipids against oxidative stress. This is achieved predominantly by two ways: /i/ by quenching of singlet molecular oxygen, and /ii/ by scavenging of oxygen and lipid radicals /Niki et al., 1985/. These specific functions of alpha-tocopherol are mostly provided for by its chromanol nucleus, whereas the phytyl side chain /Fig.1/ facilitates the incorporation and retainment of the molecule in the lipid bilayer /Niki et al., 1985/. However, the effect of the phytyl side of alpha-tocopherol on these functions has not been clearly established experimentally /Fragata et al., 1985; Niki et al., 1985; Perly et al., 1985; Villalain et al., 1986/. Recently it was shown that alpha-tocopherol can play a structural role in the membrane restricting the molecular motions of acyl lipids /Erin et al., 1984/. In order to deter-mine more precisely the role of the alpha-tocopherol phytyl chain in its structural effects in thylakoid membrane, the effects of alpha-tocopherol and its derivatives with different chain length /Fig.1/ on the parameters of induction curves of the millisecond delayed fluorescence of chloroplasts were studied.

MATERIAL AND METHODS

The pea chloroplasts were isolated as in Yordanov et al. /1987/. Ethanol solutions of AT and its derivates were added to chloroplst stock suspension /200 μg chlorophyll/ml/, vigorious shaken for 1.5 min, after which the suspension was diluted to a final concentration of 10 μg chlorophyll/ml.

The induction kinetics of millisecond delayed fluorescence was measured by a method described by Goltsev et al./1987/.

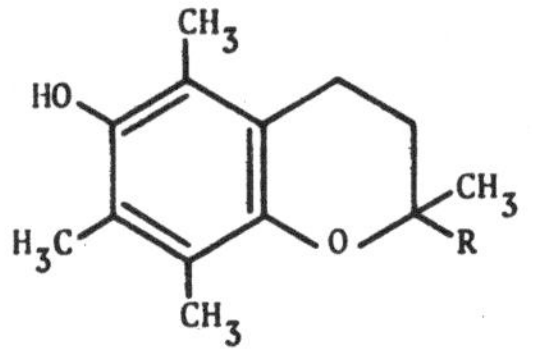

2,5,7,8-tetramethyl-6-hydroxy-2-alkyl-chromane (C_n)

C_1 R= -CH_3

C_3 R= -CH_2-CH_2-CH_3

C_6 R= -CH_2-CH_2-CH_2-$\underset{\overset{|}{CH_3}}{CH}$-$CH_3$

C_{11} R= -CH_2-CH_2-CH_2-$\underset{\overset{|}{CH_3}}{CH}$-$CH_2$-$CH_2$-$CH_2$-$\underset{\overset{|}{CH_3}}{CH}$-$CH_3$

α-tocopherol(C_{16}) R= -CH_2-CH_2-CH_2-$\underset{\overset{|}{CH_3}}{CH}$-$CH_2$-$CH_2$-$CH_2$-$\underset{\overset{|}{CH_3}}{CH}$-$CH_2$-$CH_2$-$CH_2$-$CH$-$CH_3$

Fig. 1   Chemical structure of alpha-tocopherol and its derivatives

RESULTS AND DISCUSSION

Figure 2 depicts the effect of AT and its derivates on one of the parameters of induction kinetics of delayed fluorescence of dark-adapted /3 min/ pea chloroplasts. It is known that delayed fluorescence intensity at the first induction maximum /I/ is proportional to the rate of excitation energy migration from the light-harvesting complexes to photosystem II reaction centre chlorophyll /Goltsev et al.,1987/. ATV and its derivatives

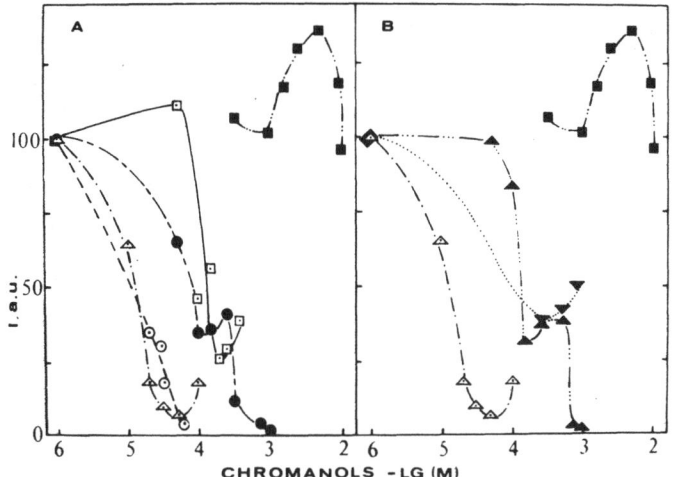

Fig.2 Effect of AT derivatives /A/ and AT-acetate derivatives
/B/ on the phase I of induction kinetic of thylakoid
membrane delayed fluorescence. Each sample contains
10 mM Tricine buffer, pH 7.8,5 mM $MgSO_4$ chloroplasts
with chlorophyll concentration 10 μg Chl/ml and different concentrations of chromanols. ■ -AT /C16/; □ -C11;
▲ -C16 acetate; ● -C6; ◔ -C3; △ -C1; ▼ C1 acetate

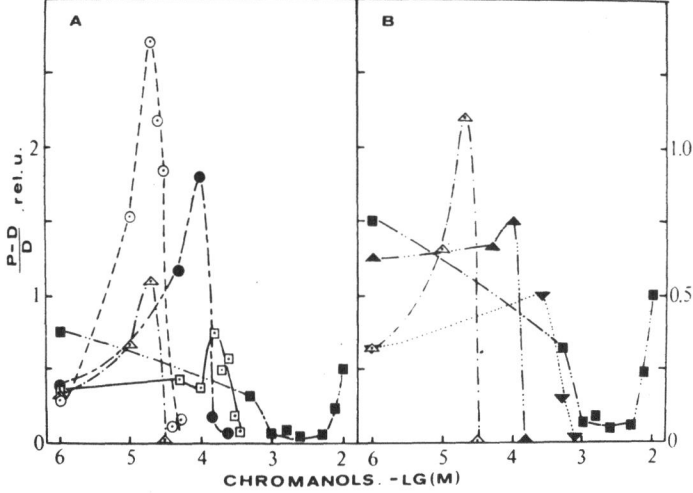

Fig.3 Effect of AT derivatives /A/ and AT-acetate derivatives
/B/ on the phase P-D/D of induction kinetic of thylakoid
membrane delayed fluorescence. Each sample contains 10 mM
Tricine buffer, pH 7.8,5 mM $MgSO_4$, chloroplasts with
chlorophyll concentration of 10 μg Chl/ml and different
concentrations of chromanols. ■ -AT /C16/; ▲ C16 acetate;
□ -C11; ● -C6; ◔ -C3; △ -C1; ▼ -C1 acetate.

caused concentration-dependent suppression of I , while short-chain AT derivates / Cl, C3, C6/ were by far more efficient than AT itself and its long-chain derivate Cll. The concentrations of AT and Cl which produced 50 % inhibition differed by more than two orders of magnitude. The efficiency of energy migration to the reaction cenre is determined by its distance from the light-harvesting complexes in the thylakoid membrane and is strongly dependent on the lateral diffusion rate, i.e. on the fluidity of the membrane lipid matrix /Barber,1984/. Thus, it can be concluded that short chain derivates of AT enhance the lateral mobility of light-harvesting complexes in the membrane due to "fluidization" of the lipid bilayer. AT proved to be uncapable to produce this "fluidizing" effect up to concentration of 10 mM. The OH-group of the chromanol nucleus of AT and its derivates is of great importance in these effects, since both AT-acetate and Cl-acetate suppress energy migration in quite different concentration ranges /Fig.2B/.

The slow phase of the delayed fluorescence induction curve P-D/D reflects the formation and maintenance of photo-induced pH transmembrane proton gradient /Wraight and Crofts, 1971/. It is evident that the Δ pH value depends on the electron flow rate and on the thylakoid membrane integrity. The concentration dependence of AT derivates on P-D/D is biphasic /Fig.4/. In the low concentration range the increase of Δ pH is observed. Further rise in the concentrations of the added AT derivatives leads to a sharp decrease of the proton gradient. The positions of the maxima on the concentration curves differ for AT derivates with different chain length, the short chain derivates having maxima at lower concentrations. Comparing these data with the data on Fig.2, it should be pointed out that the maximal values of the proton gradient correspond to the concentrations of AT derivates causing most pronounced suppression of I. This probably means that enhanced diffusion of light-harvesting complexes from appressed to non-appressed regions of the thylakoid membrane in the presence of AT derivates results in more efficient energy supply to photosystem 1 and consequently in acceleration of cyclic electron transport. In the absence of endogenous electron acceptors this would lead to a pronounced increase of the proton gradient. Further increase in the concentrations of AT derivatives in the membrane is responsible for the excessive "fluidization" of the lipid bilayer, disturbance of the barrier properties and disappearance of the proton gradient. As in case of I measurements, acetylation of chromanol nucleus of AT and Cl caused changes in the effects on the proton gradient, indicating the essential role of the hydroxy-group /Fig.3B/.

In conclusion, both phytyl chain and the 6-hydroxy group of AT are essential for its proper orientation and functioning in the thylakoid membrane. Modification of the AT molecule leads to its damaging effects on the thylakoid membrane, more pronounced for the shorter chain derivatives of AT. This explains to some extent the existence of long hydrocarbon chains in oder neutral lipids of photosynthetic membraens, e.g. ubiquinols, plastoquinones, carotenoids, etc. /Ondarroa and Quinn, 1986, Niki et al., 1985/.

REFERENCES

1. Barber, J., 1984, Lateral heterogeneity of proteins and
    lipids in the thylakoid membrane and implications for elec-
    tron transport, in: "Advances in Photosynthesis Research",
    vol.III, C.Sybesma, ed., M.Nijhoff/Dr.W.Func publs., The
    Hague.
2. Erin, A.N., Spirin, M.M., Tabidze, L.V. and Kagan, V.E.,
    1984, Formation of   -tocopherol complexes with fatty acids.
    A hypothetical mechanism of stabilization of biomembranes
    by vitamin E, Biochim.Biophys.Acta,774:96.
3. Fragata, M., El-Kindi, M. and Bellemare, F., 1985, Mixing
    of single-chain amphiphiles in two-chain lipid bilayers.
    2.Characteristics of chlorophyll A and $\mathcal{L}$-tocopherol incor-
    poration in unilamellar phosphatidylcholine vesicles, Che-
    mistry and Physics of Lipids, 37:117.
4. Goltsev, V.N., Yordanov, L., Stoyanova, Ts., and Popov, O.,
    1987, Effect of mono- and divalent cations and pH on the
    temperature sensitivity of some functional characteristics
    of chloroplasts isolated from heat-acclimated and non-accli-
    mated plants, Planta, 170:478.
5. Gounaris, K., Barber, J. and Harwood, J.L., 1986, The thyla-
    koid membranes of higher plant chloroplasts, Biochem.J.,237:
    313.
6. Niki, E., Kawakami, A., Saito, M., Yamamoto, y., Tsuchiga
    J. and Kamoga, Y., 1985, Effect of phytyl side chain of
    vitamin E on its antioxidant activity, J.Biol.Chem., 260:
    2191.
7. Ondarroa, M. and Quinn, P.J., 1986, Proton magnetic reso-
    nance spectroscopic studies of the interaction of uniqui-
    none-10 with phospholipid model membranes, Eur.J.Biochem.,
    155:353.
8. Perly, B., Smith, I.C.P., Hughes, L., Burton, G.W. and In-
    gold, K.U., 1985, Estimation of the location of natural
        -tocopherol in lipid bilayers by ^{13}C-NMR spectroscopy,
    Biochim.Biophys.Acta, 819:131.
9. Villalain, J., Aranda, F.J. and Gomes-Fernandez, J.C., 1986,
    Calorimetric and infrared spectroscopic studies of the in-
    teraction of $\mathcal{L}$-tocopherol and $\mathcal{L}$-tocopherol acetate
    with phospholipid vesicles, Eur.J.Biochem., 158:141.
10. Williams, W.P., Gounaris, K. and Quinn, P.J., 1984, Lipid-
    -protein interactions in the thylakoid membranes of higher
    plant chloroplasts, in: "Advances in Photosynthesis Research,"
    vol.III, C.Sybesma, ed., M.Nijhoff/Dr.W.Func publs., The
    Hague.
11. Yordanov, I., Goltsev, V., Stoyanova, T. and Venediktov, P.,
    1987, High - temperature damage and acclimation of the pho-
    tosynthetic apparatus. I.Temperature sensitivity of some
    photosynthetic parameters of chloroplasts isolated from
    acclimated and non-acclimated bean leaves, Planta, 1970:471.

METABOLIC REGULATION DISTURBANCE MECHANISMS OF NERVE

CELL Na-PUMP IN RADIATION PATHOLOGY

Anatoliy I. Dvoretsky, Alla M. Shainskaya,
Elena G. Yegorova, and Sineric N. Ayraoetyan

Research Institute of Biology, Dniepropetrovsk
State University, Institute of Experimental
Biology, Academy of Sciences of the Armenian SSR
Yerevan, USSR

INTRODUCTION

The change of the biological membrane permeability is considered to be one of the earliest effects of ionizing radiation /IR/.[1] Nevertheless, the radiation disturbance mechanisms of the systems regulating the ions homeostasis /including the main monovalent cations - sodium and potassium/ have not been investigated yet. Studies of these effects are of great interest, especially for the nerve tissue, taking into consideration the interpreting of the neurologic disturbance development mechanisms in the case of radiation pathology. In connection with this, the IR action mechanisms on the transport systems of Na^+ and K^+ in the model experiments on the giant neurons of Helix pomatia were investigated.

MATERIAL AND METHODS

Experiments were carried out on ganglion cells of Helix pomatia. The animal and the isolated ganglion were irradiated with a dose equal to 200 Gy and an hour later they were investigated. The velocity of 22-Na efflux from the nerve cells under different conditions was the main factor to judge about the functional activity of the Na,K - pump. The Na,K - ATPase activity was determined in the crude fractions of the plasmatic membranes prepared as described earlier.[2] The ATPase was assayed by measuring the amount of inorganic phosphate released from ATP colorimetrically.[3] Differences in the inorganic phosphate generated during the reaction with and without ouabain were calculated and used as Na,K - ATPase activity. [3]H - ouabain /Amersham Radiochemical Center, England, sp.act. 35 Ci/mmol/ was used for the estimation of the number of membrane functionally active units. The calculation was made using the described method earlier[4] by means of the SL-4221 Liquid Scintillation counter /Intertechnique, France/. The fractionation of the individual lipids was performed by the method of one-dimensional chromatography in the thin layer of silicagel of the KSK type, using such solvents as chlorophorm, methanol, ammonia-

255

- 65:35:5. The mineralization of lipid phosphorus was performed with sulphur and nitrous acids, with subsequent calculation of Pi on 1 mg of the dry fraction weight. The content of the free fatty acids was determined according to the previously described method.[5] The ATP concentration was calculated by means of the Lumenometer-1250 /LKB, Sweden/. The passive permeability of the membranes was determined according to the measuring of ^{22}Na diffusive input into the cell under conditions which exclude the ^{22}Na active efflux from the cells /cold potassium-free solution/. All the data was treated statistically.

RESULTS

The data obtained /Fig.1a/ show the increase of sodium ions content in the nerve ganglion after irradiation as a result of their diffusive input into the cells. The recording of the current voltage characteristics during the 3 hours /Fig.1b/ showed increasing conductivity of the irradiated cell membrane, which was accompanied by gradual membrane depolarization.

The efflux of the sodium ions from the cell is known to take place with the help of two mechanisms: ouabain-sensitive /Na, K-pump/ and ouabain-nonsensitive /Na:Ca-exchange/. The investigation of ^{22}Na efflux in the absence of ouabain /Fig.2a/, with two processes taking place in the cell, showed that IR causes the supression of ^{22}Na efflux. The ^{22}Na efflux determined by the Na/Ca-exchange dose differs from that of normal cells under the conditions of potassium-free solution and in the presence of ouabain /Fig.2b/ when the Na,K - pump was fully out of action. The determination of the activity of Na,K - ATPase /the functional unit of the Na,K - pump/ proves its increase by 76 % in relation to the normal cells /0.58 ± 0.007 and $1.02\pm0.05\mu$M P2 per 1 mg of protein per hour in the preparation of normal and irradiated animals/ an hour after the irradiation.

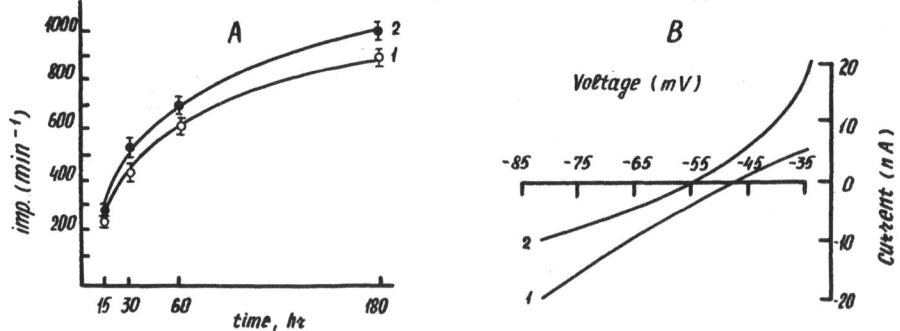

Fig.1   The effect of ionizing radiation on the passive input of ^{22}Na/A/ and current-voltage characteristics /B/ of <u>Helix pomatia</u> giant neurons /1 - control, 2 - after radiation/

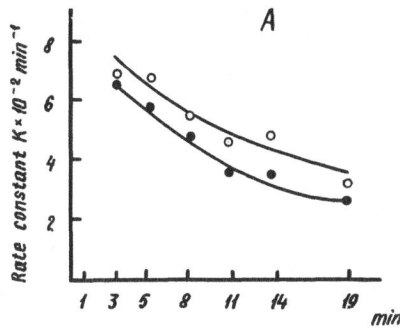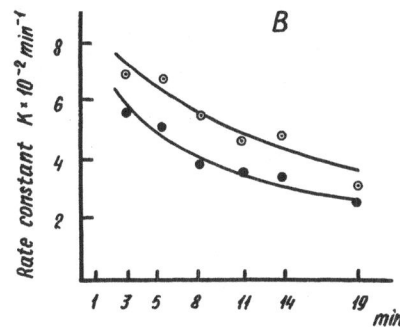

Fig.2　The effect of ionizing radiation on ouabain-sensitive
/A/ and ouabain-insensitive /B/ efflux of ^{22}Na from
Helix pomatia giant neurons /o - control: o - control
+K_o+ouabain: o - after radiation/

Thus, IR leads to strongly marked disturbance of the
functioning machanisms of the Na, K-pump of Helix pomatia
giant neuros. But plasmatic membranes obtained from the irradiated animals possess a heighthened enzyme activity. Such
differently directed effect of the radiation factor can be
explained by the increase the number of pumps in the membrane
because of the reserve ones which can be caused by the pump-
-dependent increase of the neuron's volumes and by the level-
ling of the invagination of the cell surface.[4] However, the
energy starvation and the other factors do not allow them to
function actively in the irradiated neuron.

The results of the experiments used for the investigation
of the ^{3}H-ouabain binding process and determination of the ATP
level in the irradiated cell proved the assumption mentioned
above. The data obtained proved not only the increase of the
number of functionally active pump units in the membrane but
also the essential modification of the active centres'. affinity to ouabain. The number of receptors having high affinity
to ouabain /K_m - $3x10^{-10}$/ is essentially increased, while the
second group of receptors having lesser affinity /K_m -$3x10^{-9}$/
become analaqeus according to the third linear component
which provokes the inhibitory action on the Na, K-pump activity.

The results of our experiments on the influence of IR on
the nerve ganglion's ATP content showed that within 15 minutes
after the exposure the ATP level is lowered by 28,55 %. The
maximum lowering by 28,45 % was observed an hour later.
It should be stressed that the same picture of the binding
of the tracer ouabain molecules and inhibiting of the pump
were observed in the case of the neuron's treatment by de-
cilenic acid,[6] which increases the fluidity of the lipid phase
of the membrane. In this connection the reorganization in the
lipid bilayer of the neuron membrane under the IR action may be
one of the initiating processes in the development of the disturbance in the membrane ion transport system.
Investigations undertaken in this connection have proved
the essential changes of the phospholipid content of Helix
pomatia nerve ganglia. The lowering of the level of phospha-

tidylcholine /by 46 %/, sphingomyelin /by 42 %/, phosphatidylethanolamine /by 28%/is accompanied by an increase of the lisoforms of phosphatidylcholine and phosphatidylinozitol. Accumulation of the phospholipid lisoforms points to activation of
lipid peroxidation. The changes in the phospholipid spectrum
are accompanied by an increase in the free fatty acids content
by 20 % an hour after irradiation. This fact demonstrates the
activation of endogeneous phospholipid hydrolisis.

DISCUSSION

Summing up everything written above, one can come to the
conclusion that IR induces essential changes of the transport
function of biomembranes. It is expressed in disturbance of the
passive permeability processes of the mebranes as a result of
the structural reorganization taking place because of the activation processes of lipid peroxidation and phospholipid hydrolysis. The radiation disturbances lead to strong input of Na
and Ca ions into the neuron and pump-dependence increase of the
cell volume.
An increase on the number of pump units takes place in
response to the radiation effect. These units were in a screened
form earlier. However, the decrease of the general energy level
of the cell, the change of the physico-chemical enzyme properties and its active centres caused the disturbance of the regu-
lative mechanisms of the Na, K-pump and in the functioning of
the enzyme's oligomeric complex, as the result of post-irradiation modification of the lipid surrounding and of structural
reorganization of biomembranes.

REFERENCES

1. Bacq, Z. and Alexender, P., 1983, Fundamental of radiobio-
    logy, Foreign Languages Publishing House, Moscow.
2. Daily, I.W. and Partington, C.R., 1977, Adrenergic recep-
    tors and adenylate cyclase in membrane preparations from
    the central nervous system, Catecholamines Basic and Cli-
    nical Transfers 1·
3. Taussky, H. and Shorr, E., 1953, A microcolorimetric method
    for the determination of inorganic phosphorus, J.Biol.
    Chem. 2:2.
4. Airapetyan, S.N., Dadalyan, S.S., Marikyan, G.G., Avetisyan
    and Gevorkyan, A.M., 1981, On the "reserve" molecules of
    Na, K-ATPase in cell membranes, J.Reports of Academian
    Science of USSR, 4:258.
5. Voronko, V.A., Nikushkin, E.V., Kryzhanovsky, G.M. and
    Germanov, C.V., 1982, Endogenic phospholipase hydrolysis
    in the brain hemisphere under epileptic activity, J.Exper.
    Biology and medicine, Moscow, 12:55.
6. Saghiyan, A.A., Dadalyan, S.S., Takenaka, T., 1986, The
    effect of short-chain fatty acids on the neuronal membrane
    functions of Helix pomatia, J.Cell.Moll.Neurobiol., 6:2

LUMINOMETRIC TEST FOR HUMAN ERYTHROCYTE ATP MONITORING AFTER GAMMA-IRRADIATION AND EXOGENEOUS ATP ADMINISTRATION

B. Galutzov, S. Ivanov, M. Ratcheva-Kantcheva, and S. Todorov

Department of Biophysics and Radiobiology, Biological Faculty, University of Sofia
Dr.Tzankov 8, 1421 Sofia

INTRODUCTION

The essential intracellular intermediate ATP which functions in the storage and transfer of cellular energy is generally accepted as an important cell constituent.

Under steady-state conditions /i.e.d [ATP] / dt = 0 / the ATP turnover rate is equal to either the rate of ATP formation or the rate of ATP utilization and could be estimated if either process was suspended without affecting the rate of the other.

The red blood cell ATP content is known to decrease in vivo and in vitro ageing /1,2/. Radiation-induced erytrocyte alterations are related to a decrease in the intracellular ATP content /3/. The red blood cell ATP pool in nutrient-free suspending medium is limited and corresponds to cell viability and stability, and could be used as an indication of radiation-induced membrane damage. The ageing process and radiation damage of membranes, their repair and prevention, are recently discussed /4/.

The aim of the present study is to evaluate the effect of exogeneous ATP administration on radiation-induced alterations of the red blood cells, using the ATP-monitoring technique.

MATERIAL AND METHODS

Human red blood cell suspension was used for in vitro gamma-irradiation. Exogeneous ATP administration /10 μM and 50 μM/ was performed 30 min, before or 1 and 5 h after radiation treatment. Radiation was applied with a ^{60}Co gamma-ray source in a dosage of 500 Gy and with a dose rate of 24,5 Gy.min^{-1}. Nonirradiated control cell suspensions and post-irradiation incubation were carried out at 4°C.

A modification of the bioluminiscent assay /5/ of ATP in erytrocytes with LUMINOMETER 1250 LKB was used. The amount of ATP in test samples was calculated by using the mV values measured upon addition of the internal standards. The luciferin/luciferase reagents give a time-independant light output /e.g. ATP monitoring reagent LKB-Wallac/, the light output being proportional to ATP concentration over the $10^{-11}- 10^{-6}$ M range.

RESULTS AND DISCUSSION

Cell suspension ATP content in control sample /C/ versus radiation-treated sample content /P/ presented as C/P ratio was monitored in post-irradiation incubation for 1, 5, 25 and 50 hours /Fig.1/. The C^X/P^X ratio indicates ATP-treated control and irradiated red blood cell suspension /Fig2./. The small decrease in the C/P ratio registered on the 1st and 5th hours of post-irradiation incubation is indicative about radiation--induced changes in the erythrocyte metabolic activity /Fig.1/. It could be suggested that initiation of radiation damage activated reactions is related to an increase of cell ATP-content. There is experimental evidence ATP pools in mononuclear leukocytes damaged by 30 Gy gamma-radiation are indistinguishable /6/.

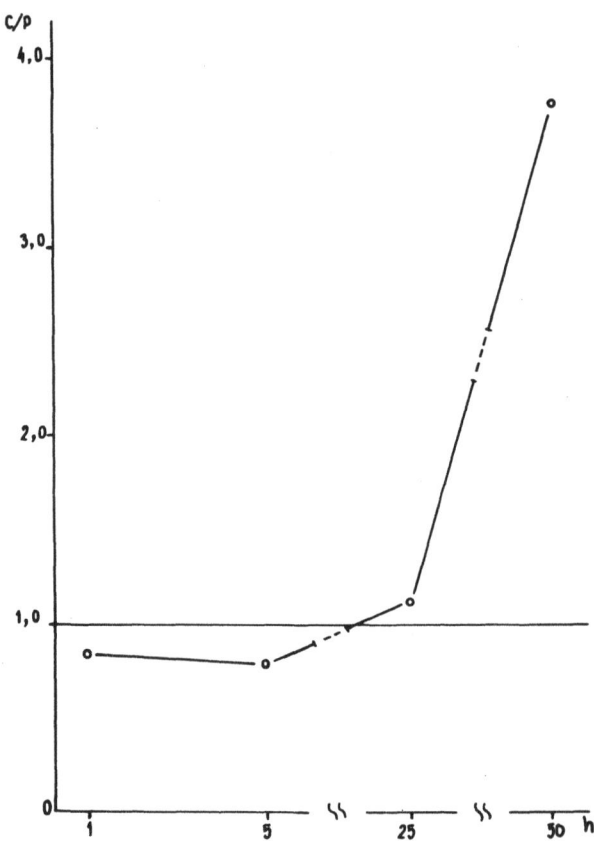

Fig.1   Post-irradiation time dependence of the C/P ratio

The enhanced C/P value between the 25th and 50th hour of incubation could be related to significant reduction of red blood cell ATP-pool and/or radiation-induced haemolysis, including hydrolysis of extracellular ATP.

Free radical-mediated peroxidation effects play an important role in the mechanism of membrane damage provoked by ionizing radiation /7/. The formation of highly reactive oxygen species, such as superoxide anion radical $/\cdot O_2^-/$, hydrogen peroxide $/H_2O_2/$ and hydroxil radical $/\cdot OH/$, may be of significance not only in the initiation of erythrocyte membrane damage but also for ATP molecule destruction.

In exogeneous ATP-treated samples the radiation-induced ATP decrease was monitored as related to the Increase in the C^X/P^X ratio during post-irradiation incubation /Fig.2/.

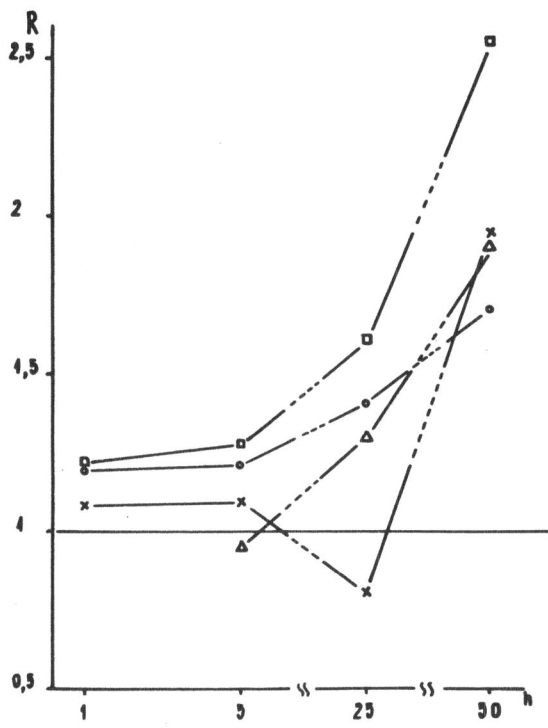

Fig. 2   Post-irradiation time dependence of the C^X/P^X on ATP administered   □ - 30 min before irradiation /10 M/ and o 50 M; 1 hour after irradiation and △ - 5 hours after irradiation

Exogeneous ATP administration results in some decrease in the ATP content of the irradiated cell suspension, registered 1-5 hours after the beginning of the post-irradiation incubation. It could be suggested that initiation of radiation-induced haemolysis and modification of extracellular ATP are evoked in the C^X/P^X ratio increase.

In the case of post-irradiation administration of ATP the low value of the parameter under investigation is indicative of the significance of exogeneous ATP presence during gamma-irradiation. The further increase in the registered

C^X/P^X value provides information about the development of fundamental cellular destructive processes, which are most pronounced for exogeneous ATP treatment /10 M/ 30 min before irradiation. The lower rate of the C^X/P^X increase, compared to C/P, could be attributed to some modification activity of exogeneous ATP administration, when red blood cells suffer radiation damage in nutrient-free suspending medium.

Dynamic observation of C/P and C^X/P^X ratio may be considered as a quantitative presentation of cell ATP-pool and cell membrane destruction. The examination of the maximum value registered for the C/P ratio - 4 as compared to $/C^X/P^X/_{max}$ about 2 - is indicative about the effect of exogeneous ATP as a supressor of destructive cellular responses to gamma-radiation.

More detailed information could be obtained by considering red blood cell integrity parameters, monitoring haemolysis during post-irradiation incubation.

Application of the dynamic observation of cell ATP-content ratios may be useful as a rapid test for monitoring different native and externally induced destabilization processes, as well as for evaluation of modification effects on the membrane structure and functional activity.

REFERENCES

1. Palmer, F.B., 1985, Polyphosphoinositide metabolism in aging human erythrocytes, Can.J.Biochem.Cell Biol.,63, 927-931.
2. Hoegman, C.F., Hedlung, K., 1985, Storage of red cells in a CPD/SAGM system using teruflex PVC, Vox Sanguinis, 49, 177-180.
3. Pantev,T.T., 1984, D.Sc.Dissertation, Med.Acad., Sofia, 240.
4. Petkau, A., 1986, Protection and repair of irradiated membranes, In: "Free radicals, aging and degenerative diseaser", Alan R.Liss Inc. 481-508.
5. Ataullakhov, F.I., Pichugin, A.V., 1981, Modification of luciferase method of determining ATP concentration in erythrocytes, Biofizika, 26,1,86-88.
6. Kjellen, E., Jonsson, G., Pero, W.R., Christensson, P., 1986, Effects of hyperthermia and nicotinamide on DNA repair synthesis, ADP-ribosyl transferase activity, NAD^+ and ATP pools and cytotoxicity in gamma-irradiated human mononuclear leukocytes, Int.J.Radiat.Biol.,49, 1, 151-162.
7. Bielsk, B.H., Gebicki, J.M., Application of radiation chemistry to biology, In: Free radicals in biology, 3, 1-51, Ed.Pryor, W.A., Academic Press, New York, 1977.

ELECTROROTATION OF LIDOCAINE-TREATED HUMAN ERYTHROCYTES

Radostina Georgieva[*] and Roland Glaser[**]

[*]Medical Institute - Stara Zagora
Department of Physics and Biophysics
[**]Humboldt-Universitaet zu Berlin, GDR
Department of Biologie

INTRODUCTION

During the past years electrorotation wins recognition as an interesting method of investigation of different biological objects. The method provides a possibility for an assessment of the biological membrane properties.[1,2,3,4,5] Also there is a possibility to detect the alterations of the membrane, caused by different drugs, disease agents, etc.

Local anaesthetic agents are membrane-active substances, which can interact, due to their chemical structure, both with the lipid and with the protein phase of the cell membrane. For that reason they are especialy intersting objects of investigation. The aim of the present study is to find the effect of lidocaine, a typical and widely applied local anaesthetic, on the electrorotation of human erythrocytes, and, if possible, to obtain information about possible alterations of the erythrocyte membrane properties.

Lidocaine is an amide-type anaesthetic. Its feature are: relatively high potency = 4, low pK_a value = 7.7, relatively high lipid solubility = 2.9 and very high protein-binding coefficient.[6]

MATERIAL AND METHODS

We studied human erythrocytes from a blood bank /0 Rh +/ not older than 3 days.

Lidocaine /Pharmachin, Bulgaria, 2 % solution per injection/ was added to the whole blood samples in concentrations resp. 0.2; 0.9, 2.1, 4.3 mmol/l and were incubated in vitro for 5 min at room temperature /23-28°C/. These concentrations were selected near the value necessary for 50 % antihaemolysis; for lidocaine it is 1 mmol/l.[6]

The erythrocyte separation was performed by 10 min centrifugation at 2000 x g.

The electrorotational measurements were performed in 300 mosm sucrose solution and 1 mmol/l phosphate buffer /pH=7.4/ with specific electroconductivity 0.0172 ± 0.0188 $S.m^{-1}$. About

5 µl erythrocytes were resuspended in 10 ml solution to produce a suitable value of the haematocrit.

The measurements were made in 4-electrode chamber with 10 V power and frequency range between 32 kHz and 10 MHz. At the same time we considered the incubation time in sucrose solution of every cell.

The results were evalued applying the single-shell-cell model according to FUHR et al.[7]

Non linear regression analysis was performed in order to find the most appropriate shape of the theoretical curves with the experimental spectrum of the cell.

RESULTS AND DISCUSSION

The results from the electrorotation measurements of one of the four blood samples are shown in a diagram /Fig.1/. The results for the other three samples are similar. The diagram shows the change of the first characteristic frequency /f_{c1}/ and the maximum negative rotation /R/ of the erythrocytes according to the time which the red cells spend in suspension in the control sample and after incubation in four different concentrations of lidocaine. /$R=w/E^2$/, where w - angular velocity of the cells, E - intensity of the electrical field./ Obviously, f_{c1} of the cells being for less than 10 min in the medium with extremely low ionic strenght, is significantly higher than in the control with same incubation. The shift of f_{c1} to higher frequencies corresponds obviously to the increase of the lidocaine concentration. We observed significant increase of the maximum rotation without concentration dependence.

The change of the f_{c1} range and the extent of the torque number N_{min} of the cell at the mentioned frequency according to the single-shell-model may result from the following parameter change: cell radius - r, membrane conductivity - G_m, intracellular conductivity - G_i, membrane permitivity - ε_m, resp. membrane capacity - C_m. The relationship between the above parameters and f_{c1} and N_{min} is shown on Fig.2.

Table 1. Values of some fundamental electrical paremeters of erythrocytes incubated in vitro by different lidocaine concentrations calculated under the conditions: $G_m=1.10^{-7}$ S.m^{-1}, d=8 nm, $\varepsilon_i=50$, $\varepsilon_e=80$

a/ all measured cells
b/ Cells measured between 2 and 7 min in sachharose solution

Table 1

	Lidocaine concentration				
	0	0.2	0.9	2.1	4.3
f_{c1} kHz	169.2	200.2	212.8	223.2	236.2
G_i S.m^{-1}	0.1747	0.2242	0.1923	0.2228	0.2334
ε_m	8.98	7.365	7.094	7.105	6.43
C_m [mF.m^{-2}]	9.94	8.145	7.925	7.861	7.12

Table 1 (Continued)

Table 1 (Continued)

	Lidocaine concentration				
	0	0.2	0.9	2.1	4.3
f_{c1} kHz	186.7	230.6	249.1	246.9	269.6
G_i S.m^{-1}	0.1601	0.2416	0.297	0.2503	0.3131
$\mathcal{E}_m$	7.81	6.789	6.55	6.24	5.78
C_m mF.m^{-2}	8.643	7.648	7.25	6.91	6.397

Since the electrorotation solution does not contain lidocaine, in order to avoid the change of environmental conductivity G_e, the observed alteration of the f_{c1} during the incubation in sucrose solution probably results from extraction of lidocaine from the cells to the external medium. Nearly 15 min later erythrocytes incubated in lidocaine have the same electrorotation behaviour as the control sample. That is why, conclusions about the influence of lidocaine on the erythrocyte membrane may be reached for the measurements between 2 and 7 min after placing them in the sucrose.

The nonlinear regression of the theoretical electrorotation spectra corresponding to the single-shell-cell model was performed at G_m=const=1×10^{-7} S.m^{-1} /G_m - membrane conductivity/. The values of G_i - intracellular conductivity; $\mathcal{E}_m$ - membrane permitivity, resp. C_m - membrane capacity, calculated under these conditions, are shown in Table 1.

It can be seen that the rise of G_i and decrease of $\mathcal{E}_m$ and C_m correlate well with the increase of the lidocaine concentration in the incubation solution.

The obtained lower values of the membrane permitivity and capacity may be seen in two aspects. First, we can assume that $\mathcal{E}_m$ decrease because of the structural changes mainly of membrane prozeins. The high protein-binding coefficient of the lidocaine confirms this assumption.[6,8,9] Second, an increase in the average cell membrane thickness /d/ is lickely to occur because of the lidocaine molecules including in the lipid double layer. In fact in that case C_m decreases and the values calculated for $\mathcal{E}_m$ are lower, because a constant membrane thickness /d=8nm/ is accepted.

The problem with the G_i changement is more complex. The frequency range, within which the electrorotation is now measured, does not still include the second characteristic frequency f_{c2} which is dependent by G_i. For that reason G_i is still calculated not so precisely. It could be supposed that the erythrocyte intracellular conductivity under physiological conditions is considerably higher than the calculated one of the control cells. This is in accordance with the data[10], who reach G_i values from 0.455 to 0.715 S.m^{-1} by means of other method.

REFERENCES

1. Arnold, W.M. and Zimmermann, U., 1982, Rotating-field- induced rotation and measurement of the membrane capacitance of single mesophyll cells, Z.Naturforsch. 37 c, 908:915.
2. Fuhr, G., Gimsa, J. and Glaser, R., 1985, Interpretation of electrorotation of protoplasts, I.Zheoretical considerations, Studia Biophysica 108, 149:164.
3. Gimsa, J., Fuhr, G. and Glaser, R., 1985, Interpretation of electrorotation of protoplasts, II. Interpretation of experiments, Studia Biophysica, IO9, 5:14
4. Glaser, R. and Fuhr, G., Gimsa, J. and Hagedorn, R., 1985, Electrorotation - capabilities and limitations, Studia Biophysica 110, 43:50.
5. Glaser, R., Fuhr, G. and Gimsa, J., 1983, Rotation of erythrocytes, plant cells and protoplasts in an outside rotating electric field, Studia Biophysica, 96, 11:20.
6. Seeman, P., 1972, The membrane action of anaesthetic and tranquilazers, Pharmacol.Rev. 24, 583:655.
7. Glaser, R. and Fuhr, G., 1986, Electrorotation of single cells - a new method for assessment of membrane properties, in: Blank, M. /Editor/: Electric double layers in biologie, Plenum Press, New York.
8. Chan, D and Wang, H.H., 1984, Local anaesthetic can interact electrostatically with membrane proteins, Biochim.Biophys. Acta 770, 55:64.
9. Yamaguchi, T., Watanabe, S. and Kimoto, E., 1985, ESR spectral changes induced by chlorpromazine in spin-labeled erythrocyte ghost membranes, Biochim.Biophys.Acta 820, 157:164.
10. Pilwat, G. and Zimmermann, U., 1985, Determination of intracellular conductivity from electrical breakdown measurements, Biochim.Biophys.Acta 820, 305:314.

SPECTRAL CHARACTERISTICS OF CHLOROPHYLL A AT DIFFERENT STATES OF SOLVATION

Radka Vladkova and Stefka Taneva

Central Laboratory of Biophysics
Bulgarian Academy of Sciences
Sofia 1113, Bulgaria

INTRODUCTION

The experimental data available suggested the monomeric nature of the cation radicals and triplets of P680 and P700, the primary electron donors in plant photosynthesis[1]. Interactions of monomeric Chlorophyll a /Chl.a/ electrostatics /for a review, see[2]/, are proposed as a factors, responsible for the observed in vivo behavior of these complexes. Chl.a-water interactions have been used for modelling of P680 and P700 /for a review, see[3]/. Recent works[4,5], have demonstrated that polar solvents can be used for modelling of protein environment and four types of specific solvation, depending on solvents' empirical parameters /electrophilicity, nucleophilicity and steric factors/, have been distingished. Thus, bisligation and hydrogen-bonding of Chl.a in the case of methanol and only bisligation in the case of pyridine have been shown[4].

A comparative study of the electronic absorption, steady--state and time-decay fluorescence of Chl.a in two different solvents /methanol and pyridine/ and in solvent-water mixtures is presented in this work.

Information on the influence of different polar environments and water on the ground and singlet excited states' properties of monomeric Chl.a is gained.

MATERIAL AND METHODS

Chl.a was puchased from Sigma Chem.Co. The solvents used were preliminary dried, distilled and kept over activated molecular sieves 4 Å /Merck/. Chl.a absorption and fluorescence spectra were recorded on a Specord UV/VIS double beam spectrophotometer and on a Perkin-Elmer /MPF-44B/ spectrofluorometer /bandpass of 4 nm and excitation at 430 nm/, respectively. Time-decay fluorescence curves were obtained using a laser impulse fluorometer LIF 200, including N_2 laser with wavelength 337.1 nm /pulse duration 0.7 ns/ in conjuction with a boxcarintegrator. The decay curves were recorded after a cut-off filter 660 nm. The method for data analysis was presented in[6].

All measurements were carried out in closed quartz cuvettes /1 cm path length/ in the same conditions and at room temperature. The concentration of Chl.a molecules was 1.8×10^{-6}M in the case of PYR and 4.5×10^{-6}M in the case of MeOH solutions.

RESULTS AND DISCUSSION

Absorption spectra of Chl.a in MeOH- and PYR-water mixtures at different mole fractions of water are presented in Fig.1. A strong red shift in the Chl.a absorption in both Soret and red regions are well documented in many papers[7]. The appearence of aggregates absorbing at 710 nm and at 746 nm /"crystalline" Chl. a form/, at the same mole fractions of water, depending on the manner of solvent and water additions is clearly seen from the Fig.1. /curves 2a and 3a/. The 710 nm absorbing form is unstable and time dependent appearence of a maximum at 746 nm was observed. These data confirm the suggestion of Dijkmans[7] that water/ MeOH ratio in the close vicinity of Chl.a molecule is the determining factor for the types of aggregates obtained.

In Chl.a-PYR-water mixtures considerable changes, strongly dependent on the quantity of water added, are observed solely in the Soret and 639 nm bands regions. There not quantitative data available on what mole fraction of water /f/ above, the band at 639 nm, attributed to the S_o-S_2 /Q_x/ transition of bis-ligated Chl.a[8,9] disappeares. The decrease of this band and its red shift are accompanied by an increase of absorption around 583 nm and at f higher than 0.5 the band at 639 nm completely disappeares. Thus the spectra observed at the former f values /f > 0.5/ are similar to those of Chl.a in MeOH. More clearlc, this simillarity is revealed from the comparison of the spectral characteristics summarized on Table 1. The spectral features ε_s /$\mathcal{E}_s$, and θ_s were considered by Renge and Avarmaa[4] as a criteria for H-bonding of Chl.a in solutions. The values obtained for f > 0.5 are within the range of the reported values for H-bond formation. The appearence of resolved peaks at 583 and 534 nm /Fig.1b-curve 3/, points out that L_1H is the more possible type of specific solvation[4]. Data obviously indicated that aggregate formation does not occur up to 0.71 mole fractions of water unlike in the case of MeOH-water mixtures. This may be attributed to the entirely different capability of these two solvents and of PYR and water to ligate Mg and to form H-bonds with Chl.a[4]. Additionally, perturbations of the ring V of Chl.a, leading to enolization, empimerization and enolate anion formation might take place[10,11]. In accordance, the observed increasing of pH of PYR-water mixtures /from 6 to 9/ when f increases from 0.5 to 0.85 could be connected with such types of Chl.a transformations[11].

On the Table 1 , the spectral characteristics obtained from the steady-state fluorescence measurements are presented as well. The comparison of these features for Chl.a in PYR-water with those for Chl.a in MeOH supports the suggestion that in the former mixtures Chl.a is additionally specific solvated through H-bonding like in MeOH.

The relative spectral chracteristics of Chl.a in PYR-water mixtures /relative to that in PYR solution/, obtained from the integral absorption and fluorescence and from the time-decay nanosecond fluorescence measurements are summarised on Table 2. The increase of A_w/A_o and I_w/I_o ratios with the increase of water content indicates that a great number of molecules are initially excited to the first excited state and a great number

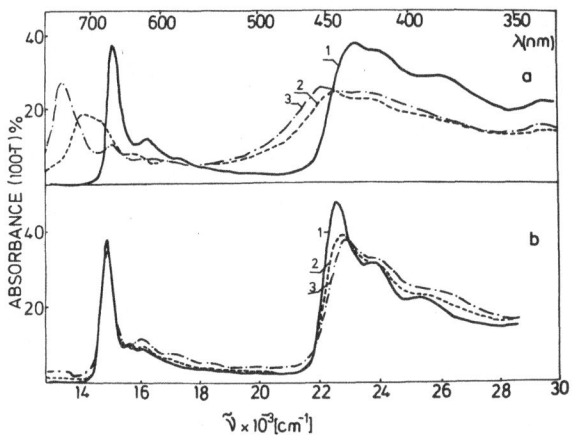

Fig. 1 Absorption spectra of Chl.a in MeOH-water /a/ and in PYR-
-water /b/ mixtures. 1a: f=0; 2a: f=0.55 3a: f=0.55;1b:
f=0; 2b: f=0.33, 3b: f=0.71.

Table 1 Spectral characteristics of Chl.a in water-solvent mixtures, f the mole fraction of water; λ_s, λ_r the absorption maxima im nm of Soret and red bands; $\mathcal{E}_s / \mathcal{E}_j$,
the ratio of the extinction coefficients of the Soret
maximum to the nearest satelite band; δ_s, δ_r and δ_f
the half-width of the Soret, red and of the main fluorescence bands; λ_1, λ_2 the wavelength in nm of the
main and the second fluorescence bands; F_1/F_2 the
ratio of their intensity.

Sol-vent	Absorption							Fluorescence			
	f	λ_s	λ_r	$\mathcal{E}_s / \mathcal{E}_s$,	δ_s	δ_r	λ_1	λ_2	F_1/F_2	δ_f
MeOH	0	431.5	665.7	1.05	85	23.5	675	730	10.5	24.5
PYR	0	443.5	671	1.69	38	20.5	676.5	735	19	20.5
PYR+	0.38	440	671	1.29	57	22.4	676	733	16	21.5
H$_2$O	0.53	438	670.7	1.26	68	23.2	676	731	13.6	23
	0.71	436.5	670.2	1.2	79	24		675.5	730	13.2	24

of molecules are capable to return back to the ground state
by fluorescence. The decrease of the ratios F_w/F_0 and τ_w/τ_0
when $I_w/I_0 > 1$ can be explained by processes taking place concomitant with the depopulation of the singlet excited state
by fluorescence from the initially excited molecules, i.e. the
quenching of their ns-fluorescence before they return to the
ground state. The relative increase in the steady-state long
wave fluorescence with the increase of the mole fractions of
water, compare with that of the main fluorescence band /Table 1 -
- F_1/F_2/, demonstrates the probable existence of species which
can act as energy traps.

Table 2. Relative spectral characteristics of Chl.a in pyri-
dine-water mixtures. A_w/A_o* the ratio of the integral
absorption within the 480-710 nm region; F_w/F_o the
ratio of the integral fluorescence within the 608-780
nm region; I_w/I_o the ratio of the intensities at the
maximum of the time-decay fluorescence curves. τ_w/τ_o
the ratio of the fluorescence lifetimes.

f	A_w/A_o	F_w/F_o	I_w/I_o	τ_w/τ_o
0	1	1	1	1
0.38	1.04	1	1.06	0.99
0.53	1.09	0.98	1.11	0.97
0.71	1.15	0.95	1.15	0.93
0.82	-	0.79	0.72	0.88

*Subscripts w and o denote the presence and absence of water
in the mixture, respectively.

For Chl.a fluorescence in MeOH-water mixtures both I_w/I_o
and τ_w/τ_o decrease in the same manner with increasing of water
content.
      All these effects can be summarized as follows: /i/ The
presence of water in the slovation shell of Chl.a molecule,
formed by the strong nucleophilic solvent PYR /which is without H-bonding capability/, gives rise to the possibility for
H-bond formation and additional perturbation of the most nucleophilic Chl.a center. /ii/ These perturbations increase the
cross section of the photoexcitation reactions of Chl.a molecule. /iii/ The heterogeneous solvation shell of Chl.a, formed
by the presence of two entirely different specific solvation
agents./pyridine and water/[4] in the immediate vicinity of Chl.a
molecule, favors intra/intermolecular processes concomitant
with nanosecond fluorescence, i.e. the existence of energy
traps.
      These data might be useful in the modelling of the primary
processes occuring after excitation of P700.

REFERENCES

1. Rutherford, A.W., 1986, How close is the analogy between
   the reaction centre of Photosystem II and that of purple
   bactria?, Biochem.Soc.Trans., 14:15.
2. O'Malley, P.J. and Babcock, G.T., 1984, Electron nuclear
   double resonance evidence supporting a monomeric nature
   for P700[+] in spinach chloroplasts, Proc.Natl.Acad.Sci. USA
   81:1098.
3. Boxer, S.G., 1983, Model reactions in photosynthesis,
   Biochim.Biophys.Acta, 726:256.
4. Renge, I. and Avarmaa, R., 1985, Specific solvation of
   chlorophyll a : solvent nucleophility, hydrogen bonding
   and steric effects on absorption spectra, Photochem.Photo-
   biol. 42:253.

5. Renge, I.V. and Bitova, E.V., 1987, Solvation of chloro-
    phylls by solvents, modelling protein environment, Biofi-
    zika in press.
6. Vladkova, R.S. and Spassov, V.Z., 1987, States of chloro-
    phyll a in mixed solvents: nonlinear least squares analysis
    of time-decay fluorescence curves, in "IV Symposium of Plant
    Metabolism Regulation. Proceedings, in press.
7. Dijkmans, H., 1973, Aggregation of chlorophylls in vitro.
    Absorption spectroscopy of chlorophyll a in water-methanol
    solutions, Eur.J.Biochem. 32:233.
8. Cotton, T.M., Trifunas, A.D., Ballschmiter, K. and Katz,
    J.J., 1974, State of chlorophyll a in vitro and in vivo
    from electronic transition spectra and the nature of antenna
    chlorophyll, Biochim.Biophys.Acta 368:181.
9. Katz, J.J., Shipman, L.L., Cotton, T.M. and Janson, T.R.,
    1978, in "The Porphyrins", D.Dolphyn ed., Plenum Press,
    London.
10. Hynninen, P.H., Wasielewski, M.R. and Katz, J.J., 1979,
    Chlorophylls VI.Epimerization and enolization of chloro-
    phyll a and its Magnesium-free derivatives, Acta Chem.Scand.
    B33:637.
11. Hynninen, P.H. and Fllfolk, N., 1973, Chlorophylls I.Sepa-
    ration and isolation of chlorophylls a and b by multiple
    liquid-liquid partition, Acta Chem.Scand. 27:1463.

ELECTROPHORETIC MOBILITY OF MYOSIN-TREATED LYMPHOCYTES

FROM PATIENTS WITH MYOCARDIAL INFARCTION

Georgi Ivanov, Stylian Stoeff, Radko Pelov* and
Milka Metodieva

Department of Physics and Biophysics
Department of Cardiology, First Cardiological Clinic*
Medical Academy, Sofia 1431, Bulgaria

INTRODUCTION

Necrotic areas in the cardiac tissue appear as a result
of many heart diseases. From the destroyed cells in these
areas the specific antigen proteins get into the organism and
cause autoallergic processes. The list of autoantigens susceptible to immune reactivity and possible injury has been expand-
ed.[1] Recently years the study of the immunopathological process
in the heart includes measurement of circulating immune complexes, immunofluorescent localisation of immune deposits, demonstration of cardiac antigen-specific cytotoxicity, etc.

In clinical practice it is generally accepted to make conclusions about the immunopathological process by determination
of delayed type hypersensitivity. This hypersensitivity can be
determined by cell reaction of passive leukocyte migration,
based on the sensitization of the lymphocytes with antigens of
endogenous origin. Recently published data show that lymphocyte sensitization is accompanied by a change in lymphocytes
surface electric charge.[2,3,4] This fact allows a rapid and
exact determination of the autoallergic processes by means of
cell electrophoresis / up to 2-3 hours after the blood samples
are taken/.

The myocardic myosin shows strongly manifested tissue
antigen specifity. The existence of antigen differences in
this protein, isolated from right and left ventricle and auri-
cles is extremely useful for determination of the localization
and size of the nectrotic zone in the heart tissue with the aid
of the cell immunological reactions discussed above.[5]

The aim of the present work is to determine the electrophoretic lymphocyte mobility of patients with infractions.

MATERIAL AND METHODS

Myosin was isolated from the left human ventricle.[5] The
protein concentration was determined spectrophotometrically
using E/278/=0.555. The obtained preparation possesses a high
degree of nativity and purity. Beffore using it in the electrophoretic study, the protein solution was dialyzed against

273

0.066 M sodium phosphate buffer /pH 7.35/ and ionic strength 0.172 M.l^{-1}.

Blood samples /approximately from 10 to 15 ml/ were taken from the patient with clinically proved infarction on the fifth day of the disease and treated with pearls in order to prevent blood coagulation. The lymphocytes were isolated from blood samples in specific medium creating density gradient /modification of Boum method/. The lymphocytes were incubated in TC 199, pH 7.35 for 90 min at 37°C ; sample 3 ml / 10^6 cells and 100 g myosin/ and control sample /10^6 cells/.

The delayed hypersensitivity of blood samples towards human heart myosin was determined by the leukocyte migration inhibition test /LMIT/.[6]

The electrophoretic mobilities were measured by microscopic cell electrophoresis with a recangular chamber at a constant electric field I=10 mA and 37°C of the measured suspension. From each sample the electrophoretic velocity of 30 lymphocyte cells was determined in two directions and the small cell drift /under 10 %/ was taken into account. The average of the lymphocyte electrophoretic mobility /LEM/ for untreated /u/ and myosin-terated /u$_A$/ samples were calculated with the aid of a computer. The standart deviation /SD/, the percentage difference /ME/ of LEM between u and u$_A$, and the statistical significance /P/ were calculated as well.

Table 1. Electrophoretic mobility of myosin-treated and untreated lymphocytes

	U$\pm$SD $10^{-8}/m^2s^{-1}v^{-1}/$	U$_A\pm$SD $10^{-8}/m^2s^{-1}v^{-1}/$	ME %	P –
Patients with registered autoallergy	1.57$\pm$0.05	1.84$\pm$0.08	17	0.001
	1.76$\pm$0.08	2.02$\pm$0.05	15	0.001
	1.65$\pm$0.05	1.89$\pm$0.06	14.5	0.001
	1.68$\pm$0.07	1.92$\pm$0.05	14.2	0.001
	1.71$\pm$0.08	1.98$\pm$0.06	15.8	0.001
Patients with unregistered autoallergy	0.86$\pm$0.06	0.92$\pm$0.07	7	NS
	1.13$\pm$0.05	1.17$\pm$0.05	3.5	NS
	1.19$\pm$0.04	1.23$\pm$0.07	3.4	NS
	1.19$\pm$0.02	1.15$\pm$0.06	– 3.4	NS
	1.95$\pm$0.01	1.89$\pm$0.03	– 3.1	NS
Controls	1.47$\pm$0.03	1.26$\pm$0.05	-14	0.001
	1.45$\pm$0.04	1.31$\pm$0.04	-10	0.001
	1.50$\pm$0.04	1.35$\pm$0.06	-10	0.001
	1.43$\pm$0.04	1.22$\pm$0.04	-14.7	0.001
	1.52$\pm$0.03	1.36$\pm$0.05	-10.5	0.001

U	–LEM of untreated samples
U$_A$	–LEM of myosin-treated samples
SD	–standard deviation
ME	–percentage difference
P	–statistical significance
NS	–statistical non-significance

RESULTS

The obtained data of the electrophoretic mobility are demonstrated on Table 1.

Three main groups are well distinquished. In the first group of patients with myocardial infraction and pronounced autoallergy determined by LMIT, the electrophoretic mobility of myosin-treated lymphocytes is increased by about 15-16 % compared to untreated lymphocytes. The control group of healthy people demonstrates remarkable effect of decreasing of the electrophoretic mobility by about 10-15 %. The statistical analysis shows value of significance / $p < 0.001$/ in both cases. The third group of patients withmiocardial infaraction, but with unregistered autoallergy, shows statistically in significant differences.

All these results suggest that the electrophoretic lymphocyte mobility measurment of samples taken from patients with myocardial infraction could be succesfully used as a method for rapid determination of the delayed hypersensitivity toward specific antigen proteins of the heart.

From the data obtained it is difficult to answer why the electric charge of the lymphocytes in the studied groups differs. The causes of immune injury to the heart are complex and multiple. Our future work involves the interplay of many other factors.

REFERENCES

1. Russel, I.J. and Persellin, R.H., 1983, Heart desease, in: "Immunology in Medicine", E.J.Holborow and W.G.Reeves, eds. Academic Press, London and New York.
2. Fike, R.M. and van Oss, C.J., 1979, Simplified cell micro-electrophoretic method applied to the macrophage electrophoretic mobility test for cancer diagnosis, Immunol.Commun., 8:173.
3. Kaplan, J.H., Lockwood, S.H., Cunningham, T.J., Springer, G.F. and Uzgiris,E.E., 1981, Reactivity of Thomson-Friedenreich/T/ antigen with lymphocytes from tumor-bearing individuales as detected by cell electrophoretic mobility, in "Cell electrophoresis in cancer and other clinical research", A.W.Preece and P.Ann.Light, eds.Elsevier/North-Holland biomedical Press, Amsterdam.
4. Uzgiris, E.E., Lockwood, S.H. and Kaplan, J.H., 1982, Early membrane events monitored by changes in cell surface charge following specific antigen-lymphocyte binding, Clin.Immunol. Immunopath., 25:264.
5. Ivanov, G.G., Zueva, M.V., Zemskov, V.M., Borisova, A.M., Alkadarski, A.S., Shekhonin, B.V., Kondalenko, V.F. and Rukosuev, V.S., 1986, Poluchenie i immunomorphologicheskii analis rasnih tipov miosina cheloveka i jivotnich, Immunologija, 3:76.
6. Zemskov, V.M., Ivanov, G.G., Borisova, A.M., Zakharkin, V. A., 1982, Issledovanie hiperchuvstvitelnosti zamedlenogo tipa u bolnih revmatismom s pomostchi miosina serdza cheloveka, Terapevticheskii archiv, 6:110.

LIPID TO PROTEIN RATIO OF RECONSTITUTED PURPLE MEMBRANE FRAGMENTS

DETERMINING THE ELECTROKINETIC BEHAVIOUR OF PROTEOLIPOSOMES

Tzvetana R. Lazarova, Maria R. Kantcheva*, and
Stoyl P. Stoylov**

Central Laboratory of Biophysics
Bulgarian Academy of Sciences
*Department of Biophysics and Radiobiology
University of Sofia, 1421 Sofia, Bulgaria
**Department of Physics and Biophysics
Medical Academy, 1431 Sofia, Bulgaria

INTRODUCTION

The purple membrane as a specialized domain of the Halobacterium halobium plasma membrane contains a single protein component bacteriorhodopsin /BR/ with protonmotive activity used
by the bacterium to synthesize ATP.[1] Reconstitution of this
active light-driven proton pump presents valuable information
about BR function as a generator of an electrochemical potential
gradient /μ_H^+/.[2] The electrokinetic studies of purple membrane
fragments has been used in the membrane structure and function
analysis, providing evidence for understanding the molecular
mechanism of the light-driven proton pump.[3,4] The microelectrophoretic technique has been successfully used for estimation of
the zeta-potential and surface charge density of reconstituted
vesicles.[5]

The present work was carried out in an effort to determine
the electrokinetic behaviour of reconstituted purple membrane
/PM/ fragments in dependence on the lipid to protein ratio.

MATERIAL AND METHODS

PM fragments were isolated by a standart procedure.[6] The
concentration of BR was determined spectrophotometrically,
using $\varepsilon_{570} = 6.3 \times 10^4 M^{-1}.cm^{-1}$. The egg phosphatidylcholine
/PC/ was purchased from "Sigma" and its purity was established
chromatographically. Phosphate analysis was performed.[7] PC/BR
vesicles were prepared using the freeze-thawing sonication technique /FTST/.[8] Thin-layer film of egg PC /20 mg/ was suspended
in 2 ml 1 mM Hepes pH=6.8 and sonicated in a bath-type sonicator
for 20 min. PM fragments were added to the sonicated vesicles
so that various weight ratios of PC to BR have been obtained.
Samples were frozen into liquid nitrogen, left to thaw temperature and sonicated for 2 min above the phase-transition temperature. The proteolyposomes, as previously reported, exhibit
a large trapping volume and size.[9]

The electrokinetic measurements were carried out on Cyto-pherometer "Opton" with TV monitoring system. The zeta potential /ζ/ was calculated according to the Helmholz-Smoluchowski equation:

$$\zeta = \frac{\eta u}{\varepsilon_r . \varepsilon_o} ,$$

/1/

where η is the viscosity of the suspending medium, ε_r - dielectric constant, ε_o - permittivity of free space and u - electrophoretic mobility.

The relation between surface charge density /σ/ and surface potential /ψ_s/, proposed by the Gouy-Chapman equation, was used for further calculations :

$$sinh\left(\frac{ze\,\psi_s}{2\,\kappa T}\right) = A.\sigma. C^{-1/2}$$

/2/

where C is the concentration of 1:1 electrolyte, z is the ionic charge, A = /1/8 $N\varepsilon_r.\varepsilon_o \kappa T/^{1/2}$ = 136.4 at 22°C.

In all experiments the temperature was maintained at 22°C.

RESULTS AND DISCUSSION

The electrophoretic mobilities and corresponding zeta potential values of liposomes reconstituted with various PC/BR weight ratios are presented in Table 1.

Table 1. Electrophoretic mobility and zeta potential of proteo-lyposomes formed from PC and BR in aqueous solution 1 mM Hepes - pH = 6.8

L/P /w/w/	u.10^{-18} u $\pm$ S.D. /m^2.V^{-1}.s^{-1}/	ζ /mV/
10	- 2.09 ± 0.05	- 26.89
20	- 1.77 ± 0.01	- 22.75
40	- 1.47 ± 0.03	- 18.96

The electrophoretic mobility of PC vesicles in 1 mM, 10 mM and 50 mM NaCl was found to be zero. The presence of negatively charged protein and lipid groups from the PM fragments at the proteoliposome surface determines the electrokinetic behaviour of PC/BR vesicles. Increasing BR content of proteoliposomes results in elevated zeta potential values.
The electrophoretic mobility data of liposomes formed from mixtures of PC and negatively charged lipids recently presented illustrated a well-defined dependence of the zeta potential on the charge lipid vesicle content[5] Experimental results

have been analysed according to predictions of the Gouy-Chapman-
-Stern theory. The zeta potential values of BR reconstituted
proteoliposomes are in a good agreement with the experimental
data cited above and could be used for evaluation and support
of the new theoretical models.[10]

Surface charge density values of PC/BR vesicles are shown
on Fig.1.

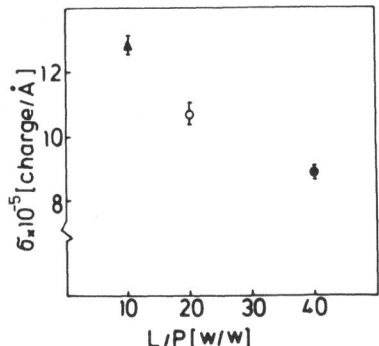

Fig. 1. Lipid to protein ratio dependence of PC/BR vesicle
surface charge density /6/

The proteoliposome surface charge density values are about an
order of magnitude smaller as compared to those of purple membrane fragments.[3] It could be suggested that the electrokinetic
behaviour of PM fragments and vesicles depends both on the
surface membrane structure and conformation.

It has to be noted that the use of the Gouy-Chapman equation involves some limitations which may interfere in the exact
evaluation of the surface charge of reconstituted vesicles. In
on attempt to minimize some errors, PM fragments were sonicated
before the reconstitution procedure in order to minimize nonuniform fragment distribution.

Ionic strength dependence of proteoliposome zeta potential
is shown on Fig.2. As expected from the Gouy-Chapman theory,
potential values become less negative with increasing the salt
concentration of the suspending medium. More significant dife-
rences in zeta potential values at various L/P ratios are registered at low ionic strenght of the suspending medium / 1 mM
NaCl/. This effect decreases with increasing the ionic strenght
value, where a large screening effect of counterions is observed.

According to our experimental results, the electrokinetic
properties of proteolyposomes as model membrane structures could
be used in an analysis which introduces hydrodynamic drag forces
related to charges located at a Debye length from the membrane-
-solution interface.Further details on the bacteriorhodopsin
surface structure and C-terminus dynamic behaviour could be
related to the electrostatic potential adjacent to purple
membrane fragments.

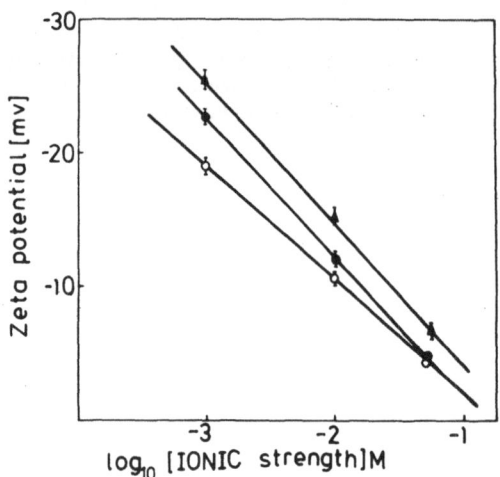

Fig.2 Ionic strength dependence of proteoliposome zeta potential

REFERENCES

1. Oesterhelt, D., Stoeckenius, W., 1971, Rhodopsin-like protein from the purple membrane of Halobacterium halobium, Nature New Biol., 233, 149-152.
2. Oesterhelt, D., Stoeckenius, W., 1973, Functions of a new photoreceptor membrane, Proc.Natl.Acad.Sci., USA, 70, 2853-2857.
3. Popdimitrova, N., Kantcheva, M.R., Dancshazy, Zs., Stoylov, S., 1985, Surface charge of purple membrane fragments, in: "Cell electrophoresis ", Eds. W.Schuett, Klinkmann, H., Walter de Gruyter, Berlin, New York, 167-173.
4. Kantcheva, M., Popdimitrova, N., Stoylov, S., 1984, Electrokinetic properties of purple membrane particles from Halobacterium halobium, Bioelectrochem.Bioenerg., 12,2-3.
5. McDaniel, R., McLaughlin, A., Winiski, A., Eisenberg, M., McLaughlin, S., 1984, Bilayer membranes containing the ganglioside G_{M_1} : Models for electrostatic potentials adjacent to biologycal membranes, Biochemistry, 23, 4618-4624.
6. Oesterhelt, D., Stoeckenius, W., 1974, Isolation of the cell membrane of Halobacterium halobium and its fractionation into red and purple membrane, Methods of Enzymology, 31, 667-678.
7. Ames,B., 1966, Assay of inorganic phosphate and phosphatases Methods in Enzymology, 8, 115-, Eds.E.F.Neufeld, V.Ginsburg
8. Casadio, R., Stoeckenius, W., 1980, Effect of protein-protein interaction on light-adaptation of bacteriorhodopsin, Biochemistry, 19, 3374-
9. Pick, V., 1981, Liposomes with a large trapping capacity prepared by freezing and thawing of sonicated phospholipid mixtures, Arch.Biochem.Biophys.212, 186-
10. McLaughlin, S., 1986, New experimental models for the electrokinetic properties of biological membranes: The location of fixed charges affects the electrophoretic mobility of model membranes., Studia biophys, in press.

EXOGENEOUS ATP ADMINISTRATION EFFECTS ON SOME BIOPHYSICAL PROPERTIES OF HUMAN ERYTHROCYTES DURING in vitro AGEING

M. Ratcheva-Kantcheva, S. Ivanov, and B. Galutzov

Department of Biophysics and Radiobiology, Biological
Faculty, University of Sofia
Dr.Tzankov 8, 1421 Sofia, Bulgaria

INTRODUCTION

A lot of biochemical and morphological data are summarized in order to improve the quality of stored red blood cells and to elucidate the mechanism of erythrocyte in vitro ageing /1-3/.
Biophysical experimental and theoretical investigations present valuable information about the processes involved in erythrcyte liquid storage destabilization and plasma membrane alterations, which results in cell destruction /4,5/.
The erythrocyte senescence process in vivo and in vitro involves: increased cell density and osmotic fragility, membrane--associated MDA-cross-linking, changes in surface IgG, cellular nucleotide production decrease, changes in antioxidative enzyme activity, as well as membrane transport alterations /6-8/.
Correlating the changes in biophysical parameters with morphological and metabolic changes, and studying the reversibility of these properties in vitro following plasma membrane stabilization, are essential.
There is evidence that exogeneous ATP modifies the cell membrane structure and function, and exhibits rejuvenation effects on stored red blood cells /9,10/. The mechanism of exogeneous ATP-induced changes in erythrocyte membrane properties is not well established. This study provides new evidence in favour of the role of ATP to enhance erythrocyte storage stability.

MATERIAL AND METHODS

Stored blood on day 1-3 was obtained from the National Blood Transfusion Centre, Sofia.
Blood samples /5 ml/ were centrifuged, the buffy coat removed, and the red cells washed and resuspended in phosphate buffered saline /PBS/ pH=7.4, to a haematocrit value of 15 %.
Hypotonic haemolysis in saline solution /0.3, 0.6 and 0.9% NaCl/ was used for osmotic fragility measurements. Aliquots of 0.05 ml were mixed with 2 ml NaCl solution. The mixtures were incubated for 10 min at $20^{\circ}C$ and adsorbtion was determined spectrophotometrically /λ =690 nm/. The rate of haemolysis value

were presented in the time 5/10 s and 5/20 s.

Cell size distribution curves were obtained with Celloscpe using standard procedures.

The erythrocyte electrophoretic mobility was measured in PBS with Cytopherometer "Opton".

The parameters under investigation were monitored on the 1st, 5th and 25th hour of in vitro incubation.

Exogeneous ATP(Phosphobion - Romania) treatment of red blood cells was performed during nutrient-free liquid storage (30 min, 1 or 5 h) in a concentration of 10 ⲭM or 50 ⲭ M.

RESULTS AND DISCUSSION

During the initial time of in vitro incubation (1 and 5 h) of erythrocytes in PBS (4°C) no significant changes in osmotic fragility were established. Statistically significant changes in erythrocyte membrane osmotic properties were registered on the 25th hour of incubation (data not shown). The osmotic fragility increase in hypotonic suspending medium (0.6 % NaCl) could be related to alterations in the membrane transport activity, and age-dependent individual cell membrane properties, which reflects the initial destabilization process in the red blood cell /RBC/ population. Monitoring of the rate of haemolysis during in vitro ageing presented evidence about reduction in membrane stability. A progressive increase of the parameter under investigation was established in the two phases of the time of haemolysis.Cell size distribution curves during in vitro incubation show changes in the erythrocyte population size /Fig.1/. On the 5th hour of liquid storage some increase /about 7 %/ in the cell population homogeneity was registered. There are indications that nondiscrete heterogeneity of human erythrocyte volume is a sensitive indicator of clinical importance /11/.

The shift of the 5th hour distribution curve to the left, indicates a decrease in the RBC volume. This effect is expected to be related to the simultaneously increased electrophoretic mobility and it could be attributed to changes in cell membrane transport.

The elevated cell population size as presented on the 25th hour of incubation may be related to matabolic energy depletion, destructive processes and alterations in the erythrocyte membrane, corresponding to the cell ageing phenomenon.

The increase in red blood cell volume is considered as a destabilization process increasing the cell osmotic fragility. There are reports concerning the decrease in the RBC deformability during in vitro storage /12/.

Exogeneous ATP administration resulted in a well pronounced preservation of cell volume as registered during in vitro incubation /Fig.2/. The cell size homogeneity increase and volume reduction on the 5th hour of incubation were found to be reduced as compared to the same effect in the nontreated cell suspension. There are no significant changes in the distribution curves on the 5th and 25th hour. This stabilization of the cell population size may be of importance in erythrocyte osmotic stability effects as presented on Fig.3.

282

The most pronounced decrease in the mean value of the rate of haemolvsis was observed when ATP (50 μM) was applied on the 1st hour of in vitro cell ageing. This effect was found to decrease in the second phase of haemolysis (5/20 s).

The osmotic fragility data presented a real support of the registered cell size distribution effects of exogeneous ATP treatment (50 μM) of RBC suspension. A decrease in osmotic fragility (0.6 % NaCl) was observed during in vitro storage. The osmotic stabilization was most pronouned after ATP treatment at the beginning of nutrient-free liquide storage /30 min/.

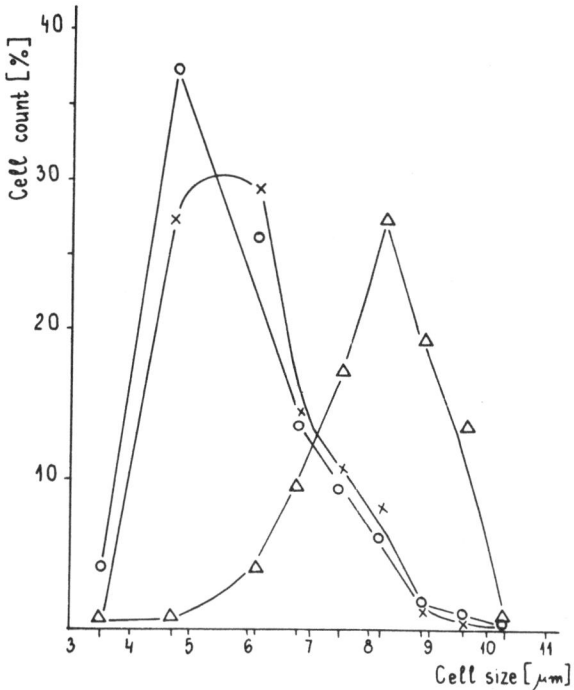

Fig. 1 Cell size distribution curves during in vitro incubation
/O - 1 hour; x- 5 hours; Δ - 25 hours/

The in vitro ageing of RBC results in changes of the cell volume and membrane properties, related to cell population integrity and osmotic stability. According to our experiments, it may be summarized that ATP administration results in a well pronounced protective effect against RBC ageing if applied at the very beginning of the storage time.

There is growing evidence that ATP modifies the cell membrane structure and properties related to permeability changes /13, 14/.

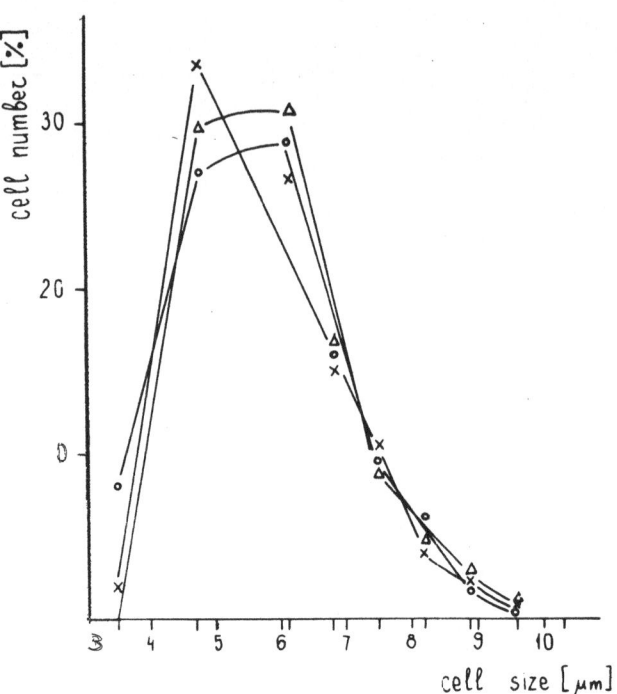

Fig.2 Cell size distribution curves after exogeneous ATP treatment (50 μM) registered during in vitro incubation:
○ - 1 hour; × - 5 hours; △ - 25 hours

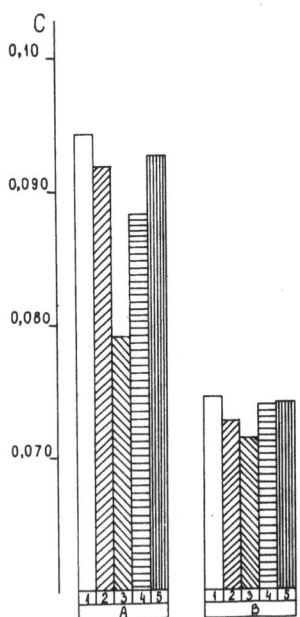

Fig.3 Exogeneous ATP effects on the RBC mean rate constant of
haemolysis value: 1 - Control; 2,3,4 - ATP treatment
(50 μM) during incubation for 30 min, 1 and 5 - ATP
(10 μM) - 30 min

The mechanism of exogeneous ATP-induced changes on erythrocyte membrane properties is not well established. There is evidence about changes in membrane transport activity /15/.

The evaluation of biophysical properties as compared to morphological and metabolic parameters should provide a convenient approach to the elucidation of processes associated with RBC stabilization.

REFERENCES

1. Hogman, C.F., Verdier, C.H. de, Ericson, A., 1985, Studies on the mechanism of human red cell loss of viability during storage at 4°C in vitro, Vox Sang. 48, 257-268.
2. Estep, T., Pedersen, R., Miller, T., Stupar, K., 1984, Characterization of erythrocyte quality during the refrigerated storage of whole blood containing di-/2-ethylhexyl/ phtalate, Blood, 64, 1270-1276.
3. Beutler, E., Kuhl, W., West, C., 1982, The osmotic fragility of erythrocytes after prolonged liquid storage and after reinfusion, Blood, 59, 1141-1147.
4. Svetina, S., 1982, Relations among variations in human red cell volume, density, membrane area, hemoglobin content and cation content, J.Theor.Biol., 25, 123-134.
5. Heindrich, R., Gaestel, M., Glaser, R., 1981, The electric potential across the erythrocyte membrane: A mathematical model, Acta biol.med.germ. 40, 765-770.
6. Hessel, E., Lerche, D., 1985, Cell surface alterations during blood-storage characterized by artificial aggregation of washed red blood cells, Vox Sang., 49, 86-91.
7. Hebbel, R.P., 1986, Autoxidation and sickle erythrocyte membrane; A possible model of iron decompartmentalization. In:"Free radicals, ageing and degenerative disseases", Alan R.Liss, 395-424.
8. Webster, N., Toolthill, C., 1986, Effects of blood storage on red cell anitoxidative systems, Acta haematol., 75, 30--33.
9. Rosengurt, E., Heppel, L.A., 1975, A specific effect of external ATP on the permeability of transformed 3T3 cells, Biochem.Bophys.Res.Commun., 67, 1581-1588.
10. Ratcheva-Kantcheva, M., Ivanov, S., Galutzov, B., Todorov, S., 1987, ATP-effects on erythrocyte during *in vitro* ageing, Biochem.Biomed.Acta, in press.
11. Bessman, J.D., Hurley, E.L., Groves, M.R., 1983, Nondiscrete heterogeneity of human erythrocytes: Comparison of Coulter principle flow cytometry and soret-hemoglobinometry image analysis, Cytometry, 3,4, 292-295.
12. Galea, G., 1986, Filtrability of ACD-stored red cells, Vox saguin, 51, 152-156.
13. Rorive, G., Kleinzeller, A., 1972, The effect of ATP and Ca^{2+} on the cell volume in isolated kidney tubules, Biochem. Biophys.Acta, 274, 226-239.
14. Horstman, D.A., Tennes, K.A., Putney, J.W., 1986, ATP-induced calcium mobilization and inositol 1, 4, 5-triphosphate formation in H-35 hepatoma cells, FEBS Lett., 204, 2, 189-192.
15. Chang, K.-J., Cuatrecasas, P., 1974, ATP-dependent inhibition of insulin-stimulated glucose transport in fat cells, J.Biol.Chem., 249, 10, 3170-3180.

RADIOPROTECTIVE EFFECTS OF ATP ON in vitro GAMMA-IRRADIATED HUMAN RED BLOOD CELLS

S. Ivanov, B. Galutzov, and M. Ratcheva-Kantcheva

Department of Biophysics and Radiobiology
Biological Faculty
University of Sofia, Bulgaria

INTRODUCTION

Studies on radiation damage to biological entities have clearly shown the importance of red blood cell /RBC/ morpho-biochemical and biophysical parameter monitoring in the study of the initiation and development of alterations after exposure to ionizing radiation.[1] The observed structural rearrangement of erythrocyte membrane components and modification in cell membrane functions result in electrophoretic mobility changes,[2] alterations in lectin-binding activity[3] and haemolysis.[4]

The elucidation of processes related to red blood stability could enhance the effective application of chemical substances for repair and reduction membrane damage.[5]

Exogeneous ATP treatment of erythrocytes presents some evidence to induce membrane alterations.[6] There is scanty information about the mechanism of ATP-membrane surface interaction.[7,8] The purpose of our study has been to investigate the exogeneous ATP effects on gamma radiation induced RBC damage.

MATERIAL AND METHODS

Human blood samples were centrifuged, the buffy coats removed, and the cells washed and resuspended in phosphate-buffered saline /pH=7.4/ to a haematocrit of 15 %.

A 60 Co gamma-source was used to irradiate red blood cell suspensions. Samples were exposed at a dose rate of 24 Gy/min at a temperature of around 20°C. Post-irradiation incubations of the irradiated cell suspensions were carried out at 4°C.

Exogeneous ATP, commercially available for clinical use /Phosphobion-Romania-10 mg/ml/, was added to the RBC suspension 30 min before irradiation, or one or five hours after gamma-irradiation in doses of 10 and 50 μM.

Erythrocyte osmotic fragility was monitored spectrophotometrically / λ =690 nm/. The rate of haemolysis was determined according to Riefkind and Ahraki.[7] Cell size distribution was obtained using a celloscope. The parameters under investigation were registered on the 1st, 5th and 25th hour of postirradiation incubation.

EXPERIMENTAL RESULTS

The osmotic behaviour of RBC during postirradiation incubation is presented on Fig.1. The spectrophotometrically detectable absorbtion A /690 nm/ corresponds to cell integrity in isotonic saline.

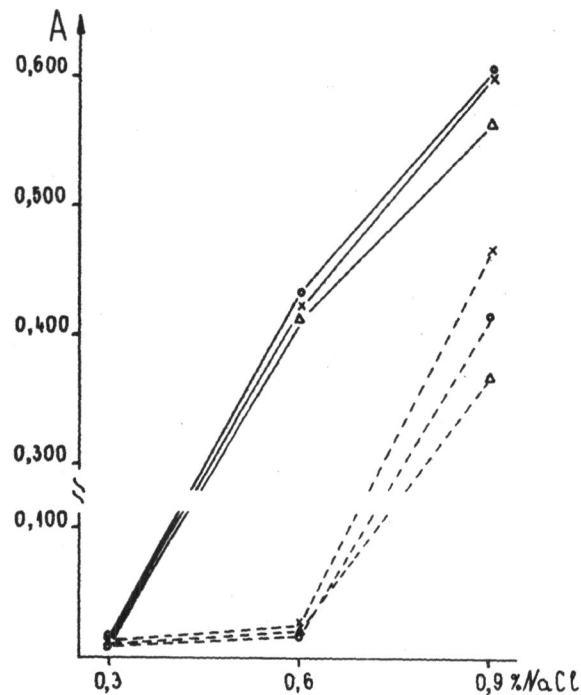

Fig.1 Osmotic fragility of human red blood cell, registered on the 25th hour of postirradiation incubation. Solid line-nonirradiated cell suspension. Dashed line-irradiated cell suspension: o - Control; ATP - added /50 μM/: x - 1st hour of postirradiation incubation; Δ - 5th hour of postirradiation incubation

RBC haemolysis after gamma-irradiation resulted in decrease of the parameter A under investigation. The cell lysis value /10 %/ registered on the 5th hour of postirradiation incubation increased to about 30 % on the 25th hour of incubation.Exogeneous ATP treatment /50 μM/ one hour after irradiation demonstrated some decrease /8 %/ in radiation haemolysis.
RBC osmotic behaviour monitored in hypotonic suspending medium /0.6 % NaCl/ showed a dramatic decrease in irradiated cell membrane stability. This effect may be considered as a result of the process of irreversible membrane damage during in vitro storage in nutrient-free suspending medium. ATP treatment in the case did not modify significantly the radiation-induced osmotic fragility increase.
The average rate constants of osmotic haemolysis during post-irradiation incubation are presented on Fig.2.

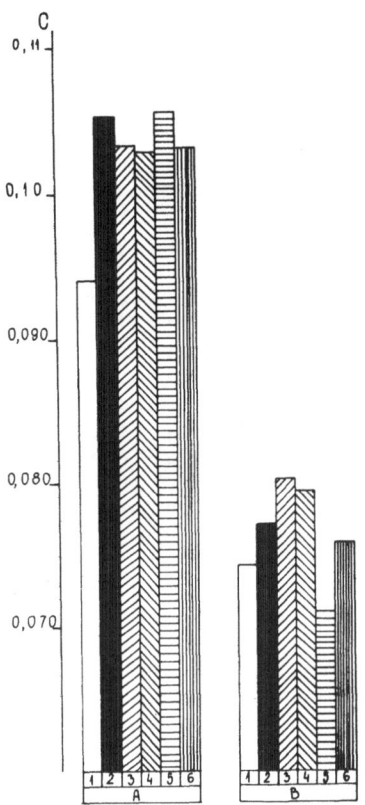

Fig.2 Mean value of RBC osmotic haemolysis rate constant
during postirradiation incubation: Panel A - 5/10 s
Panel B - 5/20 s.1 - nonirradiated sample ; 2 - irradiation /500 Gy/ 3 - ATP /50 μM/ treatment 30 min before
irradiation 4 - ATP /50μM/ 1 hour after irradiation ;
5 - ATP /50μM/ 5 hours after irradiation ; 6 - ATP
/10 μM/ 30 min before irradiation

     The elevated rate constant values of irradiated RBC samples
correlated with the increased osmotic fragility cited above.
The ATP administration /50 μ M/ demonstrated a small protective
effect in the first phase of osmotic lysis.
     Ionizing radiation may indirectly react with the protein
and lipid membrane components through the formation of free
radicals. Haemolytic damage is found to be mediated mainly by
HO radicals.[10] Cell size distribution curves of RBC monitored
during postirradiation incubation present evidence about progressive increase in RBC diameter - d . The position of d_{max}
/cell diameter value most frequently presented in the erytrocyte
population/ registered on the 5th and 25th hour is shifted to
the right. This effect could be related to erythrocyte membrane
transport alterations. Permeability changes of RBC due to ionizing radiation were previously reported as a result of biophysical studies using spin-labelled substances.[1] There are experimental results indicating radiation-induced cell volume increase
observed in rat erythrocyte volume curves.[11] The increase in
cell volume due to alterations in membrane transport enzymes
could be related to the registered RBC destabilization process

and haemolysis. The progressive cell size homogeneity decrease of about 10 %, observed during the time of postirradiatiin incubation, reflects predominantly the normal age-related cell alteration.

The molecular mechanism of cell membrane damage involves protein destruction, frequently attributed to spectrin conformation organization.[2] The electrophoretic mobility decrease registered during postirradiation incubation /data not shown/ confirm membrane surface charge density changes as previously reported.[12,13]

If we consider exogeneous ATP effects on the radiation-induced alterations of RBC stability in dependence on the time of administration, a striking difference is observed. The ATP treatment of erythrocytes before irradiation results in protective effects only as presented to reduce radiation-induced haemolysis. Osmotic fragility decrease of irradiated RBC is observed when ATP administration was performed after in vitro gamma irradiation. In the case it could be suggested that ATP treatment may be involved in processes related to repair of cell membrane damage. Changes in sodium channel conformation are found to alter the radiosensitivity of the channel.[14] One might expect that ATP-determined voltage-sensitive channel conformation is involved in the protective effects on whole-cell preparation.

ATP administration one hour after irradiation reduce the osmotic fragility of irradiated erythrocytes, but does not change the radiation-induced haemolysis. Experimental data analysis confirm the role of ATP during the radiation damage of human erytrocyte.

REFERENCES

1. Leyko, W., Bartosh, G., 1986, Membrane effects of ionizing radiation and hypertermia, Int.J.Radiat.Biol.,49,743-770.
2. Sato, C., Kojima, K., Nishizawa, K., 1977, Target of X-irradiation and dislocation of sialic acid in decrease of cell surface charge of erythrocytes, Radiation Research, 69,367-374.
3. Takahashi, K., Kaneko, I., 1986, Radiation-induced alteration in binding of concavalin A to cells and their susceptibility to agglutination, Int.J.Radiat.Biol., 49,979-986.
4. Shapiro, B., Kollman, G., 1968, The nature of the membrane injury in irradiated human erythrocytes, Radiation Research, 34, 335-346.
5. Petkau, A., 1986, Protection and repair ofirraaiated membranes. In: Free Radicals, aging and degenerative diseases, 481-508, Alau R.Liss.Iuc.
6. Sato, C., Kojima, K., Nishizawa, K., 1977, Translocation of hyaluronic acid in cell surface of cultured mammalian cells after X-irradiation and its recovery by added adenosine triphosphate, Biochim.Biophis.Acta, 470,446-452.
7. Rorive, G., Kleinzeller, A., 1972, The effect of ATP and Ca^{2+} on the cell volume in isolated kidney tubules. Biochim. Biophis.Acta, 274, 226-239.
8. Rosengurt, E., Heppel, L.A., 1975, A specific effect of external ATP on the permeability of transformed 3T3 cells, Biochem.Biophis.Res.Commun.,67,4,1581-1588.
9. Rifkind, I., Araki, K., Hadley, C., 1983, The relationship between the osmotic fragility of human erythrocytes and cell age, Archives of Biochemistry and Biophys., 222,2,582-589.

10. Bartosz, G., Leyko, W., 1981, Radioprotection of bovine erythrocytes to haemolysis, Int.J.Radiat.Biology, 39,39-46.
11. Valet, von G., Metzger, H., Kachel, V., Ruhenstroth-Bauer, G., 1972, Die Volumenverteilungs-Kurven von von Rattenery-trozyten nach Roentgenganzkoerperbestranhlung.Blut, XXIV, 274-282.
12. Ivanov, St., Kantcheva, M., Galutzov, B., 1985, The biochemical and morphologic blood analysis and electrophoretic behaviour of rat erythrocytes after gamma irradiation. In: Cell Electrophoresis, Walter de Gruyter & Co., Berlin, New York - Printed in Germany.
13. Vranska, Ts., Pantev, T., Ryzhov, N., Fedorenko, B., 1985, Electrophoretic mobility of erythrocytes exposed to accelerated helium nuclei or to gamma-rays, as modified by adeturone, Int.J.Radiat.Biol.47,285-290.
14. Freschi, J.E., Mpran, A., 1986, Effect of gamma radiation on sodium channels in different conformations in neuroblastoma cells, Biochem.Biophys.Acta, 858, 31-37.

ELECTROPHORETIC MOBILITY EFFECTS AND CELL SIZE DISTRIBUTION IN DEPENDENCE ON NITROUS ACID AND IRON CONTENT IN INTENSIVE CULTURE OF Scenedesmus acutus

K.Benderliev, M.Ratcheva-Kantcheva[x], and N. Ivanova

Institute of Plant Physiology, Bulgarian Academy of Sciences, [x]Department of Biophysics and Radiobiology Biological Faculty University of Sofia

Mass open field microalgal cultures are maintained as a source for biologically active compounds and for other products in Biotechnology /1,2/.

Nitrite accumulation in algal cultires, though well known, has long been neglected as being undesirable. It can be attributed to denitrification and to processes slowing down algal protein synthesis velocity in comparison with nitrate reduction velocity /3/.

It was shown recently that the algicide effect of nitrites is due to nitrous acid, the latter initiating oxidative stress characterized by enhanced peroxidation catalysed by iron /4/ and partly inhibited by molybdenum and urea /5/. It was also established that cell damage is minimized by addition of $\mathcal{L}$ -tocopherol, butylated hydroxytoluene, ethylenediaminetetra-acetic acid and thiourea /unpublished results/.

The particle electorphoresis technique is applied for investigation of different surface charge properties of cells and model membrane preparations /6, 7, 8/.

A considerable understanding of the mechanism of iron-enhanced nitrous acid inhibition of microalgal growth is gained by complex surface electric charge, cell size and yield gain examination.

In this paper we investigate the role of iron ions in nitrous acid poisoning using Scenedesmus acutus cells. A more comprehensive idea concerning the sequence of nitrous acid-induced events in microalgal suspension is given.

MATERIAL AND METHODS

The green microalga Scenedesmus acutus Tom. 8 was intensively cultured under laboratory conditions. The mineral nutrient medium has the following composition in mg/l : $MgSO_4.7H_2O$ - 1976; KH_2PO_4 - 680; $FeSO_4.7H_2O$ - 23; H_3BO_3 - 6.18; $MnSO_4.4H_2O$ - - 2.23; $CuSO_4.5H_2O$ - 2.49; $ZnSO_4.7H_2O$ - 2.87; EDTA Na_2 - 100. The nitrogen /1120 mg/l/ was supplied as $NaNO_3$ or different combinations of $NaNO_3+NaNO_2$ to ensure the necessary quantities of nitrous acid at the given pH value. The mixing air /100 $dm^3/h/l/$ contained 3 % CO_2. Continuous fluorecent illumination

/20 W/m^2/ was used. Samples were taken after cultivation for 24 h. The electrophoretic mobility was measured with cytopherometer "Opton". Cell size distribution was measured with celloscope. Yield gain was determined gravimetrically.

RESULTS AND DISCUSSION

Table 1 shows that 37 μg/l HNO$_2$ reduce algal yield gain about three-fold. Higher nitrous acid quantities cause zero yield gain and cell lysis.

Table 1. Effect of nitrous acid on the yield gain of <u>Scenedesmus acutus</u> /EDTA-chelated Fe=5.58 mg/l/

HNO$_2$ / μg/l/	0	37.6	122	188	263
Yield gain /mg/ml/	1.3	0.442	0.078	-0.188	-0.265

The decrease of EDTA-chelated iron concentartion through a range from 5.58 to 1.39 mg/l slightly represses growth in the absence of HNO$_2$ and diminishes the deleterious effect of the nitrous acid /Table 2/.

Tble 2. Effect of EDTA-chelated iron concentration on the yield gain of <u>Scenedesmus acutus</u> in the absence and presence of nitrous acid

	HNO$_2$ / μg/l/				
	0			188	
Fe /mg/ml/	1.39	2.79	5.58	1.39	2.79	5.58
Yield gain /mg/ml/	0.778	1.027	1.338	-0.025	-0.153	-0.188

Cell size distribution curves /Fig.1/ are indicative of the decrease in both cell size and homogeneity of the algal population after nitrous acid treatment, these effects being greater for larger HNO$_2$ and Fe quantities. Obviously a difference exists between iron-mediated nitrous acid alteration of cell growth and cell division.

A statistically significant increase in the electrophoretic mobility of algal cells up to 35 % was registered /Fig.2/ after nitrous acid treatment.

This effect shows that HNO$_2$ induces cell surface electric charge density changes due to cell membrane alterations. Accumulation of negatively charged peroxidation products cannot be excluded. Enchanced iron concentrations in the presence of nitrous acid reduce the electrophoretic mobility, most probably due to interaction of positively charged ferro-nitrous ions with negatively charged cell surface.

According to our recent results, HNO$_2$ accelerates Fe-EDTA and Fe-urea complexes destruction. The free iron ions are involved in the well-known /9/ Fenton process with O$_2$ and H$_2$O$_2$ producing the most damaging free radical species OH$^{\cdot}$. Our pre-

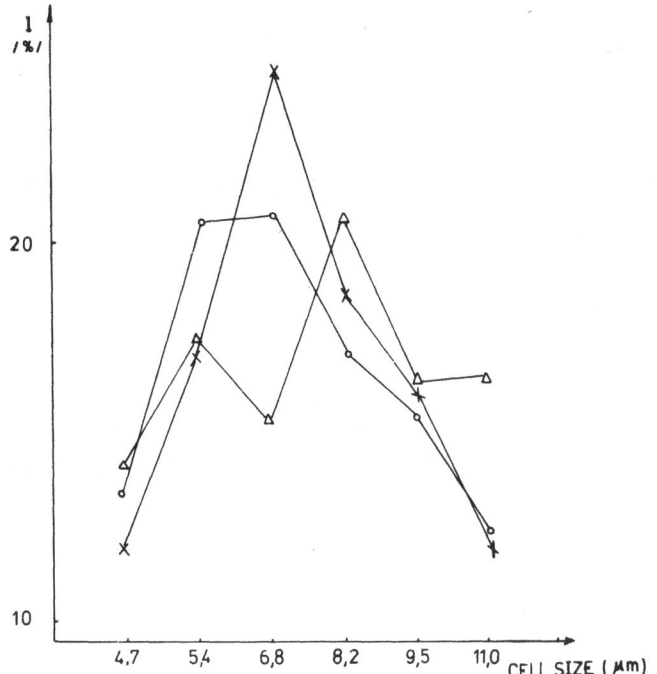

Fig.1 Cell size distribution curves of <u>Scenedesmus acutus</u>
cells after nitrous acid and iron treatment
+ control; o - HNO$_2$ - 188 μg/l, Fe-1.39 mg/l; Δ - HNO$_2$
- 263 mg/l, Fe - 5.58 mg/l

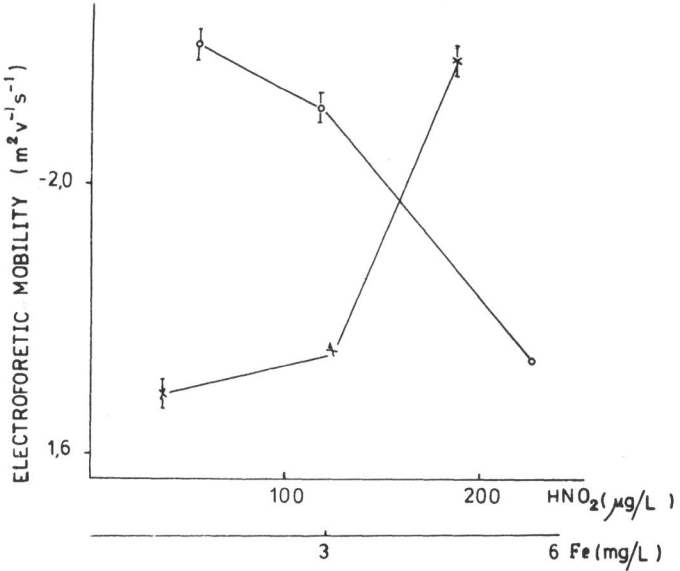

Fig. 2 Electrophoretic mobility of <u>Scenedesmus acutus</u> cells
in dependence of nitrous acid and iron concentration
x - Fe - 5.58 mg/l; o - HNO$_2$ - 188 μg/l

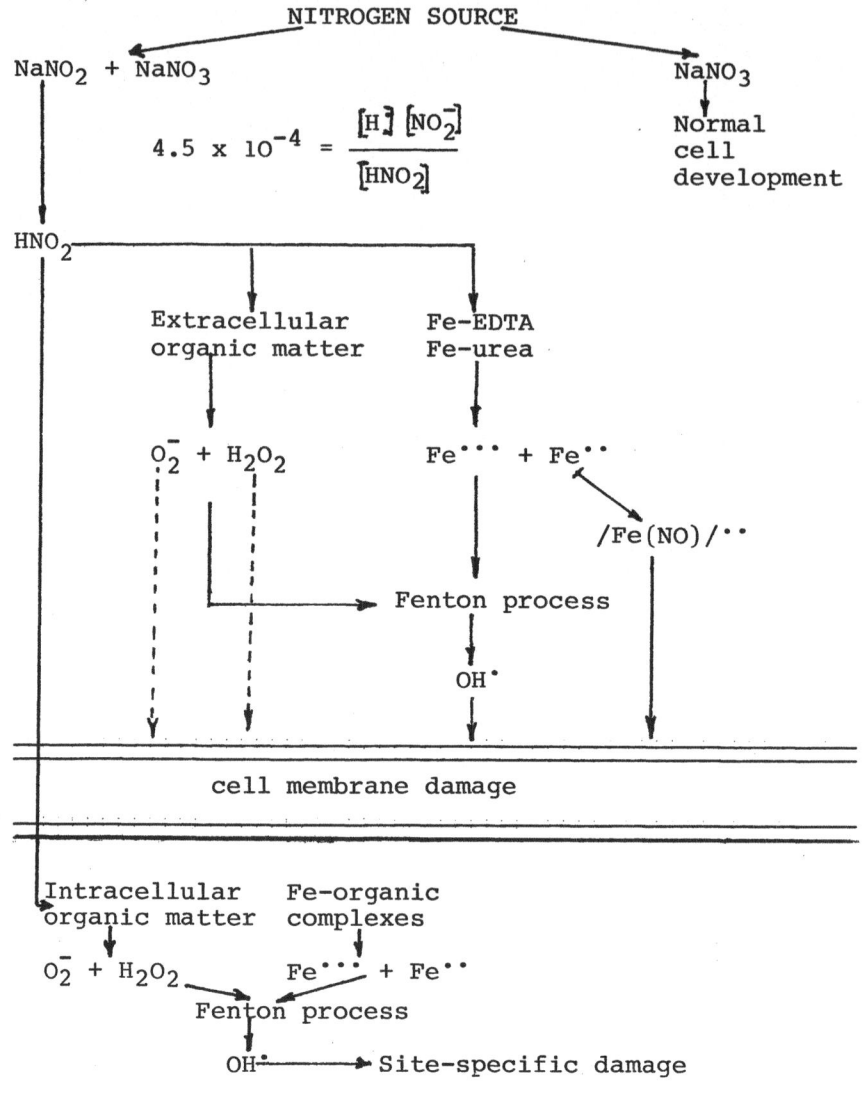

Fig. 3   Sequence of process induced by HNO_2 in green microalgal
         suspension

vious studies /4, 5, 10/ indicate that the following changes
in the cell content occur due to nitrous acid treatment: thio-
barbituric acid-reacting matter, phaeophorbides, nitrosamines
and extracellular reducing matter increase while carotene, chlorophyll a , chlorophyll b , proteins, carbohydrates and lipids
diminish. Summing up the results in our previous and the present
paper, and our unpublished results with metal chelators and
free radical scavengers, a fuller idea /Fig.3/ about the events
induced by nitrous acid in green microalgal suspension could
be presented /Fig.3/.

The electrophoretic mobility data demonstrate that the
electrokinetic approach is helpful for registration of cell
damage resulting from free radical attack and lipid peroxidation.

REFERENCES

1. Metting, B., Pyne, J., 1986, Biologically active compounds from microalgae, Enzyme Microb.Technol.,8,386-394.
2. Karuna-Karan, A., 1985, Commercial applications of large--scale culture of microalgae, Biotech.85, USA, Online Publ. Pinner, UK, 85-95.
3. Apparicio, P., Azuara, M., 1984, Wavelength dependence of nitrite release and the effects of different nitrogen sources and CO_2 tensions of Chlamydomonas reinhardii inorganic metabolism, In: Blue light effects in biological systems, ed.H. Senger, Spr.Verlag, Berlin, 196-206.
4. Benderliev, K., Ivanova, N., Puneva, I. /in print/. Effect of nitrous acid on the quality and quantity of Scenedesmus incrassatulus byomass. C.R.Acad.Bulg.Sci.
5. Benderliev, K., Ivanova, N., Puneva, I., 1987, Effect of sodium nitrate and urea on the protective action of molybdenum against nitrous acid-induced oxidative stress in Scenedesmus incrassatulus culture. Proc.Fourth Natl.Conf.Botany, Sofia.
6. Kantcheva, M., Popdimitrova, N., Stoylov, S., 1984, Electrokinetic properties of purple membrane particles from Halobacterium halobium, Studia Biophysica, 12, 626, 173-175.
7. McLaughlin, S., 1985, New experimental models for the electrokinetic properties of biological membranes: Studia Biophysica, 110, 1-3, 25-28.
8. Benderliev, K., Ratcheva-Kantcheva, M., 1987, Electrokinetic behaviour and algal flocculation in dependence on parameters of suspending medium and cell physiological activity, Bioelectrokinetics 87, Sofia University /in print/ /bulg./.
9. Halliwell, B., Gutteridge, J., 1986, Oxygen free radicals and iron in relation to biology and medicine: Some problems and concepts. Arch.Biochem.Biophys. 246, 2, 501-514.
10. Ivanova, N., Puneva, I., Benderliev, K., 1987, Effect of nitrous acid on the growth and development of asynchronous and synchronous cultures of Scenedesmus incrassatulus.IV Internat.Symp.Plant Metabolism Regulation, 1986, Varna, Bulgaria /in print/.

INFLUENCE OF CONSTANT MAGNETIC FIELD UPON CHEMILUMINESCENCE

IN WHEAT ROOTS

Ts. Gimishev and K. Tsolova

Faculty of Biology
Dept. of Plant Physiology
Sof. University "Kl. Ochridiski"

It has been found relatively late that living tissues have
the ability to generate weakest light emission in the visible
range of the spectrum, i.e. in its blue-green part. It was found
that such light is observed only in the presence of O_2 and consequently it is a result of oxidation reactions.

Oxygen as a participant in chemiluminescence, which is a
biradical reaction in the cells can take part in at least two
stages: initiation of the reaction and its continuation. The
O_2 molecule possesses two free electrons, i.e. it appears to
be biradical.

Chemiluminiscence takes place at the expence of the energy
released during the recombination of the free radicals, and the
intensity of light emission is proportional of the rate of their
recombination.

According to contemporary opinions, the superweak chemiluminescence is reflection to a definite degree of the energy
exchange taking place in the cells. There is little written
information about the influence of magnetic fields over chemiluminescence in plants.

The purpose of this investigation is to study the action
and consequence of constant magnetic field with induction B=45mT
over the chemiluminescence in wheat roots in dependence on the
exposure /t/. The generated light quanta in the roots are registered by a photometric device. The cuvette with the reactive
mixture is fixed in a definite position which makes it possible
to recolate the geometrical conditions in the measurement of the
tests. A photomultiplier, cooled by tap water, is installed in
the lower part of the camera. It is connected to an analyzer,
which magnifies the quanta transformed into an electric signal.
The magnified electronic signals are transmitted for registration
to the counting device.

For the registration of the chemiluminescence of the sample
the equipment is switched on an hour before the measurements
for the stabilization of the working regime. Each measurement
starts and finishes with a reading of the background, which by
the dark current of the photomultiplier. The measurements are
taken at an interval of 100 s.

The total quantum yield for the whole interval /100 s/ is
estimated by the elimination of the background value for the

corresponding period. The H_2O_2-induced chemiluminescence of the roots has been investigated. For this purpose 300 mg roots from six-days-old etiolated plants are used. The roots are placed in 1 ml 1% solution of hydrogen peroxide.

The experimental data for the action of constant magnetic field /CMF/ upon the chemiluminescence show strong light emission at two-hour exposure to CMF. The lowest light emission is observed at 16-hour exposure /Fig.1/. The results concerning the CMF effects upon the total quantum yield /Fig.2/ are identical with those shown on Fig.1.

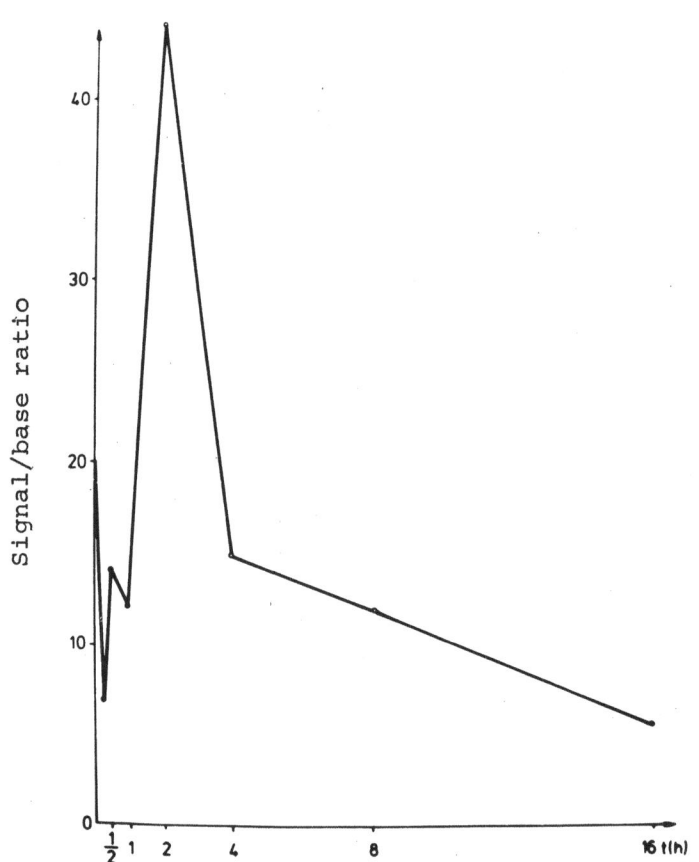

Fig.1   Influence of CMF on the chemiluminescence

From the above mentioned evidence about the action of CMF on chemiluminescence it is evident that most effective are the shortest exposures in the field /1 h and 2 h/. For the rest of the exposures the effect is strongly inhibitory and the weakest of light emission under the action of CMF is observed after 16-hour exposure.

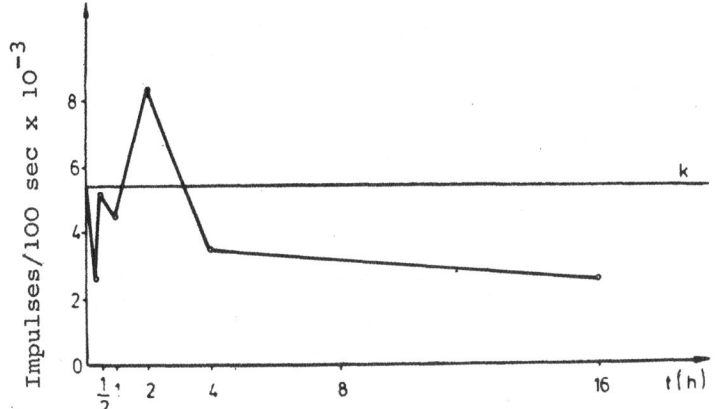

Fig. 2 Influence of CMF upon the total quantum yield

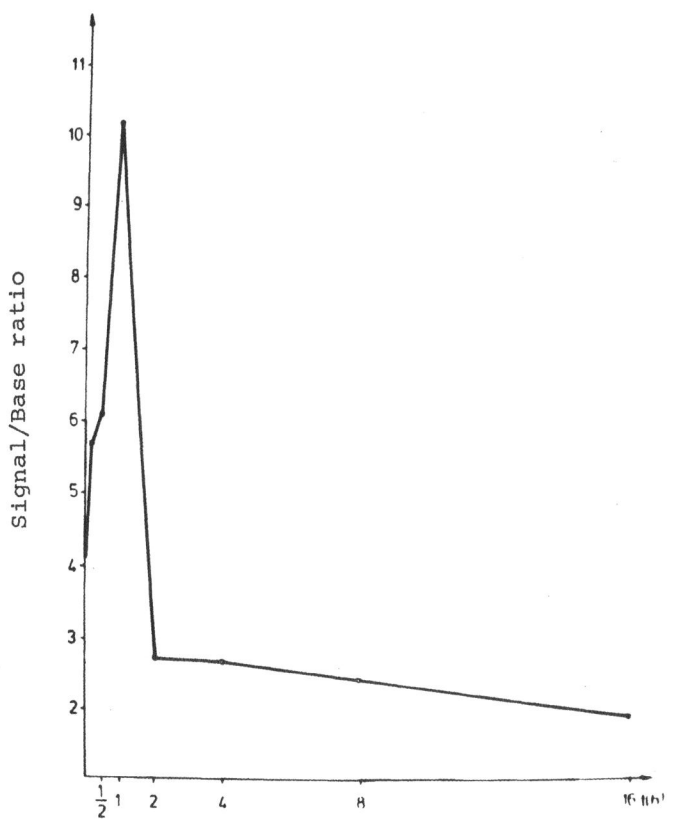

Fig. 3 Postmagnetic effect of CMF on the chemiluminescence

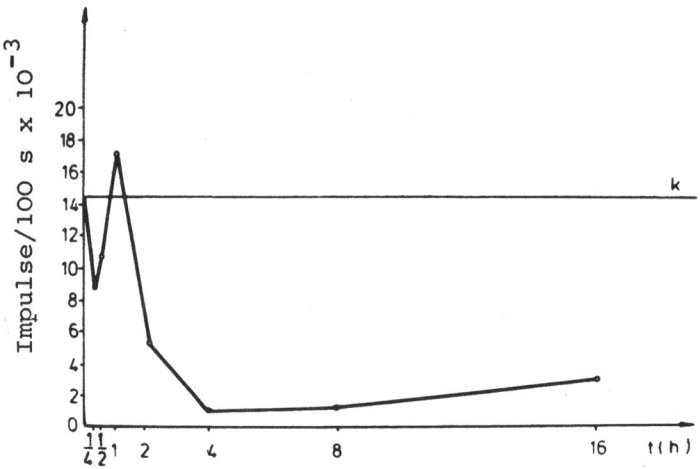

Fig. 4 Postmagnetic effect of CMF upon the total quantum yield

At present the basic biochemical reactions are considered
to take place with the participation of the free radicals
which play the   role of active semiproducts. It is established that MF is able to change the speed of the free radical
reactions /1, 2, 4/.

We can assume that the applied MF influences the electron
state of the molecules, resulting from the recombination of
the free radicals. In support of such an idea is the parallel
change of chemiluminescence and concentration of free radicals
in the tissues.

Consequently the low limit of light emission that reflects
the energy state of the molecules can serve as a source of information about processes taking place at electron level. The
low-limit light emission is basically connected with oxidative
processe.

Data on the correlation between light emission and energy
effectiveness of respiration are obtained by studying of the
influence of MF characterized with different field intensity
on the roots of broad bean /3/. The different organs of the
young plant emit amounts of light energy. The brightest light
emission is observed in the root system of plants.

The postmagnetic effect of CMF on the chemiluminescence
and on the summarized quantum yield is shown on Fig.3 and
Fig. 4 . A clearly stimulative effect after one hour exposure
time is to be seen. The four hours and the eight hours exposures are characterized by a strongly manifested inhibition
effect, while the sixteen hours exposure lead to increased
light emission,but its value remains lower than that of the
control.

The biochemical reactions proceed with the participation
of the free radicals not only with the direct action of CMF,
but also with the magnetic treatment of the imbibed wheat
seeds.

It is quite possible that the observed changes in the action of MF are result of the membrane protein conformational changes of the spatial location of the respiratory ensembles, and of the quaternary structure of the enzymes. These changes are connected with the changes of biological oxidation and with the alteration in the antioxidant acitivity of the lipoprotein membranes.

REFERENCES

1. Buchachenko, A.L., 1974, Chimitcheskaja poljarizatcija elektrono i jader, Moskwa, Nauka. /in Russian/
2. Buchachenko, A.L., Sagdeev, R.Z., Salichov, K.M., 1978, Magnitnie i spinovie effecty, Novosibirsk, Nauka.
3. Doscotch, E.V., Strekova, V.I., Tarakanova, G.A., Tarussov, B.N., 1969, Spontannaja sverhslabaja chemilumineschensija rastenii v svjazy s izmeneniem ich zhiznedejatelnosty v postojannom magnitnom pole. Phyziologija rastenii, t.16, v.2, 272-277.
4. Koretskaja, T.F., Vesselovski, V.A., Pogosijan, S.I., Jolkevitch, V.N., 1968, O sootnoshenii mezdu intenzivnostju dichanija i sverchslabim svetcheniem kornei.A.N.USSR.
5. Tarussov, B.N., Ivanov, I.I., Petrusevitch, J.M., 1967, Sverchslaboe svetchenie biologitcheskich sistem, Izd.Moskwa un-ta.

PURPLE MEMBRANE FRAGMENTS AS A MODEL FOR STUDYING THE EFFECTS

OF LIGHT INDUCED ELECTROPHORETIC MOBILITY INCREASE

M. Ratcheva-Kantcheva*, N. Popdimitrova, S. Stoylov, and D. Kovatchev**

*Department of Biophysics and Radiobiology
 Biological Faculty, University of Sofia
**Department of Biology, Medical Academy, Varna

INTRODUCTION

Electrophoretic mobility investigations of purple membrane fragments of Halobacterium halobium using the particle electrophoresis method as primarily reported in 1982 are indicative for the pH dependence of membrane surface electric charge, as well as for a reversible acid induced purple to blue colour transition /1/. The blue membrane form demonstrates a more pronounced aggregation, which correlates with the reduced surface electric charge /data not shown/.

The electrokinetic behaviour of purple membrane fragments was further monitored with the Laser Doppler velocimetry technique /2/.

A resonance Raman molecular probe is also used to measure surface potential changes of purple membrane fragments /3/.

The surface charge density value $\sigma = 8.02.10^{-4}$ charges per $Å^2$ derived in our experiments /4/ was correspondent to those calculated from Ehrenberg and Berezin /-5.3x10^{-4}/ and Packer et al./4.55x10^{-4}/ charges /$Å^2$ /3,2/. There are evidences about membrane surface charge density increase after illumination. The zeta potential calculated from our electrophoretic mobility data at light conditions presents a value of -35.4 mV in comparison to the value of -31.9 mV registered at reduced light intensity /5/.

The research reported here addresses light induced changes of purple membrane electrokinetic behaviour, as modified by lectin-membrane interaction.

MATERIAL AND METHODS

The electrophoretic mobility of purple membrane fragments was investigated using the particle electrophoresis technique with OPTON Cytopherometer. The observation light with an intensity of 22 mW/cm^2 indicated as high level of illumination /A/ was kept at the lowest possible intensity for measurements at low level of illumination /B/. Mobilities of the purple membrane fragments were calculated in m^2/V.s/±S.E.M./; each independant value was obtained by timing the movement of 15 to 20 fragments pooled from the three membrane preparations. The standard deviation for the measurements of purple membrane mobility was between 4 % and 6 %.

Cosequently repeated mobility measurements at low and high illumination levels were obtained in order to check the light induced surface charge increase.

The lecitin-membrane effects on light-induced purple membrane electrophoretic mobility shift are tested for phytochemagglutinin /PHA-P/, Difco in a concentration of 10 and 20 µg/ml.

Absorption spectra were recorded with a Shimadzu UV-260 spectrophotometer.

RESULTS AND DISCUSSION

The well pronounced electrophoretic mobility changes registered at low ionic strength /10 mM NaCl/ presented variations in the effect observed between 20 % and 70 %. The reversible light induced changes are monitored consequently during the time as shown on Fig.1.

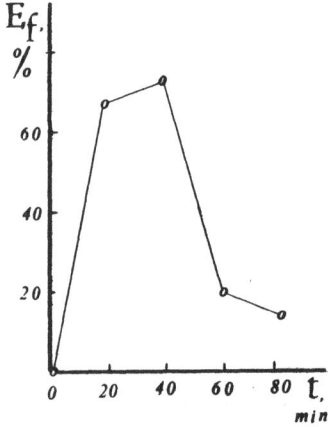

Fig.1 Light-induced electrophoretic mobility stimulation in the time of reversible change of the illumination level

The decrease in the electrophoretic mobility stimulation registered on the 60th and 80th min. could not be explained with changes in the bacteriorhodopsin functional activity. It could be suggested that the electrophoretically detectable pro-

cedure may be responsible for the effect observed after prolonged illumination in the electrophoretic chamber, under the
influence of constant magnetic field. Conformation changes of
purple membrane aggregates in the case are not excluded.

The purple membrane fragment aggregation process induced
by PHA-P treatment results in some decrease in the light induced
electrophoretic mobility stimulation /Table 1/.

Table 1   Electrophoretic mobility dependence on the light conditions in the chamber by PHA-P treatment
A and B high and low level of illumination

Light	$u \times 10^{-8} /m^2.v^{-1}.s^{-1}/$		
		PHA-P / g/	
	Control	10	20
Conditions			
A	$- 1.565 \pm 0.047$	$- 1.272 \pm 0.054$	$- 1.119 \pm 0.056$
B	$- 1.204 \pm 0.060$	$- 1.047 \pm 0.076$	$- 0.964 \pm 0.136$
Effect /%/	23.07	17.69	13.85

A progressive decrease in the electrophoretically detected
light induced surface charge effect from 5 % to 10 % was calculated. It could be suggested that PHA-P binding and clustering of purple membrane fragments determines the reduced expression of the electrokinetic light stimulation. As presented by
spectroscopic examinations PHA-P purple membrane interaction
did not change the absorption characteristics of bacteriorhodopsin /Fig.2/.There are evidences about a large scale of global structural changes of purple membrane organization after
illumination /6/. The transmembrane retinal-regulated pulsating
channels during the photocycle could be related to the surface
electric charge density modification. According to reports recently presented by Grzesiek and Dencher /1986/ there is a good evidence for proton accumulation on both faces of the membrane surface - the lipid headgroup region /7/.

The bacteriorhodopsin proton pumping activity and changes
in proton accumulation in the electric double layer may be res-
posnsible for electrokinetically detectable purple membrane
surface charge density changes.

In firther analysis of light induced surface charge effects
the PHA-P membrane interaction in dependance on the ionic
strebgth and pH value of the suspending medium have to be considered /8, 9/.

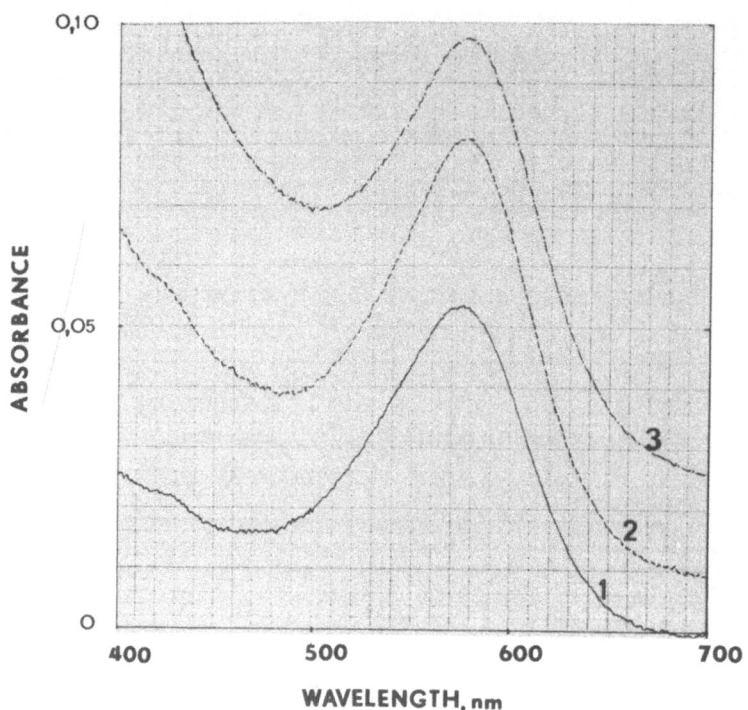

Fig. 2   Absorption spectrum of purple membrane fragments in
50 mM NaCl: Contro - 1; PHA-P - treatment - 10 µg - 2;
20 µg - 3.

REFERENCES

1. Popdimitrova, N., Stoylov, S., Kantcheva, M., 1983, Electrophoretic measurements on purple membrane particles, Electrophoresis'82, Walter de Gruyter, 507-510.
2. Packer, L., Arrio, B., Johannen, G., Volfon, P., 1984, Surface charge of purple membranes measured by laser doppler
   velocimetry, Biochem.Biophys.Res.Commun.,122, 1, 252-258.
3. Ehrenberg, B., Berezin, Y., 1984, Surface potential on
   purple membranes and its sidedness studied by a resonance
   raman dye probe, Biophys.J., 45, 663-667.
4. Popdimitrova, N., Kantcheva, M., Dancshazy, Zs., Stoylov,
   S., 1985, Surface charge density of purple membrane fragments, Proceedings of International Meeting on Cell Electrophoresis, Walter de Gruyter, 167-172.
5. Ratcheva-Kantcheva, M., Popdimitrova, N., Stoylov, S., 1984,
   Electrophoretic properties of purple membrane particles
   from Halobacterium halobium, Bioel.Bioenerget., 12, 173-175.
6. Draheim, J.E., Cassim, J.Y., 1985, Large scale global structural changes of the purple membrane during the photocycle,
   Biophys.J., 47, 497-507.
7. Kantcheva, M., Popdimitrova, N., Stoylov, S., Kovatchev, D.,
   1984, The effect of phytohemagglutinin on the electrophoretic mobility of purple membrane particles, Electrophoresis, 5, 160-161.

9. Stoylov, S., Kantcheva, M., Popdimitrova, N., Kovatchev, D., 1986, Electrostatic surface biomembrane properties - neglected variables in lectin-membrane interaction, _Studia biophysica_, 1-2, 21-25.

AUTHOR INDEX

INDEX